Infosys Science Foundation Series

Infosys Science Foundation Series in Mathematical Sciences

The **Infosys Science Foundation Series in Mathematical Sciences**, a Scopus-indexed book series, focuses on high-quality content in the domain of pure and applied mathematics, biomathematics, financial mathematics, operations research and theoretical computer science. With this series, Springer hopes to provide readers with monographs, textbooks, edited volume, handbooks and professional books of the highest academic quality on current topics in relevant disciplines. Literature in this series will appeal to a wide audience of researchers, students, educators and professionals across the world.

Lakshmi Burra · Fabio Zanolin

The Duffing Equation

Periodic Solutions and Chaotic Dynamics

Springer

Lakshmi Burra
Department of Mathematics
International Institute of Information Technology
Hyderabad, Telangana, India

Fabio Zanolin
Department of Mathematics, Computer Science and Physics
University of Udine
Udine, Italy

ISSN 2363-6149 ISSN 2363-6157 (electronic)
Infosys Science Foundation Series
ISSN 2364-4036 ISSN 2364-4044 (electronic)
Infosys Science Foundation Series in Mathematical Sciences
ISBN 978-981-97-8300-7 ISBN 978-981-97-8301-4 (eBook)
https://doi.org/10.1007/978-981-97-8301-4

Mathematics Subject Classification: 34C25, 34C28, 34C60, 37B10, 37B55, 47H10, 47H11, 54H20

This Springer imprint is published by the registered company Springer Nature Singapore Pte Ltd.
The registered company address is: 152 Beach Road, #21-01/04 Gateway East, Singapore 189721, Singapore

Preface

This book provides a self-contained introduction to some aspects of the Duffing equation, investigated by the authors in recent years with special emphasis on the existence and multiplicity of periodic solutions, subharmonic solutions and different approaches to prove rigorously the presence of chaotic dynamics. In 1918, Georg Duffing introduced a nonlinear oscillator model with a cubic stiffness term in the differential equation of a linear harmonic oscillator to describe the hardening spring effect observed in many engineering, mechanical and physical problems. This is a simple but important model. In spite of its apparent simplicity, the equation has many interesting analytical aspects. Just to mention the interest this equation generated in mathematicians, Littlewood in the 60s focused his attention in the periodically forced case and analyzed the equation for the existence of bounded solutions. Subsequently, with the research of Ueda and others and especially with the classical book of Guckenheimer and Holmes, the presence of chaotic dynamics in this equation was detected.

We start this book with a study of the autonomous equation that represents a simple model of dynamics of a mechanical system with one degree of freedom (Chap. 1). In this special case, we have an associated energy function because the system is conservative and in the case of a center, we can give a precise formula for the period of the orbits. Yet detecting the qualitative properties of the period is not an easy task and it is a topic of current investigation. In this area of research, one must mention the classical works of Urabe, Opial, Chicone among many more other important contributors. Since then there have been several achievements till date. This chapter contains basic information about the autonomous Duffing equation, including the analysis in phase-plane and the properties of the time-mappings and some main results for the monotonicity of the period in a center. This has been independently elaborated from different sources.

Chapter 2 deals with the problem of existence of periodic solutions. An important operator, the Poincaré map, is introduced here which assigns to an initial point the value of the solution after a fixed time T. The fixed points of the Poincaré map are the periodic solutions of the given differential equation. On the other hand, periodic points of the Poincaré map correspond to subharmonic solutions. Finally, as a map of the plane into itself, more complicated structures and patterns can arise giving numeric evidence of the presence of the so-called chaotic dynamics. In this connection, we introduce some tools to detect and rigorously prove the existence of fixed points and periodic points as well as analytically prove the existence of chaotic dynamics according to some classical definitions introduced in the last decades. More precisely, we shall introduce and discuss applications of the Brouwer fixed point theorem and the Brouwer and Leray–Schauder topological degree, as well as the Poincaré–Birkhoff theory. This chapter is based on classical methods, which, nevertheless are quite relevant till today. As a source material, we use [19] for the first part of the chapter and [1] for an application of the Poincaré–Birkhoff theorem. The remaining part of the chapter is based on the unpublished lecture notes given by the second author (F. Zanolin) for a course delivered at the ICTP in Trieste.

Starting with Chap. 3, we introduce the concept of chaotic dynamics with respect to the trajectories of different classical versions of the Duffing equation. At the beginning, this concept is analyzed from an heuristic point of view, by means of some numerical simulations which involve the map $t \mapsto (x(t), x'(t))$ in the phase-plane, as well as the Poincaré section and bifurcation diagrams with respect to some relevant parameters. This chapter is mainly a preparation of the subsequent sections of the book. Here we give evidence, by means of some original simulations, of the complex nature of the solutions to periodically forced Duffing-type equations.

In Chap. 4, we discuss different notions of chaos from an analytical and topological point of view. We also introduce a method (named *stretching along the paths*, synthetically *SAP*), to prove rigorously the presence of chaotic dynamics and to be applied in the rest of the book. More emphasis is devoted to some special configuration, namely, the topological horseshoe-type geometry and the geometry associated with the *linked twist maps* (in short LTM). The topological nature of our approach does not require the more complicated analytical conditions of hyperbolicity, usually required in the classical Smale's horseshoe dynamics. This chapter is the main core of the book. The SAP technique was originally introduced in [12–14] and then extended and applied in subsequent articles. A detailed description of this method can also found in the book of the first author [2], in the PhD dissertations (under the supervision of the second author) of M. Pireddu [16] and O. Gjata [9], as well as in some chapters of the Ph.D. thesis of A. Pascoletti [15] and E. Sovrano in [17]. In our presentation, we recall and prove the main theorems of the SAP technique as well as its applications to LTM. We conclude the chapter with a section concerning the Melnikov method and its applications. We take the basic introduction of this technique from the classical book of Guckenheimer and Holmes [11], also following [10]. A second application of this method is taken from [18].

The remaining chapters of the book are devoted to show some applications of our method to a few selected examples which illustrate different mathematical models in which the periodically forced Duffing equation plays a significant role. Actually, the SAP method in connection with the linked twist maps has proved to be very effective and several applications can be found in the recent literature. The examples in Chaps. 5 and 8 are partly inspired from our previous works, while the content of Chaps. 6 and 7 deals with brand new applications.

More in detail, we start with Chap. 5, dealing with the most classical Duffing equation with cubic nonlinearity. This is a special case of a superlinear equation, already studied in the fundamental work of Littlewood. We show that our method can be applied not only to the classical case but can also be easily adapted to more recent models investigated by other authors, such as the Duffing equation involving a power term with fractional order and singularities as well. This chapter contains some applications to the "classical" Duffing equation with a periodic perturbation, including the celebrated model studied by Ueda, where we provide a rigorous proof of chaotic dynamics for a small friction. The second part of the chapter considers a variant of this equation, which includes a singularity at the origin. We refer, respectively, to [4] and to [5] as main sources for this chapter. Chaotic dynamics in periodically perturbed Duffing-type equations with singularities at the origin are studied also in other recent contributions [3, 6, 8], which are not considered in the book. We refer the interested reader to the above-cited articles for more information on this topic.

Chapter 6 is devoted to a Duffing equation arising in the study of modulated laser dynamics. We prove also the presence of subharmonic solutions of arbitrarily large order.

In Chap. 7, we provide new results to the well-known and extensively studied pendulum equation. There is a rich and vast literature about chaotic dynamics for the periodically perturbed nonlinear pendulum, involving both theoretical and numerical results. Nonetheless, we consider a new situation not previously analyzed, where the periodic perturbation is due to an alternating constant torque. This is a very interesting model which motivated the pioneering studies of Tricomi on synchronous electrical machines and more recently reconsidered and improved in the works of Mawhin.

Chapters 6 and 7 present some new applications to a Duffing equation arising in a model of Laser dynamics and to a pendulum equation. The content of these chapters has never been published before.

A final application is given in Chap. 8 to a diverse application involving a Helmholtz resonator. This is a peculiar phenomenon, which, for instance, is manifested when air resonates in a cavity. The same model has been applied more recently in the study of forced tides, like in the case of a closed basin connected to the open sea with a narrow channel. We reconsider an equation previously studied by Maas and Doelman from a different point of view and provide a further application of our method. In this chapter, we follow loosely an outline from our article [7], without entering in all the technical details which can be found in [7].

Last but not the least, in our analysis of the above-considered multifaceted occurrences of the Duffing equation, some crucial steps are due to the rigorous proof of the monotonicity of the period map of an associated autonomous system. These results, for which we provide either a direct proof or a new demonstration based on Chicone's theorem, are new and of independent interest.

The topics in the book are presented at an expository level without entering too much into technical detail. The book is addressed to researchers in the field of chaotic dynamics as well as graduate and undergraduate students with a basic knowledge of topology, analysis, ordinary differential equations and dynamical systems.

Hyderabad, India — Lakshmi Burra
Udine, Italy — Fabio Zanolin

References

1. A. Boscaggin, F. Zanolin, Pairs of nodal solutions for a class of nonlinear problems with one-sided growth conditions. Adv. Nonlinear Stud. **13**(1), 13–53 (2013)
2. L. Burra, Chaotic Dynamics in Nonlinear Theory, Springer, New Delhi, (2014)
3. L. Burra, F. Zanolin, Non-singular solutions of a Rayleigh-Plesset equation under a periodic pressure field. J. Math. Anal. Appl. **435**(2), 1364–1381 (2016)
4. L. Burra, F. Zanolin, Complex dynamics in a planar hamiltonian system for an equation of the duffing type. EPAM J. **2**(1), 3–24 (2016)
5. L. Burra, F. Zanolin, Periodically perturbed superlinear Duffing type equations with singularities: subharmonic solutions and complex dynamics. Dyn. Contin. Discrete Impuls. Syst. Ser. A Math. Anal. **26**(5), 349–364 (2019)
6. L. Burra, F. Zanolin, Monotonicity of the period function and chaotic dynamics in a class of singular ODEs. J. Math. Anal. Appl. **507**(2), Paper No. 125814, 17 (2022)
7. L. Burra, F. Zanolin, A topological approach to the problem of chaotic tides. Nonlinear Anal. Real World Appl. **68**, Paper No. 103699, 14 (2022)
8. L. Burra, F. Zanolin, Analysis of the period map for a singular ODE related to the Liebau phenomenon. J. Math. Anal. Appl. 537, no. 2, Paper No. 128317, 26 pp., (2024)
9. O. Gjata, Periodic solutions and chaotic dynamics in a Duffing equation model of charged particles. Ph.D. Thesis, University of Udine, (2020)
10. O. Gjata, F. Zanolin, An application of the Melnikov method to a Piecewise Oscillator. Contemp. Math. **4**(2), 249–269 (2023)
11. J. Guckenheimer, P. Holmes, Nonlinear oscillations, dynamical systems, and bifurcations of vector fields. Pergamon, (1988)
12. D. Papini, F. Zanolin, Periodic points and chaotic-like dynamics of planar maps associated to nonlinear Hill's equations with indefinite weight. Georgian Math. J. **9**(2), 339–366 (2002)
13. D. Papini, F. Zanolin, Fixed points, periodic points, and coin-tossing sequences for mappings defined on two-dimensional cells. J. Fixed Point Theory Appl. **2004**(2), 126568, (2004)
14. D. Papini, F. Zanolin, On the periodic boundary value problem and chaotic-like dynamics for nonlinear Hill's equations. Adv. Nonlinear Stud. **4**(1), 71–91 (2004)
15. A. Pascoletti, Fixed points for planar twist-maps. Ph.D. Thesis, University of Udine, (2013)
16. M. Pireddu, Fixed points and chaotic dynamics for expansive-contractive maps in Euclidean spaces, with some applications. Ph.D. Thesis, arXiv:0910.3832, (2009)
17. E. Sovrano, Nonlinear differential equations having non-sign-definite weights. Ph.D. Thesis, University of Udine, (2018)
18. E. Sovrano, F. Zanolin, The Ambrosetti-Prodi periodic problem: different routes to complex dynamics. Dyn. Syst. Appl. **26**(3–4), 589–626 (2017)
19. F. Zanolin, Continuation theorems for the periodic problem via the translation operator. Rend. Sem. Mat. Univ. Politec. Torino, **54**(1), 1–23 (1996)

Contents

Notations Used in the Monograph

We introduce some basic notation that will be used throughout this work.

$\mathbb{N}, \mathbb{Z}, \mathbb{Q}, \mathbb{R}$ and $\mathbb{C}$ represent the sets of *natural*, *integer*, *rational*, *real* and *complex numbers*, respectively. In particular $\mathbb{N}$ is the set of positive integers, (excluding 0). The sets of nonnegative and positive real numbers will be indicated by $\mathbb{R}^+ := [0, +\infty)$ and $\mathbb{R}_0^+ := (0, +\infty)$. Accordingly, the first quadrant and the open first quadrant in the plane will be denoted by $(\mathbb{R}^+)^2$ and $(\mathbb{R}_0^+)^2$, respectively. Open intervals are denoted interchangeably as (a, b) or $]a, b[$.

We denote by I or by Id the *identity map* and by I_W or by Id_W the identity on the space W, when we need to specify the domain. Given a function f, we indicate by *dom*(f) its *domain* and by *Im*(f) its *range*. By $f|_M$ we mean the *restriction* of f to a subset M of its domain.

For a subset S of a topological space W, we denote by $\bar{S}$, Int(S) and ∂S the *closure*, the *interior* and the *boundary* of S, respectively. If $S \subseteq W$, we indicate by $W \backslash S$ the *complement* of S in W. In the case that $A, B \subseteq W$, we denote by $A \backslash B$ the *relative complement* of B in A. By $|A|$ we mean the *cardinality* of the set A.

A continuous map $f : W \to W$ is called a *self-map* of the space W. The *iterates* of f are defined recursively with the convention $f^0 = Id_W$, $f^1 = f$ and $f^n = f \circ f^{n-1}, \forall n \geq 2$. We say that $\bar{w} \in W$ is a *fixed point* for f if $f(\bar{w}) = \bar{w}$. We say that $\bar{w} \in W$ is a *periodic point* for f if there exists an integer $l \geq 1$ such that $f^l(\bar{w}) = \bar{w}$.

If $\bar{w} \in W$ is a *periodic point* for f which is not a fixed point, then we define the period of $\bar{w}$ as the least integer $l \geq 2$ such that $f^l(\bar{w}) = \bar{w}$.

The N-dimensional Euclidean space $\mathbb{R}^N$ is endowed with the usual scalar product $\langle .,. \rangle$, norm $\|.\|$ and distance dist$(.,.)$. In the case of $\mathbb{R}$, the norm $\|.\|$ will be replaced with the *absolute value* $|.|$. If X is a metric space different from $\mathbb{R}^N$, we denote by d_X the distance defined on it. The distance between $M_1, M_2 \subseteq X$ is indicated by $d_X(M_1, M_2) := inf\{d_X(x, y) : x \in M_1, y \in M_2\}$.

In a normed space $(X, \|.\|_X)$, we denote by $B(x_0, r)$ and $B[x_0, r]$ the *open* and *closed balls* centered in $x_0 \in X$ with radius $r > 0$, i.e. $B(x_0, r) := \{x \in X : \|x - x_0\|_X < r\}$ and $B[x_0, r] := \{x \in X : \|x - x_0\|_X \leq r\}$.

For $M \subseteq X$, we set $B(M, r) := \{x \in X : \exists w \in M with \|x - w\|_X < r\}$. The set $B[M, r]$ is defined accordingly.

Given an interval $[a, b] \subseteq \mathbb{R}$, and a real number $p \geq 1$ we define the function space $L^p([a, b])$ as the set of functions $f : [a, b] \to \mathbb{R}$ such that f is Lebesgue measurable and $\int_a^b |f(t)|^p dt < +\infty$, where the integral is the Lebesgue integral.

List of Figures

Chapter 1
The Autonomous Duffing Equation

1.1 Introducing the Equations

The Duffing equation has been widely investigated in literature because of the significance of its applications as well as for the intrinsic interest in it. Indeed, this equation is also a good model for testing the effectiveness of various technical tools of nonlinear analysis. The Duffing equation or a reference to it invariably crops up in any study involving nonlinear dynamics. Such is the impact of this equation. Actually, the Duffing oscillator is one of the prototypical systems of nonlinear dynamics. It can be used to describe diverse systems. Georg Duffing (1861–1944), a German engineer and inventor, in 1918 introduced a periodically nonlinear oscillator with a cubic stiffness term to describe the hardening spring effect detected in many mechanical systems. The original equation takes the form

$$\frac{d^2x}{dt^2} + \alpha x + \beta x^2 + \gamma x^3 = k \sin \omega t$$

as an approximation of the forced pendulum equation

$$\frac{d^2x}{dt^2} + \alpha \sin x = k \sin \omega t,$$

so that the natural choice of the coefficients was $\alpha > 0$, $\beta = 0$ and $\gamma < 0$. On the other hand, interesting dynamics appear for different signs these coefficients. Furthermore, physical models suggest that friction terms, in the form of $\delta \dot{x}$, may be inserted. Of course, we can write the nonlinear term $\alpha x + \beta x^2 + \gamma x^3$ in a more compact form as a function $g(x)$ and a forcing term as a generic periodic function $p(t)$. This leads to the study of a periodically perturbed generalized Duffing equation of the form

$$\ddot{x} + g(x) = p(t)$$

L. Burra and F. Zanolin, *The Duffing Equation*, Infosys Science Foundation
Series in Mathematical Sciences, https://doi.org/10.1007/978-981-97-8301-4_1

as well as its damped counterpart

$$\ddot{x} + c\dot{x} + g(x) = p(t).$$

Last but not the least, in analogy with the study of the pendulum equations with moving support, one can also imagine a periodic perturbation in the form of

$$\ddot{x} + a(t)g(x) = 0$$

(and again the variant of this latter equation with a friction term). These equations represent the typical way in which the Duffing equation may be considered nowadays by several authors in the literature.

Even in its classical form, involving only a cubic term, this equation is intrinsically quite complex. The different physical systems range from the pendulum, to the nonlinear behaviour of certain electrical and electronic circuits, beam-buckling, stiffening springs, Josephson amplifiers, cables, waves, and plasmas, to name a few. The Duffing oscillator first came into the forefront for studying oscillations and then chaotic nonlinear dynamics. Despite its simplicity, it depicts extremely rich dynamical behavior and research on this equation is ongoing. See, for example, the more or less recent work including chapters on the Duffing equation [2, 8, 20, 31, 36, 48, 51, 52].

The Duffing Equation

The Duffing equation is a relevant example of a simple nonlinear differential equation exhibiting many fascinating properties (to quote J.K. Hale [32, Ch. V.5]). Since then this equation has become, together with van der Pol's equation, one of the commonest examples in nonlinear oscillation texts and research articles. As already explained in the Introduction, the periodically forced case deals with an equation of the form

$$\ddot{x} + f(x) = e(t) \tag{1.1}$$

(in the undamped case) or

$$\ddot{x} + c\dot{x} + f(x) = e(t), \quad c > 0, \tag{1.2}$$

when a friction term is included. The function $e(\cdot) : \mathbb{R} \to \mathbb{R}$ represents an external perturbation which we will assume to be a (nonconstant) periodic function of a fixed period $T > 0$. Typical forms of the nonlinearity $f(x)$ (representing the spring characteristic) are given by

$$f(x) = \alpha x + \beta x^3, \tag{1.3}$$

with $\alpha, \beta \in \mathbb{R}$ and $\beta \neq 0$. The Duffing equation with the cubic term can be written out as

$$\ddot{x} + c\dot{x} + \alpha x + \beta x^3 = e(t), \quad c > 0. \tag{1.4}$$

This is usually considered to represent a model of a structural system which includes nonlinear restoring forces (for example, springs). When $\alpha > 0$ the Duffing oscillator may be considered to be a forced oscillator with a spring whose restoring force is given by $F(x) = -\alpha x - \beta x^3$. When $\beta > 0$, this equation represents a "hard spring", and for $\beta < 0$, it represents a "soft spring". If $\beta < 0$, the phase portrait curves are closed. For $\alpha < 0$, the Duffing oscillator describes the dynamics of a point mass in a double well potential (see Fig. 3.1, Chap. 3) and it can be seen as an example of a periodically forced steel beam which is deflected towards the two magnets. In this case, chaotic motions may be observed.

In particular, Moon and Holmes [34, 41] showed that the Duffing equation in the form $\ddot{x} + c\dot{x} - x + x^3 = \gamma \cos \omega t$ gives a simple model for the forced vibrations of a cantilever beam placed in the field of two permanent magnets. In order to start our analysis, we confine ourselves to the unperturbed Duffing equation, namely, we study the autonomous equation

$$\ddot{u} + g(x) = 0.$$

This system is conservative, and therefore we start with a brief survey of conservative systems.

1.2 Conservative Systems

In order to have a qualitative understanding of the Duffing oscillator and its many applications, we would like to introduce some mathematical concepts involved. The first of these is the concept of conservative systems [32].

Let $\Omega \subset \mathbb{R}^n$ be an open, nonempty set. Let $F : \Omega \to \mathbb{R}^n$ be a continuous vector field. We assume that for any x_0 the differential system (1.5) defined below has a unique solution. We recall that the uniqueness of solutions for initial-value problems is guaranteed if F is locally Lipschitz continuous. The solution x is defined on a maximal open interval of existence $I_{x_0} = (\alpha_{x_0}, \beta_{x_0})$ with $-\infty \leq \alpha_{x_0} < 0 < \beta_{x_0} \leq +\infty$. For any $t \in (\alpha_{x_0}, \beta_{x_0})$ the unique solution of the differential system

$$\dot{x} = F(x) \tag{1.5}$$

is denoted by $x(t) = x(t, x_0)$ with $x(0) = x_0$. A *first integral* of (1.5) is a function $E : \Omega \to \mathbb{R}$ which is constant along the solutions of (1.5). To avoid trivialities we suppose that E is not constant on any open subset of Ω. Typically we assume that the integral E is continuous along with its first partial derivatives. In this case, the condition that the integral $E(x(t)) =$ constant is equivalent to $\frac{d}{dt}E(x(t)) = \langle \nabla E(x(t)), \dot{x}(t)\rangle = 0$ and hence[1] $\langle \nabla E(x), F(x)\rangle = 0\ \forall\, x \in \Omega$. Given this condition, the system (1.5) is called *conservative*. The orbits in Ω thus lie on the level sets of the first integral E.

As a consequence of the properties of a conservative system, some important and relevant results follow. (We state the following results without proof, for detailed explanations see [32].)

Result1: An equilibrium point of a conservative system can never be asymptotically stable.

Result2: For $n = 2$, all orbits in a neighbourhood of a stable isolated equilibrium point of a conservative system must be periodic orbits with the equilibrium point being a *center.*

Conservative force expressed as a gradient ... an example from mechanics

The problems associated with differential equations usually come from mechanics, electrical engineering, or even quantum mechanics. These are typically areas where the laws of conservation are applicable. These laws relating to energy state that the total energy of the system is a constant. In physics, many force fields usually come about in the following way: From Newton's laws of motion, which asserts that a particle X, in a force field $\mathcal{F}(X)$, moves in such a way that the force vector at the location X of the particle, at any instant, equals the acceleration vector of the particle times the mass m. We have $m\ddot{X} = \mathcal{F}(X)$.

This type of a system is often called a mechanical system with n degrees of freedom. Suppose that there is a smooth function $U : \mathcal{A} \subset \mathbb{R}^n \to \mathbb{R}$ such that

$$\begin{aligned} \mathcal{F}(X) &= -\left(\frac{\partial U}{\partial x_1}(X), \frac{\partial U}{\partial x_2}(X), \ldots \frac{\partial U}{\partial x_n}(X)\right) \\ &= -\ \nabla U(X). \end{aligned} \tag{1.6}$$

The negative sign is used by convention. Here A represents an open domain.

Such a force field is called a conservative force and U is called the potential energy of the system. To show the connection with the first integral, we introduce the total energy E defined as $E = K + U$, K being the kinetic energy. The total energy function is important because it is constant along any solution curve of the system, agreeing with the definition in the previous section. To derive that the total energy of such a system is a constant, consider the associated system of equations, with $V = V(t)$, the velocity of the particle,

$$\dot{X} = V$$

[1] $\dot{E} = \frac{d}{dt}E(x(t))$.

$$\dot{V} = -\frac{1}{m}\nabla U(X) \tag{1.7}$$

and show that this is indeed a conservative system. This is a system of equations in $\mathcal{A} \times \mathbb{R}^n$. To prove that the Law of Conservation of Energy applies to this system: Let $x(t) = (X(t), V(t))$ be a solution curve of the system (1.7). Then, differentiating the total energy E with respect to t, we have

$$\begin{aligned}\dot{E} &= \frac{d}{dt}\left(\frac{1}{2}m\|V(t)\|^2 + U(X(t))\right) \\ &= m\langle V, \dot{V}\rangle + \nabla\langle U, \dot{X}\rangle \\ &= \langle V, (-\nabla U)\rangle + \langle \nabla U, V\rangle \\ &= 0.\end{aligned}$$

Hamiltonian Systems

Newton's second law gives rise to systems of second-order differential equations in $\mathbb{R}^n$ and so to a system of first-order equations in $\mathbb{R}^{2n}$, an even-dimensional space. When the forces are derived from a potential function, the equations of motion of the mechanical system have many specific properties, most of which follow from the fact that the equations of motion can be put in the form of a Hamiltonian system. The Hamiltonian formalism is the natural mathematical formalism in which to develop the theory of conservative mechanical systems [55]. Such a system of $2n$ ordinary differential equations of this form would be

$$\dot{q} = H_p, \qquad \dot{p} = -H_q,$$

or equivalently

$$\dot{q}_i = \frac{\partial H}{\partial p_i}(q, p), \qquad \dot{p}_i = -\frac{\partial H}{\partial q_i}(q, p), \qquad i = 1, \cdots, n. \tag{1.8}$$

The function H, is called the Hamiltonian and is a smooth, real-valued function defined for $(q, p) \in \mathcal{O}$, an open set in $\mathbb{R}^n \times \mathbb{R}^n$. The vectors $q = (q_1, \cdots, q_n)$ and $p = (p_1, \cdots, p_n)$ are called, traditionally, the position and momentum vectors, respectively, because, in the classical examples, that is what these variables represent. The integer n represents the number of degrees of freedom of the system.

To give a current "symplectic" representation of a Hamiltonian system, we introduce the $2n$ vector z, the $2n \times 2n$ skew symmetric matrix J, and the gradient by

$$z = \begin{bmatrix} q \\ p \end{bmatrix}, \quad J = J_n = \begin{bmatrix} 0 & I \\ -I & 0 \end{bmatrix}, \quad \nabla H = \begin{bmatrix} \frac{\partial H}{\partial z_1} \\ \vdots \\ \frac{\partial H}{\partial z_{2n}} \end{bmatrix},$$

here 0 is the $n \times n$ zero matrix and I is the $n \times n$ identity matrix. Using this formalism Equation (1.8) maybe written as

$$\dot{z} = J\nabla H(z).$$

Such a vector field would be a conservative system with the energy function E replaced by the Hamiltonian H. Hamiltonian systems have a special role in the applications to mechanics because the Poincaré map associated with such systems is area (measure) preserving [1, 39].

Integral curves of conservative systems in $\mathbb{R}^2$.

One can give a little more specific information about the nature of integral curves of conservative systems with one degree of freedom. Let us consider a general second-order scalar equation. Let $g : \mathbb{R} \to \mathbb{R}$, be a continuous function. We consider the associated scalar second-order equation:

$$\ddot{u} + g(u) = 0 \tag{1.9}$$

(namely, the autonomous Duffing equation) or its equivalent system

$$\dot{u} = v \qquad \dot{v} = -g(u). \tag{1.10}$$

The assumption is that a uniqueness theorem holds for (1.10). For convenience in the exposition, we assume $g : \mathbb{R} \to \mathbb{R}$, but the argument can be easily be adapted for g defined on an open interval, of $\mathbb{R}$, for instance, $]0, \infty)$. This is a Hamiltonian system with the total energy or the Hamiltonian function given by $E(u, v) = v^2/2 + G(u)$ where $G(u) = \int_0^u g(s)ds$. The orbits of the solutions of (1.10) in the (u, v)-plane must lie on the level curves of the function $E(u, v)$, that is, the curves represented by $E(u, v) = h$, a constant. The equilibrium points of (1.10) are where $\dot{u} = v = 0$ and $\dot{v} = -g(\bar{u}) = 0$, that is, $(\bar{u}, 0)$. If $G(u)$ has a strict local minimum at $\bar{u}$, then $(\bar{u}, 0)$ is a stable equilibrium point, with all the orbits in its neighborhood periodic, that is, it is a center. Solving $E(u, v) = h$, a constant, $v = \pm\sqrt{2(h - G(u))}$, it can be stated that any equilibrium point $(\bar{u}, 0)$, where $\bar{u}$ is not a minimum of the potential function $G(u)$ has to be unstable. Thus the point $\bar{u}$ is a strict local minimum of $G(u)$ if $g(u) < 0$ for $u < \bar{u}$ and $g(u) > 0$ for $u > \bar{u}$, with u in a neighbourhood of $\bar{u}$. In a symmetric fashion one could state the inequalities for an absolute maximum of $G(u)$. More formally these could be stated as follows:

The stable equilibrium points of (1.10) given by the strict local minima of $G(u)$ are centres and all the unstable equilibrium points corresponding to the strict local maxima of $G(u)$ are saddle points.

These concepts can be explained very clearly from some real-life examples.

1.2.1 A Conservative System from Mechanics Following Newton's Laws

As an example of the autonomous Duffing equation in (1.9), we consider the equations of Newton's laws from mechanics. According to the law of conservation of energy, the total energy is a constant, ideally speaking. In real life, systems cannot be totally conservative because of the presence of dissipative forces like friction, resistance, and other fluctuations.

A mechanical system with one degree of freedom following Newton's laws could be represented by

$$\mathscr{F}(x) = m\,\ddot{x}, \qquad \ddot{x} = \frac{d^2x}{dt^2},$$

where the force $\mathscr{F}(x)$ is a function of the displacement x only. This equation in the phase-plane in $\mathbb{R}^2$ can be written as the system

$$\begin{aligned} \dot{x} &= y \\ \dot{y} &= \frac{\mathscr{F}(x)}{m}. \end{aligned} \tag{1.11}$$

Eliminating t, and solving by separating the variables, we obtain, taking at $t = t_0,\ \ x = x_0$ and $y = y_0$

$$\frac{1}{2}my^2 - \frac{1}{2}my_0^2 = \int_{x_0}^{x} \mathscr{F}(x)dx \tag{1.12}$$

$\frac{1}{2}my^2$ is the kinetic energy of the system and $V(x) = -\int_0^x \mathscr{F}(x)dx$ is the potential energy

$$\frac{1}{2}my^2 + V(x) = \kappa, \tag{1.13}$$

where κ is the constant obtained after adjusting the limits in the integral. So, for a given value of κ, the curve traced out by the equation is the trajectory of the particle (or a level line) or a line of constant energy, in the phase-plane. The equilibrium points of (1.11) system are the points with coordinates $(x_e,\ 0)$, x_e are the roots of $\mathscr{F}(x) = 0$. Equation (1.13) clearly indicates that the trajectories are symmetric with respect to the x-axis. For these paths

$$y = \pm\sqrt{\frac{2}{m}\{\kappa - V(x)\}}. \tag{1.14}$$

At the equilibrium points (also called critical points) of system (1.11), $(x_e, 0)$, $\mathscr{F}(x_e) = 0$, and, given the choice of $V(x)$, that is, $\mathscr{F}(x) = -V'(x)$, $V'(x_e) = 0$ so that $V(x)$ has a stationary point at x_e. We now elaborate these cases for $V(x)$.

1. A strict relative minimum for $V(x)$ at $x = x_e$ Choose $V(x_e) = \kappa_0$. For the choice of $\kappa_1 > \kappa_0$, in a suitable neighbourhood of κ_0 $\frac{2}{m}\{\kappa_1 - V(x)\} > 0$ in an open interval containing x_e. For such a choice, y is real, for $b_1 \le x \le b_2$ and $y = 0$ at b_1 and b_2 (see Fig. 1.1) $y = 0$, at b_1 and b_2. Therefore the path is a closed ellipse-type trajectory, with the equilibrium point inside it. In this case, the equilibrium point is a centre and is stable.
2. A strict relative maximum for $V(x)$ at $x = x_e$.
   Let $V(x)$ have a relative maximum = κ_0 at $x = x_e$. We choose another value of $V(x) = \kappa_1 < \kappa_0$, in a suitable neighbourhood of κ_0. For $\kappa = \kappa_1$, from (1.14) y is real for all $x \le b_1$ for $x \ge b_2$ and 0 at $x = b_1$ and $x = b_2$ (see Fig. 1.2). The equilibrium point $(x_e, 0)$ is a saddle point and unstable, and the trajectory for these values of x, is a hyperbola.

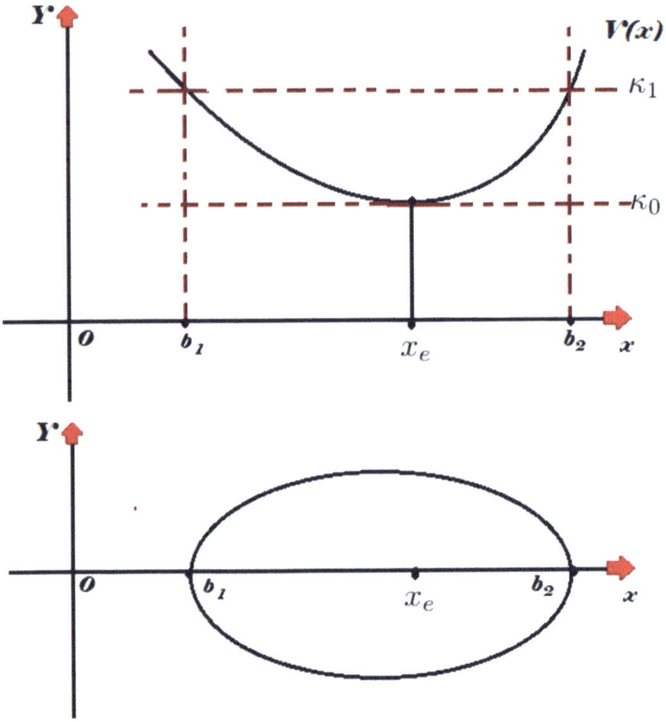

Fig. 1.1 A strict relative minimum for $V(x)$

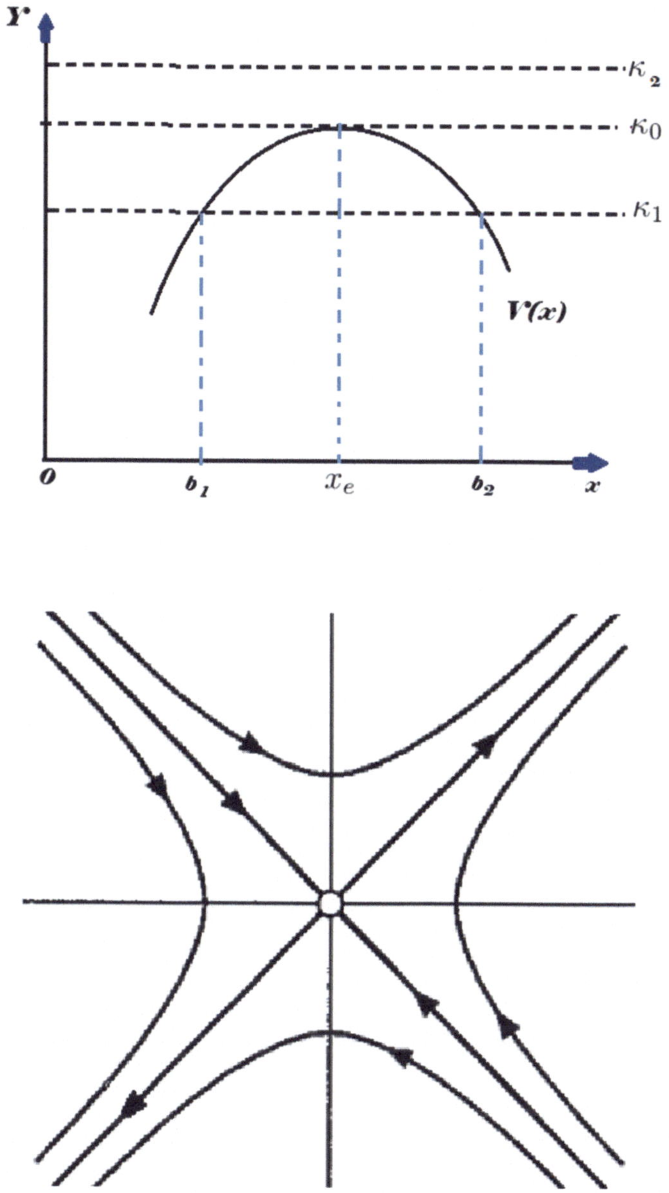

Fig. 1.2 A strict relative maximum for $V(x)$

1.2.2 The Simple Pendulum

In the chronicle of nonlinear dynamics, including as well as in Duffing's book, the pendulum plays a predominant role, and so it is befitting to start the narrative with Galileo. It was Galileo who identified the pendulum as the first example of synchronous motion. This led to the first successful clock developed by Huygens. In addition to horology (the science of measuring time), pendulums have important applications in gravimetry (the measurement of the specific gravity) and inertial navigation.

A simple pendulum consists of a mass (m) suspended from a fixed point (hinge). If all the mass is assumed to be concentrated at a point, we obtain the idealized simple pendulum. The pendulum is suspended by a string of negligible weight and of a length l. The string is attached to a point so that the bob may oscillate freely in the vertical plane. Let x be the angle which the string make with the vertical line from S at time t. We neglect frictional forces due to the air present and the only forces which the mass m is subjected to are the tension in the string and the acceleration due to gravity g. The differential equation of motion representing the angular displacement x of a pendulum of length l and g, the acceleration due to gravity is given by

$$\frac{d^2x}{dt^2} = -\frac{g}{l}\sin x$$

Let $\frac{g}{l} = L$, then we have $\frac{d^2x}{dt^2} = -L \sin x$.

In the phase-plane, this is

$$\begin{aligned}\frac{dx}{dt} &= y \\ \frac{dy}{dt} &= -L \sin x\end{aligned}$$

It is evident that the equation of the pendulum has the same form as (1.10) with $g(x)$ as $L \sin x$. The equilibrium points of this system are $(x_e, 0)$ where x_e are the roots of $\sin x = 0$. Integrating after eliminating t the equation of the energy level lines is

$$\frac{1}{2}y^2 + L(1 - \cos x) = E(x, y)$$

. Thus the potential energy in this equation is (as before) $V(x) = -\int_0^x F(s)ds =$ $\int_0^x L \sin(s)ds = L(1 - \cos x)$. As before we plot the graph of $V(x) = L(1 - \cos x)$. The derivative of the potential energy, $V'(x) = L \sin x$ and its second derivative, $V''(x) = L \cos x$. Thus the potential energy function has a relative minimum at $x = 2n\pi$ for every even integer $2n$ and a relative maximum at $x = (2n + 1)\pi$ for every odd integer $2n + 1$. Hence, we can state that $(2n\pi, 0)$ are centres and stable and $(2n + 1, 0)$ are the saddle points and unstable. Figure 1.3 shows the minima and the maxima

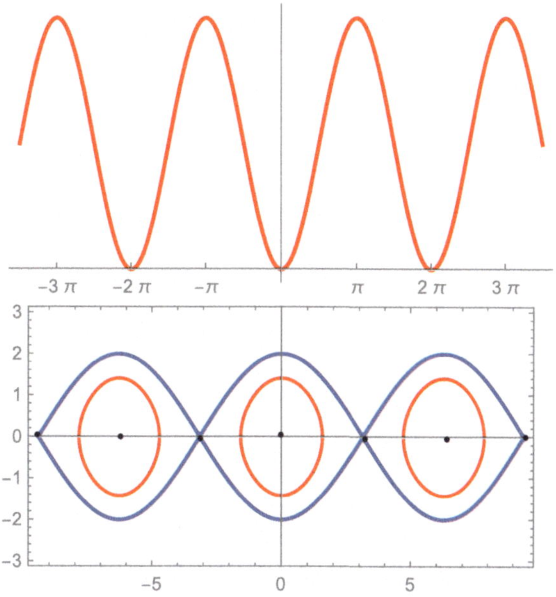

Fig. 1.3 Relative maxima for $V(x)$ as well as relative minima, the case of the simple pendulum

in the potential function as well as the corresponding trajectories for these values of the potential. More precisely, the picture shows the potential function $V(x) = L(1 - \cos x)$ with $L = 1$ (above) and the solution curves with stable equilibrium points (surrounded by periodic orbits) corresponding to the minima in the potential function and the saddle points, the unstable equilibrium points corresponding to the relative maxima in the potential function (below).

In fact, we can do a little more with the information that we have, we can derive the time period of the periodic orbits corresponding to the minima in the potential function. Since we consider a closed orbit, it must intersect the x-axis in two points, say $(x_1, 0)$ and $(x_2, 0)$, $x_1 < x_2$ and symmetric with respect to the x-axis. If T is the period of the periodic solution, $T = 2\omega$, where ω is the time required by the solution to move in the positive part of the solution that is where $y > 0$. We have $y = \pm\sqrt{2(h - G(x))}$, and a simple separation of variables gives

$$T = 2\int_{x_1}^{x_2} \frac{du}{\sqrt{2(h - G(u))}}. \tag{1.15}$$

We come back to this concept of the time-mapping formula in Eq. (1.15) in Sect. 1.4.

Current research on the pendulum and pendulum-type equations

There is a huge volume of work on the pendulum-type as well as the pendulum equations. In [35], Hubbard shows that a differential equation modelling a damped forced pendulum that is $\ddot{x} + 10^{-1}\dot{x} + \sin(x) = \cos(t)$ can exhibit extraordinarily complicated and unstable behaviour. Such results are vital in robotics: the forced pendulum is a basic subsystem of any robot. Pendulum-type equations have applications in unexpected areas like the librations of satellites. In [27], the authors study a nonlinear, second-order ordinary differential equation that models the longitudinal librations of the longest axis L of a satellite with respect to the planet-satellite centre line C. Combining theoretical arguments and numerical evidence they prove that, in the case of Hyperion, a satellite of Saturn, the angle A between L and C, can change in a chaotic manner at the moments when the distance from Hyperion to Saturn reaches its minimum value. Also in [29], the authors, under the assumption that the spin axis remains perpendicular to the orbit plane, investigate the behaviour of the longitudinal librations of any satellite. They also show that they are possibly chaotic depending on two parameters: a constant k related to the principal moments of inertia of the satellite, and the eccentricity e of its orbit. In [28], the authors show that a pendulum, whose pivot oscillates vertically in a periodic fashion, has uncountably many chaotic orbits. In [5], the authors use computational methods to investigate Hubbard's equation $\ddot{x} + 10^{-1}\dot{x} + \sin(x) = \cos(t)$, its chaotic properties and the induced dynamics. In [37, 38] Mawhin, amongst other things, obtains the existence of T-periodic solutions of the pendulum equation $u''(t) + \sin u(t) = f(t)$, for $f(t)$, a general function. Some authors study the class of equations representing the "inverted pendulum" [9]. The inverted pendulum with small parametric forcing given by $f(t)$, for example, $-u''(t) + L\sin u(t) = f(t)$, is considered as an example of a wider class of parametrically forced Hamiltonian systems. The qualitative dynamics of the Poincaré map corresponding to the central periodic solution is studied via an approximating integrable normal form in [9]. In this context, structural stability was proved under generic conditions.

1.2.3 The Duffing Oscillator

The pendulum was the main focus of Duffing's book and more or less the whole of Duffing book was devoted to it.

Consider the pendulum equation $\ddot{x} + L\sin x = 0$. The sine of the angular displacement x can be written as the series expansion

$$\sin x = x - \frac{x^3}{3!} + \frac{x^5}{5!} + \cdots ,$$

if this series is truncated at the third order, and substituted in the pendulum equation, we have

$$\ddot{x} + L(x - \frac{x^3}{3!}) = 0.$$

This is the unforced Duffing's equation with the negative cubic stiffness term.

To give a clearer picture of the dynamics, we include a simulation of a form of this equation, namely, $\ddot{x} + x - \rho x^3 = 0$ or equivalently the associated system $\dot{x} = y,\ \dot{y} = -x + \rho x^3$. From Fig. 1.4, the similarity with the phase curves of the pendulum equation (see Fig. 1.3) is very clear. The picture shows the potential function $V(x)$ with the two maxima and the minimum, the corresponding solution curves with a saddle point on either side of the origin corresponding to the relative maxima in the potential function, with the (periodic) orbits around the minimum of the potential function.

The more usual form of the Duffing equation one encounters is (1.2) with $f(x)$ as in (1.3).

Let us, for the sake of comparison, consider a simulation of the potential as well as the phase portrait of a simple form of the undamped, autonomous Duffing equation

$$\ddot{x} - x + \rho x^3 = 0$$

where ρ is a positive constant. This is shown in Fig. 1.5. The picture shows the double well potential function $V(x)$ (above) and the corresponding solution curves with a saddle point at the origin corresponding to the relative maxima in the potential function, surrounded by (periodic) orbits around the minima of the potential function (below).

The Duffing equation simulated in Fig. 1.5 shows its potential function $V(x) = -x^2/2 + x^4/4$ with three extrema, one relative maximum (at the origin) and two relative minima at ± 1. Its phase portrait shows the unstable equilibrium at the origin, a saddle point where there are two homoclinic orbits on either side of it and periodic orbits corresponding to the relative minima of the potential function.

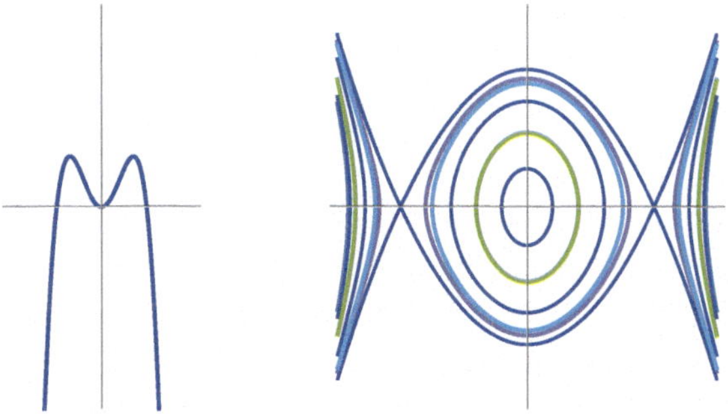

Fig. 1.4 Relative minima and maxima for the potential function $V(x) = x^2/2 - x^4/4$ as well as the phase curves for the system $\dot{x} = y, \quad \dot{y} = -x + x^3$

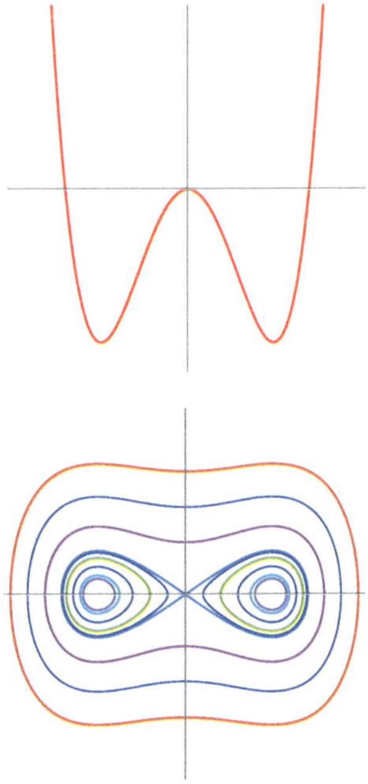

Fig. 1.5 Relative maxima for the potential function $V(x) = -x^2/2 + x^4/4$ as well as relative minima in the case of the undamped, autonomous Duffing equation $\ddot{x} - x + \rho x^3 = 0$ where $\rho = 1$

1.3 Jacobi Elliptic Functions

We now consider an interesting classical example. Elliptic functions were introduced in the nineteenth century as an exercise to extend the table of integrable functions. The basic properties of these functions follow from elementary ideas of differential equations. Let us consider the following example.

Let k be a constant with $0 < k < 1$ and $\text{sn}(t, k)$ the solution of

$$\ddot{x} + (1 + k^2)x - 2k^2x^3 = 0, \qquad x(0) = 0, \ \dot{x}(0) = 1. \tag{1.16}$$

The function $\text{sn}(t, k)$ is called the Jacobi elliptic sine function. With the usual method followed setting $\dot{x} = y$, the Hamiltonian or integral may be obtained as

$$2H = y^2 + (1 + k^2)x^2 - k^2x^4 \tag{1.17}$$

and on the solution curve $\text{sn}(t, k)$, $2H = 1$, so $\dot{\text{sn}}^2 = (1 - \text{sn}^2)(1 - k^2\text{sn}^2)$. The phase portrait of this system is similar to the one in Fig. 1.4. Here we have three equilibrium points, namely, the origin, a centre and the saddle points at $\pm\Big(\sqrt{\frac{1+k^2}{2k^2}}, 0\Big)$. The saddle points have a heteroclinic connection as can be seen in Fig. 1.4. The open region bounded by the heteroclinic level lines is filled with periodic orbits surrounding the centre which is the origin. These periodic orbits are defined by $C = y^2 + (1 + k^2)x^2 - k^2x^4$, where $0 < C < (\frac{1+k^2}{2k})^2$. Therefore, it is evident that sn is a periodic function. Moreover, the function $u(t) = -\mathrm{sn}(-t, k)$ satisfies the same differential equation with the same initial conditions and therefore by the uniqueness of the solutions we have that $\mathrm{sn}(t, k) = -\mathrm{sn}(-t, k)$ showing that $\mathrm{sn}(t, k)$ is odd. If we focus our attention on the part of the periodic level line $2H = 1$, in the first quadrant we find that

$$\dot{x} = \sqrt{1 - (1 + k^2)x^2 + k^2x^4}$$

and with $(x(t), \dot{x}(t))$ connecting $(0, 1)$ at the time $t = 0$ to $(1, 0)$ at some time $t = \kappa$ with

$$\kappa = \kappa(k) = \int_0^1 \frac{dx}{\sqrt{(1 - x^2)(1 - k^2x^2)}}. \tag{1.18}$$

By symmetry we immediately find that sn is a periodic function of period 4κ. Notice that for $k \to 0^+$ we have that $\kappa(k) \to \displaystyle\int_0^1 \frac{dx}{\sqrt{(1 - x^2)}} = \frac{\pi}{2}$. Whilst, for $k \to 1^-$ we have that $\kappa(k) \to \displaystyle\int_0^1 \frac{dx}{(1 - x^2)} = \infty$. As observed in [40] the functions $\mathrm{sn}(., k)$ have a similar behaviour as the trigonometric sine function with the period $4\kappa(k)$ playing the role of 2π. The elliptic sn function and its cosine counterpart are widely studied with numerical tables included in classical handbooks of mathematics. These functions, interestingly, are implemented in computer algebra systems (Fig. 1.6).

Another expression of the period function can be obtained as follows: from Eq. (1.18) using the substitution $x = \sin\theta$ we have

$$\kappa(k) = \int_0^{\pi/2} \frac{d\theta}{\sqrt{1 - k^2 \sin^2\theta}}. \tag{1.19}$$

It is evident from both the formulas (1.18) and (1.19) that $\kappa(k)$ is a strictly monotone increasing function with respect to k. Jacobi as well as Weierstrass elliptic functions play significant roles not only in the ODEs but also in PDEs. Elliptic functions find useful applications in the particular solutions of partial differential equations. See, for instance, [18, 19, 46].

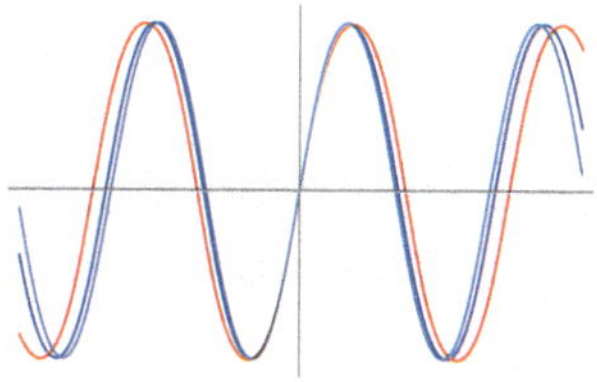

Fig. 1.6 Jacobi Elliptic sine functions for $k = 1/2, 1/3, 1/4$

Existence of harmonic and subharmonic solutions in the Duffing equation

Several other authors (see, for example, [17] and the references therein) proved the existence of infinitely many harmonic and subharmonic solutions for the second-order scalar nonlinear Duffing's equation $\ddot{x} + g(x) = p(t, x, x')$ where g and p are continuous functions with $p(t)$ bounded and periodic in the first variable and g satisfying the assumption $g(x)sgn(x) \to +\infty$ as $|x| \to \infty$. Fučik and Lovicar in [26] prove the existence of at least one T-periodic solution of (1.23) for any T-periodic forcing term $p(t)$, assuming the superlinear growth condition, $g(x)/x \to \infty$ as $|x| \to \infty$. The proof was extended by Struwe in [49] to $\ddot{x} + g(x) = p(t, x, x')$ where p could be unbounded but with utmost linear growth in x'. A generalization of the theorems of Fučik and Lovicar and Struwe has also been obtained by Capietto, Mawhin, and Zanolin in [11], using a continuation approach which allows extensions of the existence results to systems as well. On the other hand, in his pioneering work [33], Hartman, using the Poincaré-Birkhoff fixed point theorem proved the existence of infinitely many T-periodic solutions for $\ddot{x} + f(t, x) = 0$ with $f(t, 0) \equiv 0$, f_x continuous and $\frac{f(t,x)}{x} \to \infty$ as $|x| \to \infty$, uniformly in t. Further work in the lines of [11] and [33] has been done in the past 20 years, see, for example, [6, 15, 21, 30, 44, 45, 54].

Another relevant area of research involves the extension of these results to higher dimensions. The first important contributions in this area were made by Castro and Lazer in [12] and Bahri and Berestycki in [3, 4]. Further, more recent results have been obtained in [7, 10, 22, 50, 53]. Some of these papers apply a recent extension of the Poincaré-Birkhoff fixed point theorem to Hamiltonian flows in higher dimensions due to Fonda and Ureña, [23–25].

Usually, when we speak of harmonic solutions we mean solutions of a fixed period (say) T. Subharmonic solutions of order $m \geq 2$ (m an integer) are solutions of period mT with m being the minimal positive integer. This means the solutions are mT−periodic and not kT−periodic with $k = 1, \cdots, m-1$.

The aim of this book is to study the complex dynamics which arise from the Duffing-type equations with a periodic forcing term. The chaotic dynamics which arise from these equations are studied from the point of view of the "topological horseshoes", amongst other interesting phenomena in these type of equations.

1.4 Time Mapping and Some Results Related to Its Monotonicity

Let us go back to the system

$$\begin{aligned} \dot{x} &= y \\ \dot{y} &= -g(x), \end{aligned} \tag{1.20}$$

where $g : \mathbb{R} \to \mathbb{R}$ is a continuous function. System (1.20) is a Hamiltonian system, with the Hamiltonian function or the total energy given by $E(x, y) = y^2/2 + G(x)$ where $G(x) = \int_0^x g(s)ds$. The orbits of the solutions of (1.20) in the $(x, y)-$ plane lie on the level lines of the function $E(x, y)$, that is, where $E(x, y) = h$, a constant. Also, let

$$x\, g(x) > 0, \quad \forall x \neq 0 \tag{1.21}$$

and

$$\lim_{|x|\to\infty} G(x) = +\infty. \tag{1.22}$$

We elaborate on here the notion of the time map and some important results by various authors related to its monotonicity [16, 43, 47]. We observe that condition (1.21) implies that the "potential energy" G is bounded below and the map $G : \mathbb{R}^+ \to \mathbb{R}^+$ satisfies $G(0) = 0$, and is strictly increasing. For every $\alpha > 0$, the solution $(x(.), y(.))$ of (1.20) satisfying the initial conditions $(x(0), y(0)) = (\alpha, 0)$ is unique and defined on the entire real line. (x_α, y_α) indicates the dependence of the solution $(x(t), y(t))$ on the parameter α. The corresponding trajectory would then be

$$\frac{1}{2}y^2 + G(x) = G(\alpha). \tag{1.23}$$

Transposing the terms we get $y(t) = \dot{x}(t) = \sqrt{2(G(\alpha) - G(x(t)))}$ for all t in the maximal interval, $[c, 0]$ where both $x(t) \geq 0$ and $\dot{x}(t) = y(t) \geq 0$. Integrating on $[c, 0]$ we can determine a formula for the time period (as before) τ_g, given by

$$\tau_g(\alpha) := 2\left|\int_0^\alpha \frac{ds}{\sqrt{2(G(\alpha) - G(s))}}\right|, \tag{1.24}$$

where $\tau_g(\alpha)$ is the time needed for the solution to transit between two consecutive zeros.

If we use the hypothesis

$$\lim_{|x|\to+\infty} g(x)\mathrm{sign}(x) = +\infty. \tag{1.25}$$

Indeed whilst (1.21) and (1.22) guarantee that the origin is a global centre, assumption (1.25) ensures only that the level lines of large energy are closed/periodic trajectories surrounding the origin. For every $\alpha \in \mathbb{R}$, $\alpha > 0$, the function $\alpha \mapsto \tau_g(\alpha)$ is continuous.

In [43], Opial proved the following theorems that illustrate the asymptotic behaviour of the time maps:

Theorem 1.1 ([43], Theorem 8) *Let $g(x)$ be a function satisfying the conditions*

$$xg(x) > 0,$$

when $x \neq 0$ and $g(0) = 0$. Let us suppose further that we have

$$\lim_{|x|\to\infty} G(x) = +\infty.$$

1. *If $g(u)/u$ is increasing (decreasing) in $(0, +\infty)$, then $\tau_g(x)$, the time mapping is decreasing (increasing) for $x > 0$.*
2. *If $g(u)/u$ is decreasing (increasing) in $(-\infty, 0)$, then $\tau_g(x)$ is decreasing (increasing) for $x < 0$.*

Proof We prove the theorem for the interval $(0, +\infty)$, the proof for the interval $(-\infty, 0)$ may be obtained in a similar fashion.

Let the first condition be assumed, that is, $g(u)/u$ is decreasing in $(0, +\infty)$, then we prove that $\tau_g(x)$ is increasing for $x > 0$.

Consider $r \in (0, b)$ where $b \leq \infty$ then, we have

$$\tau_g(r) = \sqrt{2}\int_0^r \frac{1}{\sqrt{G(r) - G(s)}}ds.$$

Taking $p = \frac{r'}{r}$, where $0 < r < r' < b$. Define the new function

$$h_p(x) = \frac{g(px)}{p}.$$

Since $\frac{g(x)}{x}$ is decreasing, for any $x \in (0, \frac{b}{p})$, we have

$$g(x) = \frac{g(x)}{x}x > \frac{g(px)}{px}x = h_p(x).$$

Define $I(x) = \int_0^x h_p(u)du$, then $G(r) - G(x) > I(r) - I(x)$.

$$\tau_g(r) = \sqrt{2}\int_0^r \frac{dx}{\sqrt{G(r) - G(x)}} < \sqrt{2}\int_0^r \frac{dx}{\sqrt{I(r) - I(x)}} \tag{1.26}$$

$$I(r) = \int_0^r \frac{g(ps)}{p}ds = \int_0^{pr} \frac{g(s)}{p^2}ds = \frac{G(pr)}{p^2}.$$

From our choice $p = \frac{r'}{r}$, where $0 < r < r' < b$, we find that $I(r) = \frac{G(pr)}{p^2} = \frac{G(r')}{p^2}$.

$$\sqrt{2}\int_0^r \frac{dx}{\sqrt{I(r) - I(x)}} = \int_0^r \frac{dx}{\sqrt{G(r')/p^2 - G(px)/p^2}}$$
$$= \int_0^{\frac{r'}{p}} \frac{pdx}{\sqrt{G(r') - G(px)}} = \int_0^{r'} \frac{dx}{\sqrt{G(r') - G(x)}} = \tau_g(r'). \tag{1.27}$$

From (1.26) and (1.27) we have $\tau_g(r) < \tau_g(r')$ or the time map is increasing monotonically on $(0, b)$. □

Theorem 1.2 ([43], Theorem 6) *Let $g(x)$ be a function satisfying the conditions*

$$xg(x) > 0,$$

when $x \neq 0$ and $g(0) = 0$. Let us suppose further that we have

$$\lim_{|x|\to\infty} G(x) = +\infty.$$

Then, if the function $\tau_g(x)$ is increasing (strictly increasing) in the interval $(0, +\infty)$ and strictly decreasing in the interval $(-\infty, 0)$, the function $\frac{G(x)}{x^2}$ is strictly decreasing in the interval $(0, +\infty)$ and strictly increasing in $(-\infty, 0)$.

Proof (We prove only one case for the monotonicity, that is, for the case in which $\tau_g(x)$ is increasing. The other case may be proved in a similar fashion.)
In the interval $(0, +\infty)$,by hypothesis $\tau_g(x)$ is increasing. Therefore for all $\beta > 1$ and all $x > 0$ we can say that the following inequality holds good:
$\tau_g(x) \leq \tau_g(\beta x)$.

The time-mapping formula for $\tau_g(\beta x)$ is

$$\tau_g(\beta x) = \sqrt{2}\int_0^{\beta x} \frac{du}{\sqrt{G(\beta x) - G(u)}}$$
$$= \sqrt{2}\int_0^{x} \frac{du}{\sqrt{G(\beta x)/\beta^2 - G(\beta u)/\beta^2}}$$

Set $h_\beta(x) = g(\beta x)/\beta$, $(0 \leq x < +\infty,)$, then from the result that $\tau_g(x) \leq \tau_h(x) \implies H(x) \leq G(x)$ where $H(x) = \int_0^x h(s)ds$ (see Theorem 1, [43]) we have $\tau_g(x) \leq \tau_{h_\beta}(x) \implies \int_0^x h_\beta(s)ds \leq G(x)$. Integrating we get

$$G(\beta x)/\beta^2 \leq G(x)$$

dividing on both sides of this inequality with x^2 we get

$$\frac{G(\beta x)}{(\beta x)^2} \leq \frac{G(x)}{x^2}.$$

This shows that $\dfrac{G(x)}{x^2}$ is a decreasing function in the interval $(0, +\infty)$. □

Remark 1.1 We remark that Opial's condition of monotonicity for the period function is valid even when $g''(0) = 0$. Chicone's condition which follows in the next section is not valid when $g''(0) = 0$.

1.4.1 Chicone's Condition for the Monotonicity of the Time Map

In case neither of the results of the preceding sections apply, we describe a condition due to Chicone [13]. This condition is independent of the results due to Opial [43] and Obi [42].

Consider the Hamiltonian system in the plane $H(x, y) = y^2/2 + G(x)$, where G is a smooth function and has a non-degenerate relative minimum (see Fig. 1.1). Recall that $G(0) = 0$. As mentioned earlier, in this case, the orbits are closed and periodic. That is, the phase portrait consists of a centre at the origin surrounded by closed orbits, each of these orbits is an energy level line of H with energy $E \in (0, E_*)$. Let the period, $T : (0, E_*) \to \mathbb{R}$ of each periodic orbit of energy E, be the minimum period of each orbit. Following the work of Chicone, Chow and Wang, Yagasaki and others [13, 14, 56], we state a result on the monotonicity of the time map. Let $K(E) = \{x \in \mathbb{R} : G(x) \leq E\}$ and set

$$\begin{aligned} N(x) &= 6\,G(x)(G''(x))^2 - 3(G'(x))^2 G''(x) - 2G(x)G'(x)G'''(x) \\ &= (G'(x))^4 \left(\frac{G(x)}{(G'(x))^2} \right)''. \end{aligned} \tag{1.28}$$

Then the following theorem is valid:

Theorem 1.3 *If* $N(x) \geq 0$ *for all* $K \in K(E)$*, then* T *is monotone increasing on* $(0, E)$*. If* $N(x) \leq 0$ *for all* $K \in K(E)$*, then* T *is monotone decreasing on* $(0, E)$*.*

The author, in [13], observes that the monotonicity in the period function may be attributed to the convexity of the function $\dfrac{G(x)}{(G'(x))^2}$.

The proof of the theorem requires that the function $\dfrac{G(x)}{G'(x)^2}$ be convex, that is, the second derivative of $\dfrac{G(x)}{G'(x)^2}$ is ≥ 0. An outline of the proof as given by Chicone [13] is given here.

Proof Using the symmetry of each trajectory with respect to the $X-$ axis, the period of a closed orbit with energy E is given by the usual time-mapping formula:

$$T = \frac{2}{\sqrt{2}} \int_{x_0}^{x_1} \frac{dx}{\sqrt{E - G(x)}},$$

where x_0 and x_1 are the left and right intersections of the curve $H(x, y) = y^2/2 + G(x)$ with the $x-$axis. The idea is to compute the derivative $\frac{dT}{dE}$ and prove that T is monotone increasing if $N(x) \geq 0$ and decreasing if $N(x) \leq 0$. An auxiliary function $h(x)$ is defined for the purposes of computation, as follows:

$$h(x) = \begin{cases} G(x)^{1/2}, & x \in K(E) \setminus \{0\} \\ 0 & x = 0. \end{cases}$$

Since, by hypothesis $G(x)$ has a local minimum at the origin, $h(x)$ is a smooth function, and it is apparent that $h'(x) > 0$, for all $x \in K(E_*)$. Using a change of variable, setting $r = h(x)$ in the time-mapping formula we get

$$T = \sqrt{2} \int_{-\sqrt{E}}^{\sqrt{E}} \frac{dr}{h'(x)\sqrt{E - r^2}}.$$

To simplify further, another change of variable is used, $r = \sqrt{E} \sin\theta$. Then

$$T = \sqrt{2} \int_{-\frac{\pi}{2}}^{\frac{\pi}{2}} \frac{d\theta}{h'(h^{-1}(\sqrt{E} \sin\theta))}.$$

Differentiating the time T with respect to E, and then integrating the resulting integral by parts

$$\frac{dT}{dE} = \frac{1}{\sqrt{2}} \int_{-\frac{\pi}{2}}^{\frac{\pi}{2}} \frac{3(h''(x))^2 - h'(x)h'''(x)}{(h'(x))^5} \cos^2\theta d\theta. \tag{1.29}$$

Reverting back to $h^2(x) = G(x)$ one obtains

$$h' = \frac{G'}{2h} \qquad h'' = \frac{2GG' - (G')^2}{4h^3}$$

and

$$h''' = \frac{4G^2G''' - 6GG'G'' + 3(G')^3}{8hG^2}$$

From this, one obtains

$$3(h'')^2 - h'h''' = \frac{6G(G'')^2 - 3(G')^2G'' - 2GG'G'''}{8G^2}. \tag{1.30}$$

The numerator of (1.30) is $N(x)$ (see Eq. (1.28)) which by hypothesis, if positive, should imply that the period T is monotone increasing. The left-hand side of (1.30) is the numerator of the integrand in (1.29). If this quantity is positive the rest of the terms in the integrand are positive, making $\frac{dT}{dE} \geq 0$ and proving the theorem. There is strong monotonicity if the strict inequality $\frac{dT}{dE} > 0$ holds. A symmetric argument may be given of the case when $N(x) \leq 0$. □

1.4.2 Some Examples Illustrating the Condition of Chicone

(1) $x'' - b(e^x - 1) = 0$
The equation

$$x'' - b(e^x - 1) = 0 \qquad b < 0$$

has been studied by Chicone [13] as well as Chow and Wang [14] to prove the monotonicity of its time map. Written in the phase-plane this is

$$\begin{aligned} \dot{x} &= y \\ \dot{y} &= b(e^x - 1). \end{aligned} \tag{1.31}$$

The corresponding Hamiltonian for (1.31) is

$$H(x, y) = \frac{1}{2}y^2 - b(e^x - x - 1). \tag{1.32}$$

The potential function is $-b(e^x - x - 1)$, and a simple calculation shows that $N(x) = -b^3 e^x \left(-4e^x(x-1) + e^{2x} - 2x - 5\right)$. To verify that this expression is monotone increasing we need to show that $f(x) = \left(-4e^x(x-1) + e^{2x} - 2x - 5\right)$ is positive. It is readily seen that $f(0) = 0$ as well as $f'(0) = 0$, and $f''(x) = 4e^x(-x + e^x - 1) \geq 0$. Therefore $N(x) \geq 0$ and the time period T is monotone increasing.

Figure 1.7 shows the periodic level lines related to (1.31) as well as the potential function with a minimum at the origin.

An interesting fact needs to be pointed out here with respect to the function $g(x) = e^x - 1$, $g(x)/x = e^x - 1/x$ is increasing for $x > 0$. From the theorem of Opial [43], when applied to the half-period τ_g gives that it is monotone decreasing. The nature of the nonlinearity of $e^x - 1$ shows that the Chicone criteria applies and gives that the period map is monotone increasing. But the Opial criterion which applies to the half-period map shows that it is monotone decreasing.

On the other hand, if we take the example of the equation $\ddot{x} + x^3 = 0$, $g(x) = x^3$, and $G(x) = x^4/4$, here $N(x) = (3/2)x^8$, Opial's theorem can be applied and this gives that the time map is monotone decreasing, as well as the period of the orbit in fact, $\tau_g = T/2$ (as $g(x)/x$ is monotone increasing). But Chicone's theorem would

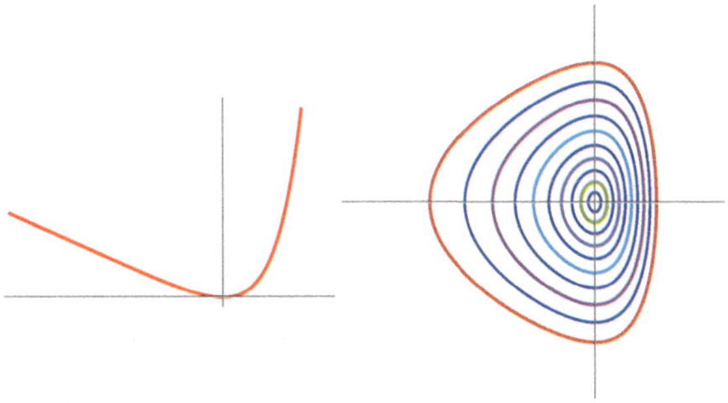

Fig. 1.7 Graph of the potential function corresponding to the Hamiltonian (1.32), $V(x) = (e^x - x - 1)$ (left) and the energy level lines corresponding to (1.31), $b = -1$ (right)

apparently provide a false conclusion that the period is increasing since $N(x) = (3/2)x^8$, which is positive for $x \neq 0$. This is because the Chicone theorem cannot be applied as 0 is a degenerate minimum ($g''(0) = 0$).

(2) The next example satisfying the condition of monotonicity is that of the pendulum equation $\ddot{x} + L \sin x = 0$ where L is a positive constant.
The corresponding Hamiltonian is

$$H(x, y) = \frac{1}{2}y^2 + L(1 - \cos x). \tag{1.33}$$

The potential energy function is $L(1 - \cos x)$.

In each open interval $(-\pi + 2\kappa\pi, \pi + 2\kappa\pi)$, with $\kappa \in \mathbb{Z}$, the level lines of energy $E \in (0, 2)$ are periodic orbits. This can be seen to be consistent with Fig. 1.8.

$N(x)$ can be calculated to be

$$\begin{aligned} N(x) &= L(2 - 5\cos x + 4\cos^2 x - \cos^3 x) \\ &= L(\cos x - 1)^2(2 - \cos x) \geq 0 \end{aligned}$$

and the inequality is strict in the given intervals for $x \neq 2\kappa\pi$. Therefore, the period function is strictly monotone increasing.

Figure 1.8 shows the periodic level lines related to the pendulum equation $\ddot{x} + L \sin x = 0$. An alternative proof of the monotonicity of the period map of the pendulum equation may also be seen by the use of Jacobi elliptic sine functions. Indeed starting from the pendulum equation

$$\ddot{x} + L \sin x = 0. \tag{1.34}$$

The time period of the periodic orbits may be obtained by integration as

$$4 \int_0^{\alpha} \frac{ds}{\sqrt{2L(\cos s - \cos \alpha)}} \tag{1.35}$$

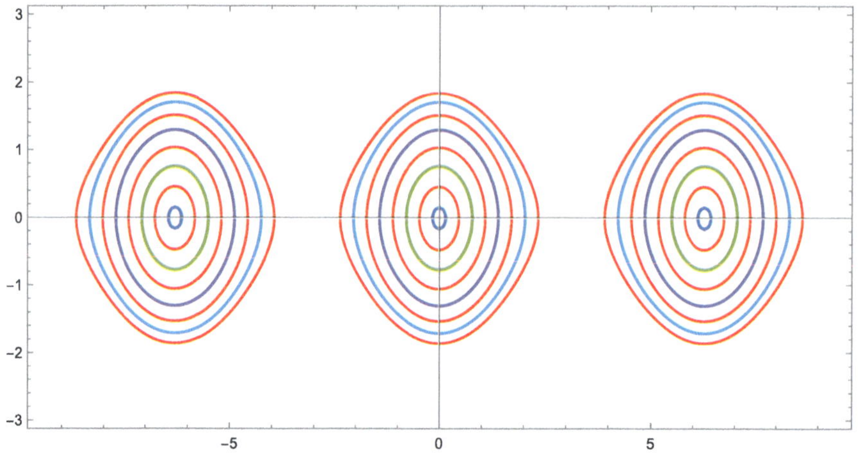

Fig. 1.8 The periodic, monotone orbits related to the pendulum equation

where $(\alpha, 0)$ is the point where the periodic orbit cuts the $x-$axis. Equation (1.35) may be simplified as

$$\frac{2}{\sqrt{L}}\int_0^{\alpha}\frac{2dx}{2\sin(\alpha/2)\sqrt{1-\frac{\sin^2(x/2)}{\sin^2(\alpha/2)}}}. \tag{1.36}$$

Setting $\frac{\sin^2(x/2)}{\sin^2(\alpha/2)} = \sin\theta$, we have $0 \leq x \leq \alpha$, $0 \leq \sin\theta \leq 1$ with $0 \leq \theta \leq \pi/2$. Equation (1.36) simplifies to

$$\frac{4}{\sqrt{L}}\int_0^{\pi/2}\frac{d\theta}{\sqrt{1-k^2\sin^2\theta}} = \frac{4}{\sqrt{L}}\kappa(k)$$

, where $k = \sin\alpha/2$ and $\kappa(k)$ as in Eq. (1.19). Since the function $\kappa(k)$ is strictly monotone with respect to k, we have another proof for the monotonicity of the period map of the pendulum equation.

(3) As another example, consider the function $g(x) = x + x^3$. Hence, $G(x) = \frac{1}{2}x^2 + \frac{1}{4}x^4$ and we obtain

$$\frac{G(x)}{g(x)^2} = \frac{1}{4}\frac{2+x^2}{(1+x^2)^2}.$$

(A graph of this function is shown in Fig. 1.9.)

Moreover, $N(x) = \frac{3}{2}x^4(-1+4x^2+x^4)$ has no constant sign.

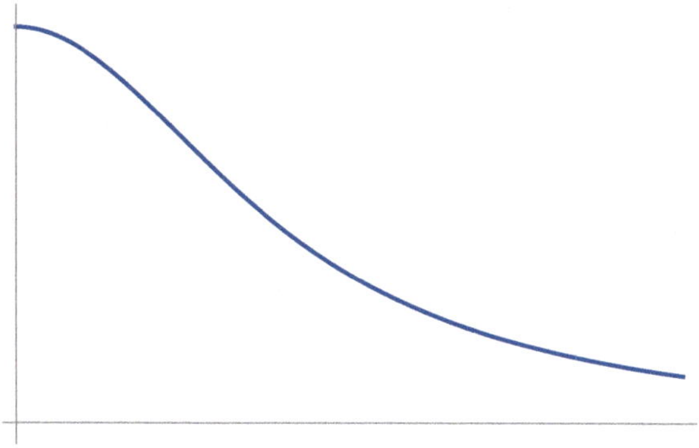

Fig. 1.9 A graph of $\frac{G(x)}{g(x)^2} = \frac{1}{4}\frac{2+x^2}{(1+x^2)^2}$

On the other hand, $g(x)/x = 1 + x^2$ is strictly increasing on $(0, +\infty)$ and strictly decreasing on $(-\infty, 0)$, and therefore, according to Opial's theorem, the period map is strictly decreasing.

1.4.3 Some Further Conditions of Monotonicity of the Time Mappings

We now prove the monotonicity of the time mapping associated to a system of the form:

$$\begin{cases} \dot{x} = y \\ \dot{y} = -g(x), \end{cases} \tag{1.37}$$

where $g : \mathbb{R} \to \mathbb{R}$ is a continuously differentiable function such that

$$\lim_{|x| \to +\infty} g(x)\text{sign}(x) = +\infty. \tag{1.38}$$

Condition (1.38) is stable with respect to bounded perturbations. Namely, if a function f satisfies $\lim_{|x| \to +\infty} f(x)\text{sign}(x) = +\infty$, and $m(x)$ is a bounded function (in particular, if m is constant) then $g(x) := f(x) - m(x)$ satisfies (1.38). We also define

$$G(x) := \int_0^x g(s)\,ds.$$

From (1.38), we know that there exists $R > 0$ such that $G(x)$ is strictly decreasing on $(-\infty, -R)$, strictly increasing on $(R, +\infty)$ and, moreover, $G(x) \to +\infty$ as $x \to \pm\infty$. Due to the conservative nature of system (1.37) we know that the associated orbits lie on the energy level lines

$$E(x, y) = \frac{1}{2}y^2 + G(x) = k \tag{1.39}$$

which are symmetric with respect to the x-axis in the phase-plane. Such level lines are closed curves (hence periodic orbits) provided that

$$k > k^* := \max_{x \in [-R, R]} G(x).$$

For every $k > k^*$ there exists a unique pair (c_k^-, c_k^+) with

$$c_k^- < -R < 0 < R < c_k^+$$

such that

$$G(c_k^-) = G(c_k^+) = k.$$

For any fixed $k > k^*$, the period of the orbit

$$\Gamma^k := \{(x, y) \in \mathbb{R}^2 \,:\, E(x, y) = k\}$$

is given by

$$\tau(k) := \tau^-(k) + \tau^+(k),$$

where

$$\tau^-(k) := 2\int_{c_k^-}^{0} \frac{dx}{\sqrt{2(G(c_k^-) - G(x))}}$$

and

$$\tau^+(k) := 2\int_{0}^{c_k^-} \frac{dx}{\sqrt{2(G(c_k^+) - G(x))}}.$$

Recall that $\tau^-(k)$ and $\tau^+(k)$ have an equivalent representation, as introduced earlier in (1.24). More precisely, we have that $\tau^+(k) = \tau_g(c_k^+)$ and $\tau^-(k) = \tau_g(c_k^-)$.

Hence, in order to prove the monotonicity of the time mapping, it is sufficient to show that the map

$$\phi(c) := \int_0^c \frac{dx}{\sqrt{(G(c) - G(x))}}$$

is strictly monotone in a neighbourhood of infinity, namely, for $|c|$ sufficiently large.

For simplicity, we discuss only the case in which c is positive and large. Our analysis follows the lines of [42].

Via a rescaling, we write $\phi(c)$ as

$$\phi(c) = \int_0^1 \frac{d\xi}{\sqrt{(G(c) - G(c\xi))/c^2}}.$$

Our goal is to show that for $c < c'$ with $c > R$ and sufficiently large, it follows that

$$(G(c) - G(c\xi))/c^2 < (G(c') - G(c'\xi))/c'^2, \quad \forall 0 \le \xi < 1.$$

Hence, by differentiating with respect to c, we get that the above condition holds true provided that for the auxiliary function

$$h(c) := cg(c) - 2G(c)$$

we have that

$$h(c) > h(\xi c), \quad \forall \xi \in [0, 1[.$$

This latter condition holds good for every $c > 0$ sufficiently large, if we suppose that $h(c) \to +\infty$ as $c \to +\infty$ and $h'(c) > 0$ for $c > 0$ large enough.

If we apply the above considerations to the case

$$g(x) := f(x) - m, \quad \text{for } m \in \mathbb{R},$$

we obtain the following result.

Lemma 1.1 *Let $f : \mathbb{R} \to \mathbb{R}$ be a continuously differentiable function such that*

$$\lim_{|x| \to +\infty} f(x)\text{sign}(x) = +\infty \tag{1.40}$$

and

$$\lim_{|x| \to +\infty} (xf'(x) - f(x))\text{sign}(x) = +\infty. \tag{1.41}$$

Then, for every $m \in \mathbb{R}$ there exists k_m^* such that for every $k > k_m^*$, (see (1.39)) the orbits of (1.37) (for $g(x) := f(x) - m$) are periodic with a strictly monotonic decreasing period.

Proof We consider the function $h(c)$ for the special choice of $g(x) = f(x) - m$. Then we have

$$h'(c) = cf'(c) - f(c) + m.$$

Clearly, $h'(c) \to +\infty$ as $c \to +\infty$ if $xf'(x) - f(x) \to +\infty$ as $x \to +\infty$. Similarly, one can check the symmetric condition for $c << 0$. □

References

1. V.I. Arnold, *Mathematical Methods of Classical Mechanics*, vol. 60 (Springer, 2013)
2. J. Awrejcewicz, *Ordinary Differential Equations and Mechanical Systems* (Springer, Cham, 2014)
3. A. Bahri, H. Berestycki, Existence of forced oscillations for some nonlinear differential equations. Comm. Pure Appl. Math. **37**(4), 403–442 (1984)
4. A. Bahri, H. Berestycki, Forced vibrations of superquadratic Hamiltonian systems. Acta Math. **152**(3–4), 143–197 (1984)
5. B. Bánhelyi, T. Csendes, B.M. Garay, L. Hatvani, A computer-assisted proof of σ_3-chaos in the forced damped pendulum equation. SIAM J. Appl. Dyn. Syst. **7**(3), 843–867 (2008)
6. A. Boscaggin, Periodic solutions to superlinear planar Hamiltonian systems. Port. Math. **69**(2), 127–140 (2012)
7. A. Boscaggin, R. Ortega, Monotone twist maps and periodic solutions of systems of Duffing type. Math. Proc. Camb. Philos. Soc. **157**(2), 279–296 (2014)
8. T. Bountis, H. Skokos, *Complex Hamiltonian Dynamics*. Springer Series in Synergetics (Springer, Heidelberg, 2012). With a foreword by Sergej Flach
9. H.W. Broer, I. Hoveijn, M. Van Noort, G. Vegter, The inverted pendulum: a singularity theory approach. J. Differ. Equ. **157**(1), 120–149 (1999)
10. A. Capietto, W. Dambrosio, Multiplicity results for systems of superlinear second order equations. J. Math. Anal. Appl. **248**(2), 532–548 (2000)
11. A. Capietto, J. Mawhin, F. Zanolin, A continuation approach to superlinear periodic boundary value problems. J. Differ. Equ. **88**(1), 990 (1990)
12. A. Castro, A.C. Lazer, On periodic solutions of weakly coupled systems of differential equations. Boll. Un. Mat. Ital. B (5) **18**(3), 733–742 (1981)
13. C. Chicone, The monotonicity of the period function for planar Hamiltonian vector fields. J. Differ. Equ. **69**(3), 310–321 (1987)
14. S. Chow, D. Wang, On the monotonicity of the period function of some second order equations. Časopis pro pěstování matematiky **111**(1), 14–25 (1986)
15. F. Dalbono, F. Zanolin, Multiplicity results for asymptotically linear equations, using the rotation number approach. Mediterr. J. Math. **4**(2), 127–149 (2007)
16. W. Dambrosio, Time-map techniques for some boundary value problems. Rocky Mt. J. Math. 885–926 (1998)
17. T. Ding, F. Zanolin, Periodic solutions of Duffing's equations with superquadratic potential. J. Differ. Equ. **97**(2), 328–378 (1992)
18. E. Fan, B.Y.C. Hon, Double periodic solutions with Jacobi elliptic functions for two generalized Hirota-Satsuma coupled KdV systems. Phys. Lett. A **292**(6), 335–337 (2002)
19. E. Fan, J. Zhang, Applications of the Jacobi elliptic function method to special-type nonlinear equations. Phys. Lett. A **305**(6), 383–392 (2002)
20. M. Fečkan, *Bifurcation and Chaos in Discontinuous and Continuous Systems*. Nonlinear Physical Science (Higher Education Press, Beijing; Springer, Heidelberg, 2011)
21. A. Fonda, A. Sfecci, A general method for the existence of periodic solutions of differential systems in the plane. J. Differ. Equ. **252**(2), 1369–1391 (2012)
22. A. Fonda, A. Sfecci, Periodic solutions of weakly coupled superlinear systems. J. Differ. Equ. **260**(3), 2150–2162 (2016)
23. A. Fonda, A.J. Ureña, A higher-dimensional Poincaré-Birkhoff theorem without monotone twist. Comptes Rendus Mathematique **354**(5), 475–479 (2016)
24. A. Fonda, A.J. Ureña, A higher dimensional Poincaré-Birkhoff theorem for Hamiltonian flows. Ann. Inst. H. Poincaré C Anal. Non Linéaire **34**(3), 679–698 (2017)
25. A. Fonda, A.J. Ureña, A Poincaré-Birkhoff theorem for Hamiltonian flows on nonconvex domains. J. Math. Pures Appl. **9**(129), 131–152 (2019)
26. S. Fucik, V. Lovicar, Periodic solutions of the equation $x'' + g(x) = p$. Casopis Pest. Mat **100**, 160–175 (1975)

27. M. Furi, A.S. Landsberg, M. Martelli, On the longitudinal librations of Hyperion. J. Fixed Point Theory Appl. **5**(2), 249–264 (2009)
28. M. Furi, M. Martelli, M. O'Neill, C. Staples, Chaotic orbits of a pendulum with variable length. Electron. J. Differ. Equ. (EJDE) [electronic only] 2004:Paper–No, 2004
29. M. Furi, M. Martelli, A. Vignoli, Longitudinal librations of a satellite. Electron. J. Differ. Equ. **2011**(134), 1–15 (2011)
30. P. Gidoni, Existence of a periodic solution for superlinear second order ODEs. J. Differ. Equ. **345**, 401–417 (2023)
31. J. Guckenheimer, P. Holmes, *Nonlinear Oscillations, Dynamical Systems, and Bifurcations of Vector Fields* (Pergamon, 1988)
32. J. Hale, *Ordinary Differential Equations*, 2nd edn. (R.E. Krieger Publishing Co., Inc., Huntington, N.Y., 1980)
33. P. Hartman, On boundary value problems for superlinear second order differential equations. J. Differ. Equ. **26**(1), 37–53 (1977)
34. P. Holmes, A nonlinear oscillator with a strange attractor. Philos. Trans. R. Soc. Lond. Ser. A Math. Phys. Sci. **292**(1394), 419–448 (1979)
35. J. Hubbard, The forced damped pendulum: chaos, complication and control. Am. Math. Mon. **106**(8), 741–758 (1999)
36. I. Kovacic, M.J. Brennan (eds.), *The Duffing Equation* (Wiley, Chichester, 2011). Nonlinear oscillators and their behaviour
37. L. Markus, *Lectures in Differentiable Dynamics*, vol. 3 (American Mathematical Soc., 1971)
38. J. Mawhin, Global results for the forced pendulum equation. Handb. Differ. Equ. **1**, 533–589 (2004)
39. K.R. Mayer, G.R. Hall, D. Offing, *Introduction to Hamiltonian Dynamical Systems and the N-Body Problem* (Springer, New York, N.Y., 2009)
40. K.R. Meyer, G.R. Hall, D. Offin, *Introduction to Hamiltonian Dynamical Systems and the N-Body Problem, Applied Mathematical Sciences*, 2nd edn., vol. 90 (Springer, New York, 2009)
41. F.C. Moon, P.J. Holmes, A magnetoelastic strange attractor. J. Sound Vib. **65**(2), 275–296 (1979)
42. C. Obi, Analytical theory of nonlinear oscillations. vii. the periods of the periodic solutions of the equation $\ddot{x} + g(x) = 0$. J. Math. Anal. Appl. **55**(2), 295–301 (1976)
43. Z. Opial, Sur les périodes des solutions de l'équation différentielle $x'' + g(x) = 0$. Ann. Polon. Math. **10**, 49–72 (1961)
44. D. Papini, Periodic solutions for a class of Liénard equations. Funkcial. Ekvac. **43**(2), 303–322 (2000)
45. D. Qian, P.J. Torres, P. Wang, Periodic solutions of second order equations via rotation numbers. J. Differ. Equ. **266**(8), 4746–4768 (2019)
46. A.H. Salas, L.J. Martinez H, D.L. Ocampo R, New solutions for the generalized BBM equation in terms of Jacobi and Weierstrass elliptic functions. Abstr. Appl. Anal., pages Art. ID 5513266, 10 (2021)
47. E. Sovrano, F. Zanolin, Remarks on Dirichlet problems with sub linear growth at infinity. Rend. Istit. Mat. Univ. Trieste **47**, 267–305 (2015)
48. J.C. Sprott. *Elegant Chaos* (World Scientific Publishing Co. Pte. Ltd., Hackensack, NJ, 2010). Algebraically simple chaotic flows
49. M. Struwe, Multiple solutions of anticoercive boundary value problems for a class of ordinary differential equations of second order. J. Differ. Equ. **37**(2), 285–295 (1980)
50. S. Terracini, G. Verzini, Solutions of prescribed number of zeroes to a class of superlinear ODE's systems. NoDEA Nonlinear Differ. Equ. Appl. **8**(3), 323–341 (2001)
51. P.J. Torres, *Mathematical Models with Singularities: A Zoo of Singular Creatures*, Atlantis Briefs in Differential Equations, vol. 1 (Atlantis Press, Paris, 2015)
52. Y. Ueda, Survey of regular and chaotic phenomena in the forced duffing oscillator. Chaos, Solitons Fractals **1**(3), 199–231 (1991)

53. S. Wang, Periodic solutions of weakly coupled superlinear systems with indefinite terms. NoDEA Nonlinear Differ. Equ. Appl. **29**(4), Paper No. 36, 22 (2022)
54. Z. Wang, T. Ma, Infinitely many periodic solutions of planar Hamiltonian systems via the Poincaré-Birkhoff theorem. Bound. Value Probl., pages Paper No. 102, 19 (2018)
55. S. Wiggins, *Introduction to Applied Nonlinear Dynamical Systems and Chaos*, vol. 2 (Springer, 2003)
56. K. Yagasaki, Monotonicity of the period function for $u'' - u + u^p = 0$ with $p \in r$ and $p > 1$. J. Differ. Equ. **255**(7), 1988–2001 (2013)

Chapter 2
The Periodically Forced Duffing Equation

2.1 Fixed Points Techniques Using the Poincaré Map

In this chapter we focus our attention on the existence and multiplicity of periodic solutions of the periodically forced Duffing equation

$$\ddot{x} + g(x) = e(t, x, \dot{x}). \tag{2.1}$$

The analysis of (2.1) will be performed in the phase-plane with the system

$$\begin{cases} \dot{x} = & y \\ \dot{y} = & -g(x) + e(t, x, y) \end{cases} \tag{2.2}$$

where e is $T-$periodic in the t variable. In the special case when $e \equiv e(t)$ we can split

$$e(t) = \bar{e} + \tilde{e}(t)$$

where $\bar{e} = \dfrac{1}{T}\displaystyle\int_0^T e(t)dt$. Subtracting $\bar{e}$ from both sides of the equation $\ddot{x} + g(x) = e(t)$ we obtain an equivalent equation where the forcing term has mean value zero. Thus, in the study of the periodically forced Duffing equation

$$\ddot{x} + g(x) = e(t), \tag{2.3}$$

without loss of generality, we can assume $e(t) = \tilde{e}(t)$, that is the $\int_0^T e(t)dt = 0$, and introduce the primitive $E(t) = \int_0^t e(s)ds$, which is $T-$periodic. Then, instead of studying (2.2) we can analyse the equivalent system

$$\begin{cases} \dot{x} = & y + E(t) \\ \dot{y} = & -g(x). \end{cases} \tag{2.4}$$

L. Burra and F. Zanolin, *The Duffing Equation*, Infosys Science Foundation Series in Mathematical Sciences, https://doi.org/10.1007/978-981-97-8301-4_2

Similar considerations can be adapted to the case of these equations with a friction term included. In the coming sections we describe some general methods to study systems of this kind. We develop our approach for n−dimensional systems when no special conditions are imposed for the planar case. The next section follows from the work in the article [197].

Some fixed point techniques using the Poincaré map.

Let $f = f(t, x) : \mathbb{R} \times \mathbb{R}^m \to \mathbb{R}^m$ be a continuous function which is periodic of period $T > 0$ in the t-variable. The search of T-periodic solutions for the differential system

$$\dot{x} = f(t, x) \tag{2.5}$$

is a classical topic in the theory of differential equations which has been widely studied in the literature for its significance in view of concrete applications to ODE models and also as a basic source of problems which motivate the development of new tools in nonlinear analysis. A simple observation shows that finding a T-periodic solution to (2.5) is equivalent to solving the problem

$$\begin{cases} \dot{x} = f(t, x) \\ x(0) = x(T) \end{cases} \tag{2.6}$$

Indeed, any T-periodic solution of (2.5) is also a solution of (2.6) and, conversely, any solution $x : [0, T] \to \mathbb{R}^m$ of (2.6) can be extended to $\mathbb{R}$, by T-periodicity, to a T-periodic solution of (2.5) using the fact that $f(t + T, x) = f(t, x)$. Anyhow, note that problem (2.6) is meaningful if we only assume $f : [0, T] \times \mathbb{R}^m \to \mathbb{R}^m$, without necessarily requiring $f(0, x) = f(T, x)$.

Amongst various ways to obtain the solvability of (2.6) the following main approaches have been considered:

- write (2.6) as an equivalent fixed point problem for a map in $\mathbb{R}^m$;
- write (2.6) as an equivalent fixed point problem for an operator acting in a suitable function space;
- in the case that our problem has variational structure (e.g. if Eq. (2.5) is a Hamiltonian system), write (2.6) as an equivalent critical point problem for a functional defined in a suitable function space.

The aim of this section is to discuss some typical results concerning the use of fixed point theorems and topological degree which can be applied in the first of the above approaches, and then to propose some examples for the solvability of (2.6). We refer the interested reader to [31, 33, 107, 131, 135] and the references quoted therein for various existence results and continuation theorems in the frame of the Leray-Schauder degree and its developments for the periodic problem. We also refer to [139] for periodic solutions of Hamiltonian systems using critical point theory.

The following basic list of notations is used throughout this chapter, consistent with the notation introduced in the beginning. W e denote by $\mathbb{R}^m$ the m-dimensional real Euclidean space endowed with inner product $\langle \cdot, \cdot \rangle$ and norm $|| \cdot ||$. Let I be the identity in $\mathbb{R}^m$ and let ∂D and $\overline{D}$ denote, respectively, the boundary and the closure of a set $D \subset \mathbb{R}^m$. The open and the closed ball of centre the origin and radius $r > 0$ are denoted, respectively, by $B(r)$ and $B[r]$. Given a function $V : \mathbb{R}^m \to \mathbb{R}$ and a constant $c \in \mathbb{R}$, we set $[V \le c] := \{x \in \mathbb{R}^m : V(x) \le c\}$ and $[V < c] := \{x \in \mathbb{R}^m : V(x) < c\}$. Note that if V is differentiable and for some c, $V^{-1}(c) \ne \emptyset$, with c a regular value of V, i.e. $\nabla V(x) \ne 0$, $\forall x \in V^{-1}(c)$, then $\overline{[V < c]} = [V \le c]$, and $\partial[V < c] = \partial[V \le c] = V^{-1}(c)$. By $\mathcal{R}(h)$ we denote the range of a function h, that is the image through h of the domain $\mathcal{D}(h)$ of h; $|h|_p$, with $1 \le p \le \infty$ is the L^p-norm of h and it is usually referred to the interval $[0, T]$ when h is T-periodic. Finally, we point out that some of the results we are going to present can be proved in a more general setting, namely, when $f : [0, T] \times \mathbb{R}^m \to \mathbb{R}^m$ satisfies the Carathéodory conditions that is, $f(\cdot, x)$ is measurable for all x, $f(t, \cdot)$ is continuous for almost every t and, for each $r > 0$, there is a function $\zeta_r \in L^1_{\text{loc}}$ such that $||f(t, x)|| \le \zeta_r(t)$, for a.e. t and all x with $||x|| \le r$. In this case, a solution x of (2.5) is an absolutely continuous function defined on an interval J such that $x'(t) = f(t, x(t))$, for a.e. $t \in J$. In the sequel, solutions of (2.5) are assumed to be *noncontinuable,* i.e. defined on a maximal interval of existence. This convention will be constantly used, unless when the contrary is explicitly stated.

2.2 The Translation Operator

Throughout this section, we assume that *any Cauchy problem associated to the differential systems we consider has a unique solution which is globally defined in* $\mathbb{R}$. Then, for any $s \in \mathbb{R}$ and $z \in \mathbb{R}^m$ there is a unique solution $x(t) = x(t; s, z)$ to (2.5) satisfying the initial condition $x(s) = z$. Note that the map $(t, s, z) \mapsto x(t; s, z)$ is continuous and $x(t_2; s, z) = x(t_2; t_1, x(t_1; s, z))$, $\forall t_1, t_2 \in \mathbb{R}$, so that a *process* [96] in $\mathbb{R}^m$ turns out to be defined in a natural manner. For ease in the notation, we set $x(t; z) := x(t; 0, z)$ and define the Poincaré map (translation operator at the time T)

$$\Phi : \mathbb{R}^m \to \mathbb{R}^m, \quad \Phi(z) := x(T; z),$$

which is continuous. Actually, to have a well-defined continuous Poincaré map, it is sufficient to require forward uniqueness and global existence in $[0, T]$.

Lemma 2.1 *System* (2.5) *has a T-periodic solution if and only if Φ has a fixed point.*

The proof is evident and is therefore omitted (see, e.g. [106]). Just observe that any fixed point of Φ is an initial point, for $s = 0$, of a T-periodic solution of (2.5).

In view of Lemma 2.1, our programme is that of presenting in the sequel some existence results for (2.6) based on the search of fixed points for the Poincaré map

Φ. It can be worth mentioning that, in spite of the great success of more modern functional-analytic techniques, this classical approach has been constantly used with success in a lot of old and recent books and articles, see, e.g. [9, 11, 105, 106, 155, 167], just to quote a few of references. This method is also specially suited to deal with periodic problems on manifolds [33, 34, 88, 89, 146, 180, 181], problems arising from population dynamics [1, 3, 45, 100, 179, 182, 199], or planar conservative systems where the Poincaré-Birkhoff fixed point theorem may be applied to Φ yielding subharmonic solutions as well [53, 75, 165]. Here we don't consider the wide literature about the existence of nonconstant periodic orbits of autonomous systems for which we refer to [116] and the references therein.

Lemma 2.2 *There is a continuous function $\eta : \mathbb{R}_0^+ \to \mathbb{R}^+$, with $\eta(r) > r$, for all $r \geq 0$, such that the following inference holds:*

$$||z|| \geq \eta(r) \Longrightarrow ||x(t; s, z)|| > r, \quad \forall s, t \in [0, T].$$

Proof Given $r \geq 0$, define $\eta(r) := 1 + \sup\{\, ||x(t; s, w)|| : 0 \leq s, t \leq T,\ ||w|| \leq r\,\}$. Suppose, by contradiction, that there is z with $||z|| \geq \eta(r)$ and $s_1, s_2 \in [0, T]$, such that $||x(s_1; s_2, z)|| < r$. Setting $w = x(s_1; s_2, z)$, we obtain $z = x(s_2; s_1, w)$ and therefore, by the definition of η, $||z|| \leq \eta(r) - 1$, a contradiction. □

Remark 2.1 The requirement of uniqueness for the solutions to the Cauchy problems can be removed from Lemma 2.2 with the only expense of a longer proof (see [106, Th. 6.5] for the details). We also observe that Lemma 2.2 is still valid if we consider systems of the form $\dot{x} = f(t, x, \lambda)$ (with f continuous) and $\eta(r)$ can be chosen to be independent of λ for λ in a compact set.

Remark 2.2 A comparison between the methods of studying the properties of the translation operator, which is exposed here, and other approaches based on a functional—analytic setting, shows that a major drawback in the use of this method lies in the requirements of the uniqueness and global existence for the solutions of the Cauchy problems, which, in many cases, impose some regularity and growth restrictions on the vector fields. We point out, however, that in various applications, such a drawback is more apparent than real. Indeed, as remarked in [106, pp. 3–4], it is often possible to find a sequence of smooth vector fields $(F_n)_n$ such that $F_n \to f$, uniformly on compact subsets of $\mathbb{R} \times \mathbb{R}^m$ and such that for each n sufficiently large, the F_n's satisfy all the requirements needed to apply our theory and, moreover, $\dot{x} = F_n(t, x)$ has a T-periodic solution x_n with $x_n(t) \in \mathcal{K}$, $\forall t \in [0, T]$, $\mathcal{K}$ being a compact set which can be chosen independently of n. Then, a standard use of the Ascoli-Arzelà theorem allows to pass to the limit along a subsequence of $(x_n)_n$ which converges to a T-periodic solution of (2.5) with values in $\mathcal{K}$.

Example 2.1 Let $g : \mathbb{R} \to \mathbb{R}$ and $e : \mathbb{R}^3 \to \mathbb{R}$ be continuous functions and consider the perturbed Duffing equation

$$(\mathcal{D}) \qquad \ddot{u} + g(u) = e(t, u, \dot{u}).$$

Equation $(\mathcal{D})$ represents a simple model for the motion of a system with one degree of freedom subject to the internal force g (restoring field) and to the external forcing e. We denote by

$$G(x) := \int_0^x g(s)\,ds \tag{2.7}$$

the potential of the unperturbed system. Then we have:

Clearly, equation $(\mathcal{D})$ is the same as Eq. (2.1). We have used $u(t)$ in place of $x(t)$ because $x(t)$ represents a vector function as in Eq. (2.5).

Proposition 2.1 *Suppose there is a continuous function* $q : \mathbb{R} \to \mathbb{R}_+$ *such that* $|e(t, u, y)| \le q(t)$, $\forall\, t, u, y$ *and assume* $\inf G > -\infty$. *Then, any solution of* $(\mathcal{D})$ *can be globally defined in* $\mathbb{R}$.

Proof First of all, we write equation $(\mathcal{D})$ as the equivalent system in the phase-plane $(u, \dot{u})$,

$$\dot{u} = y, \quad \dot{y} = -g(u) + e(t, u, y).$$

Let $(u(t), y(t))$ be a noncontinuable solution of $(\mathcal{D})$ defined on a maximal interval of existence $]\alpha, \beta[$, containing the initial point t_0, and consider the auxiliary function $v(t) := 2G(u(t)) + L + 1 + y(t)^2$, where $L > 0$ is a constant such that $\inf 2G \ge -L$. Differentiating with respect to $t \in]\alpha, \beta[$, we have $|v'(t)| = |2g(u(t))y(t) + 2y(t)\big(-g(u(t)) + e(t, u(t), y(t))\big)| \le 2|y(t)|q(t) \le (1 + y(t)^2)q(t) \le q(t)v(t)$ and now, Gronwall inequality yields $|u'(t)| = |y(t)| \le v(t) \le v(t_0)\exp|\int_{t_0}^t q(s)\,ds| := q_1(t)$. Hence, $|u(t)| \le |u(t_0)| + |\int_{t_0}^t q_1(s)\,ds| := q_2(t)$. Thus, in conclusion, $|u(t)| + |u'(t)| \le q_3(t) := q_1(t) + q_2(t)$, with $q_3 : \mathbb{R} \to \mathbb{R}_+$ a continuous function. The last inequality clearly prevents the possibility that $\lim_{t\to\alpha^+}(|u(t)| + |u'(t)|) = +\infty$ with $\alpha > -\infty$, or that $\lim_{t\to\beta^-}(|u(t)| + |u'(t)|) = +\infty$ with $\beta < +\infty$. Then, by the fundamental theory of ODEs, we obtain $]\alpha, \beta[= (-\infty, +\infty)$. □

The conclusion is still true if $e(t, u, y)$ is a Carathéodory function such that $|e(t, u, y)| \le q(t)$, for a.e. t and all u, y, with $q \in L^1_{loc}(\mathbb{R})$. An improved version of this result is given in [165] where it is assumed that $\inf_{\mathbb{R}}(G(x) + Kx^2) > -\infty$, for some $K \ge 0$. Periodicity solutions for equation $(\mathcal{D})$ are considered in [31, 33, 53, 62, 64, 75, 90, 130, 139, 165] and their references.

Example 2.2 The case of the Duffing Equation with singularities [75].

We consider the differential equation

$$\ddot{u} + g(u) = e(t) \tag{2.8}$$

where $g : (A, B) \to \mathbb{R}$ is a continuous function and $e : \mathbb{R} \to \mathbb{R}$ is $T-$periodic $(T > 0)$, with $e \in L^1$, A and B are such that

$$-\infty \le A \le B \le +\infty$$

and we deal with a function $g(u)$ which has superlinear growth at the boundary of its domain as explained in conditions (i_1) and (i_2) below.

Choose a point $c \in (A, B)$. We confine our attention to the open strip

$$S := (A, B) \times \mathbb{R}$$

Let us consider the equivalent system

$$\dot{u} = y + E(t), \qquad \dot{y} = -g(u) + \bar{e} \tag{2.9}$$

where $E(t) := \displaystyle\int_0^t (e(\xi) - \bar{e})d\xi$ and $\bar{e} := (1/T)\displaystyle\int_0^T e(t)dt$. Thus $E : \mathbb{R} \to \mathbb{R}$ is continuous and T−periodic with mean value 0. Moreover as previously observed, we can suppose $\bar{e} = 0$. As before $G(x) := \int_c^x g(s)ds$.

We define some conditions

(i_1) $$\lim_{x \to A^+} G(x) = \lim_{x \to B^-} G(x) = +\infty$$

(i_2) $$\lim_{x \to A^+} \frac{g(x)}{(x - c)} = \lim_{x \to B^-} \frac{g(x)}{(x - c)} = +\infty.$$

We observe that: (i_1) is always satisfied at $A = -\infty$ or at $B = +\infty$ when (i_2) is assumed. Also (i_2) is independent of the choice of the point $c \in (A, B)$. Some terminology is introduced: For $f : \mathcal{O} \to S, \ \alpha \mapsto f(\alpha)$, with $\alpha \in \mathcal{O}$, where $(\mathcal{O}, \prec)$ is a directed set, and we write

$$f(\alpha) \to \partial S$$

if for every compact set $\mathcal{K} \subset S$, there is an $\alpha_{\mathcal{K}} \in \mathcal{O}$ such that $f(\alpha) \notin \mathcal{K}$, for all $\alpha \in \mathcal{O}$ with $\alpha \succ \alpha_{\mathcal{K}}$.

As in the previous example, all the solutions are globally defined. The next proposition follows from Lemma 2.2, see also [75, Proposition 3.2]

Proposition 2.2 *Assume (i_1) and (i_2). Then for every compact set $\mathcal{K} \subset S$ and each $m \in \mathbb{N}$, there is a compact set $\mathcal{B} = \mathcal{B}(\mathcal{K}, m) \subset S$ with $\mathcal{K} \subset \mathcal{B}$ such that for each solution $z = (u, y)$ of the system*

$$\dot{u} = y + E(t), \quad \dot{y} = -g(u), \tag{2.10}$$

the following inference is valid:

$$z(0) \notin \mathcal{B} \implies z(t) \notin \mathcal{K} \ \forall t \in [-mT, mT].$$

In particular, for each $m \in \mathbb{N}$, there is a compact set $\mathcal{R}_m$ with $\mathcal{M} \subset \mathcal{R}_m \subset \mathcal{S}$ such that

$$z(0) \notin \mathcal{R}_m \implies z(t) \notin \mathcal{M} \ \forall t \in [-mT, mT] \tag{2.11}$$

holds for any solution of (2.10).

For detailed proof of this result see [75, Proposition 3.2].

A typical fixed point result in $\mathbb{R}^m$ can be seen in the following theorem, usually referred as the **Poincaré-Bohl theorem.** Since we apply the result in this section we state now the theorem without proof (For proof see Sect. 2.6).

Theorem 2.1 *Let $\Omega \subset \mathbb{R}^m$ be an open bounded set with $0 \in \Omega$, such that*

$$\phi(z) \neq \mu z, \quad \forall z \in \partial\Omega, \ \forall \mu > 1, \tag{2.12}$$

where $\phi : \overline{\Omega} \to \mathbb{R}^m$ is a continuous function. Then ϕ has at least one fixed point in $\overline{\Omega}$.

It is also easy to see that, instead of $0 \in \Omega$ and (2.12) we can suppose that there is a point $z_0 \in \Omega$ such that $\phi(z) \neq z_0 + \mu(z - z_0)$, $\forall z \in \partial\Omega$, $\forall \mu > 1$. An obvious consequence of Theorem 2.1, which generalizes the Brouwer fixed point theorem, is the following variant of Rothe's theorem (see [46, Ex. 5, p. 33]) which is also called Birkhoff-Kellogg theorem [17].

Corollary 2.1 *Let $\Omega \subset \mathbb{R}^m$ be an open bounded set such that $\overline{\Omega}$ is strictly star shaped and let $\phi \in C(\overline{\Omega})$ be such that $\phi(\partial\Omega) \subset \overline{\Omega}$. Then ϕ has at least one fixed point in $\overline{\Omega}$.*

(The proof of the corollary is given in Sect. 2.6). The recent book of Dinca and Mawhin [49] gives a comprehensive treatment of the Brouwer fixed point theorem and all the related results like Theorem 2.1 and Corollary 2.1, within a precise historical context. The interested reader would find in [49] the detailed treatment of this topic.

In 1934 a fundamental paper of Leray and Schauder [111] founded the theory of the topological degree for compact perturbations of the identity in normed spaces. We refer to [46, 49, 107, 120, 131] for a general treatment of the Leray-Schauder theory, its extension to more general classes of operator equations and applications to various boundary value problems, including the periodic one. With this respect, we particularly recommend the lecture notes [135] of Mawhin for the Montreal conference devoted to the continuation of theorems with various interesting results not covered in the present section.

2.3 Application of Brouwer—Type Fixed Point Theorems

Definition 2.1 [18, 38, 98, 147, 193] Let Q be a domain in real Euclidean space $\mathbb{R}^n$, $X(t, x)$ a function on $\mathbb{R} \times Q \to \mathbb{R}^n$ and M a closed subset of Q. We shall be concerned with trajectories of the vector field X, that is, with solutions of

$$\dot{x} = X(t, x), \quad x(t) \in Q.$$

The set M is said to be *flow invariant* or *positively invariant* for X if every trajectory $x(t)$ which meets M at a time t_0 must remain in M for $t > t_0$. Thus, in the case of flow invariance,

$$x(t_0) \in M \implies \text{the solution } x(t) \in M \text{ for } t_0 \leq t < t_1,$$

where $[t_0, t_1)$ is the maximal right interval of existence for the trajectory through the point $x(t_0)$.

In mathematics, the Bony-Brezis theorem, [20, 26, 166] due to the French mathematicians Jean-Michel Bony and Haim Brezis, gives necessary and sufficient conditions for a closed subset of a manifold to be invariant under the flow defined by a vector field, namely, at each point of the closed set M, the vector field must have a non-positive inner product with any exterior vector normal to the set, whenever such an exterior normal vector exists.We may recall that a vector $\nu \neq 0$ is an exterior normal vector at a point u to the closed set M if $u \in M$ is a point of M with minimal distance from $u + \nu$. Such a theorem of flow invariance is also referred to as the flow invariance Theorem of Nagumo [147] (see also [38, 98, 193]).

We start with a simple application of Corollary 2.1 for the solvability of (2.6).

Theorem 2.2 *Let $V : \mathbb{R}^m \to \mathbb{R}$ be a continuously differentiable function and assume that for some $p \in \mathbb{R}^m$ and $c > V(p)$, we have that $V^{-1}(c)$ is a bounded set containing at least two points, with*

$$(V_1) \qquad (\nabla V(z)|z - p) > 0, \quad \forall z \in V^{-1}(c),$$

$$(V_2) \qquad (\nabla V(z)|f(t, z)) \leq 0, \quad \forall z \in V^{-1}(c), \ \forall t \in [0, T].$$

Then problem (2.6) *has at least one solution $\tilde{x}$ with $\tilde{x}(t) \in [V \leq c]$, $\forall t \in \mathbb{R}$.*

Proof Let us set $\Omega := [V < c]$. Note that c is a regular value for V and Ω is an open set with $p \in \Omega$ and $\partial\Omega = V^{-1}(c)$, $\overline{\Omega} = [V \leq c]$.

We claim that Ω is bounded and strictly star shaped with respect to the point p. Indeed, let $z \in \overline{\Omega} \setminus \{p\}$ and consider the function $v(s) := V(p + s(z - p))$, for $s \in \mathbb{R}_+$. Since $v(0) < c$, $v(1) \leq c$ and, by (V_1), $v'(s) > 0$ for each $s > 0$ (if any) such that $v(s) = c$, it follows that $v(s) < c$ for all $s \in [0, 1)$ and hence we have that the segment $\{p + s(z - p) : 0 \leq s < 1\}$ is contained in Ω for each $z \in \overline{\Omega}$. This proves that $\overline{\Omega}$ is strictly star shaped with respect to p. With a similar argument we also see that when $z \in \partial\Omega$, then $p + s(z - p) \notin \overline{\Omega}$ for $s > 1$. Consider now the projection $\Pi : V^{-1}(c) \to \partial B(1)$, $\Pi(z) = (z - p)/||z - p||$. If we prove that Π is onto, then we can conclude that Ω is bounded. So, by contradiction, if Π is not onto, then we can find $\bar{z} \in V^{-1}(c)$ and v with $||v|| = 1$, such that $V(p + sv) < c$ for all $s \in \mathbb{R}^+$ and $\Pi(\bar{z}) \neq -v$ (use the fact that Π is one-to-one on $V^{-1}(c)$, which has at least two points). Let $M > 0$ be such that $V^{-1}(c) \subset p + B(M)$ and take $R > 1$ such that $||Q(\lambda)|| > M$ for each $\lambda \in [0, 1]$, where $Q(\lambda) := (1 - \lambda)R(\bar{z} - p) + \lambda R v$. We have $V(p + Q(0)) = V(p + R(\bar{z} - p)) > c$ and $V(p + Q(1)) = V(p + Rv) < c$. Hence, there is $w = Q(\lambda^*)$, for some $\lambda^* \in]0, 1[$, such that $V(p + w) = c$. Therefore, $z = p + w \in \partial\Omega$ with $||z - p|| = ||w|| > M$. This contradicts the boundedness of $V^{-1}(c)$ which provided the definition of M. In this manner, our claim about the shape of Ω is proved. Note that if $m \geq 2$ and $V^{-1}(c) \neq \emptyset$, then $V^{-1}(c)$ has always infinitely many points.

Using the Stone-Weierstrass approximation theorem, for any $n \in \mathbb{N}$, there is a sequence of C^1-functions $(g_{n,k})_k$ such that $\lim_{k\to+\infty} g_{n,k}(t, x) = f(t, x) - n^{-1}\nabla V(x)$, uniformly on compact subsets of $\mathbb{R} \times \mathbb{R}^m$. Then, by (V_2), for every n, we can find an index k_n such that $(\nabla V(z)|g_{n,k}(t, z)) < 0$, $\forall z \in V^{-1}(c)$, $\forall t \in [0, T]$, holds for $k \geq n_k$. Now, using standard results about flow - invariance (see, e.g. [66, Cor. 2]), we have that the set $\overline{\Omega} = [V \leq c]$ is positively invariant for the solutions of the differential systems

$$\dot{x} = g_{n,k}(t, x), \quad k \geq k_n. \tag{2.13}$$

Note that the smoothness of $g_{n,k}$ ensures the uniqueness of the solutions for the Cauchy problems associated to (2.13) whilst the compactness of the positively invariant set $\overline{\Omega}$ implies the global existence in the future for the solutions with initial values in $\overline{\Omega}$. Hence, if we denote by $\Phi_{n,k} : \overline{\Omega} \to \mathbb{R}^m$ the Poincaré operator associated to (2.13) we see that Corollary 2.1 applies to $\Phi_{n,k}$ and thus we find a solution $x_{n,k}$ to (2.13) satisfying $x_{n,k}(0) = x_{n,k}(T)$ and such that (by the positive invariance of $[V \leq c]$) $x_{n,k}(t) \in \overline{\Omega}$, $\forall t \in [0, T]$. Finally, we let $n \to +\infty$ and, by a diagonal argument, we can find a cross sequence $g_n := g_{n,j_n}$, with $j_n \geq k_n$ such that $g_n \to f$ uniformly on $[0, T] \times \overline{\Omega}$. A standard reasoning based on Ascoli-Arzelà theorem (see Remark 2.2) yields the conclusion, showing that a subsequence of the sequence of solutions $(x_n)_n := (x_{n,j_n})_n$, converges uniformly to a solution of (2.6). □

Remark 2.3 We observe that the inequality in (V_2) can be reversed as in

$$(V_3) \qquad (\nabla V(z)|f(t, z)) \geq 0, \quad \forall z \in V^{-1}(c), \ \forall t \in [0, T].$$

Indeed, (V_3) implies $(\nabla V(z)| - f(T - t, z)) \leq 0$, $\forall z \in V^{-1}(c)$, $\forall t \in [0, T]$, and thus we can apply Theorem 2.2 to obtain a T-periodic solution $\hat{x}$ for system

$$\dot{x} = -f(T - t, x),$$

with $\hat{x}(t) \in [V \leq c]$, $\forall t$. Clearly $\tilde{x}(t) := \hat{x}(T - t)$ is now a solution of (2.6) and so, our claim is proved.

Theorem 2.2 is a prototype of a wide collection of analogous results based on the search of fixed points for the Poincaré operator, by looking for a compact positively invariant set $\mathcal{M} \subset \mathbb{R}^m$ with $\mathcal{M}$ having the *fixed point property (i.e. any continuous* *self-map of $\mathcal{M}$ has a fixed point). In the above proof, we found $\mathcal{M} = \overline{\Omega}$. Note also that we could replace (V_1) by the assumption that c is a regular value and $[V \leq c]$ is a compact set with the fixed point property. Variants of Theorem 2.2 are available in case of $\mathcal{M}$ a compact convex set (see* [35, 66, 131, 188]*), (V_2) being replaced by similar inequalities concerning the scalar product between f and the outer normals to the convex set at its boundary, or, more generally, for sets with possibly nonsmooth boundary, invoking for the flow invariance, the theorem of Nagumo* [147] *or some variants of it.*

Example 2.3 Consider the Liénard equation

$$(\mathcal{L}) \qquad \ddot{u} + \psi(u)\dot{u} + g(u) = e(t),$$

where $\psi, g : \mathbb{R} \to \mathbb{R}$ are continuous functions and $e \in L^1_{\mathrm{loc}}(\mathbb{R})$, with $e(t+T) = e(t)$, for a.e. $t \in \mathbb{R}$. Solutions of $(\mathcal{L})$ will be intended in the generalized (Carathéodory) sense.

Taking the mean value on $[0, T]$ of both sides of $(\mathcal{L})$ we immediately see that a *necessary condition* for the existence of a T-periodic solution to $(\mathcal{L})$ is that $\bar{e} := T^{-1}\int_0^T e(t)\,dt \in \mathcal{R}(g)$. On the other hand, without loss of generality, possibly subtracting $\bar{e}$ to both sides of $(\mathcal{L})$ and renaming g, we can assume

$$\bar{e} = 0. \tag{2.14}$$

Then, setting $E(t) := \int_0^t e(s)\,ds$ ($E(\cdot)$ is now continuous and T-periodic) and $\Psi(x) := \int_0^x \psi(s)\,ds$, we can write $(\mathcal{L})$ as the equivalent system in the Liénard plane

$$\dot{u} = y - \Psi(u) + E(t) + k, \quad \dot{y} = -g(u), \tag{2.15}$$

where $k \in \mathbb{R}$ is an arbitrary constant.

Proposition 2.3 *Suppose that $\bar{e} = 0$ and*

$$(g_0) \qquad \liminf_{|x|\to+\infty} g(x)\mathrm{sgn}(x) > 0,$$

$$(\Psi_0) \qquad \big(\liminf_{x\to+\infty} \Psi(x)\big) - \big(\limsup_{x\to-\infty} \Psi(x)\big) > \mathrm{osc}(E) := (\max E) - (\min E).$$

Then equation $(\mathcal{L})$ has at least one T-periodic solution.

Proof First of all, we observe that it is sufficient to prove our result by assuming, in place of (Ψ_0), that

$$\exists\, K, d : \quad \Psi(x)\mathrm{sgn}(x) \geq K > |E|_\infty, \quad \text{for } |x| \geq d. \tag{2.16}$$

Indeed, splitting the free constant k in Eq. (2.15) as $k = k_1 - k_2$, we can choose k_2 such that $|E(\cdot) - k_2|_\infty = \frac{1}{2}\text{osc}(E)$ and k_1 such that

$$[k_1 - k_2, k_1 + k_2] \subset]\limsup_{x\to-\infty} \Psi(x), \liminf_{x\to+\infty} \Psi(x)[.$$

Then, setting $\Psi_1(x) := \Psi(x) - k_1$, $E_1(t) := E(t) - k_2$, so that (2.15) takes the form $\dot{u} = y - \Psi_1(u) + E_1(t)$, $\dot{y} = -g(u)$, with $\frac{1}{2}\text{osc}(E) = |E_1|_\infty$ and (2.16) satisfied by Ψ_1 and E_1 in place of Ψ and E. Thus, without restriction, we can assume (2.16) and take $k = 0$ in (2.15) just renaming Ψ and E. From (g_0) we can also assume that $g(x)\text{sgn}(x) \geq \varepsilon > 0$, for $|x| \geq d$ and observe that $\lim_{|x|\to+\infty} G(x) = +\infty$, recalling the definition of the potential G given in (2.7).

Now, following [109, p. 119], we define a $\mathcal{C}^1$-function $h : \mathbb{R} \to \mathbb{R}$ such that $h'(s) > 0$, $\forall\, |s| < 2d$ and $h(s)\text{sgn}(s) = \delta < K - |E|_\infty$, $\forall\, |s| \geq 2d$, with $\delta > 0$. Clearly, $h'(s) = 0$ for $|s| \geq 2d$. Then we define, for $z = (u, y) \in \mathbb{R}^2$,

$$f(t, z) := \big(y - \Psi(u) + E(t), -g(u)\big), \quad V(z) = V(u, y) = \frac{1}{2}(y - h(u))^2 + G(u).$$

By standard computations, we find that $(\nabla V(z)|z) \geq y^2 - \delta|y| + g(u)u - uh'(u)(y - h(u))$ and $(\nabla V(z)|f(t, z)) \leq -\big(g(u)\text{sgn}(u)\big)\big(\Psi(u)\text{sgn}(u) - \delta - |E|_\infty\big) - h'(u)(y - \Psi(u) + E(t))(y - h(u))$. Using (2.16) (g_0) and recalling the definition of $h(s)$, it is routine to verify that $(\nabla V(z)|z) > 0$ and $(\nabla V(z)|f(t, z)) < 0$ for z outside a compact subset of $\mathbb{R}^2$. Then, observing that $V(z) \to +\infty$ as $||z|| \to +\infty$, it is easy to see that (V_1) (with $p = 0$) and (V_2) of Theorem 2.2 are satisfied provided that we choose a sufficiently large constant c. □

Remark 2.4 The same argument of the above proof can be used to show that assumption (g_0) can be relaxed to

$$(G_0) \qquad \exists d > 0 : \quad g(x)\text{sgn}(x) > 0, \text{ for } |x| \geq d, \quad \lim_{|x|\to+\infty} G(x) = +\infty.$$

In this case, even if assumption (V_1) of Theorem 2.2 is not necessarily satisfied, nevertheless, we can achieve the thesis via the observation that the set $V^{-1}(c)$ is a simple closed curve for c large enough and hence the Schoenflies theorem guarantees the fact that $[V \leq c]$ is a compact closed set homeomorphic to the unit disc B. Therefore $[V \leq c]$ for c large has the fixed point property and the result easily follows from the flow invariance of $[V \leq c]$.

Variants of Proposition 2.3 were obtained in the fifties by Reuter [168], Mizohata and Yamaguti [192] (see also [167]) and subsequently refined by Graef in [93]. In [104] and [154], some extensions have been obtained for Liénard differential systems as well. In particular, we observe that (via a different proof, based on a priori bounds in function spaces and Leray-Schauder degree as in [154]), it is possible to prove the existence of T-periodic solutions to $(\mathcal{L})$ in case (2.14) by relaxing (Ψ_0) and (g_0) or (G_0) to the slightly more general conditions

$$\exists d > 0 : \quad g(x)\mathrm{sgn}(x) \geq 0, \text{ for } |x| \geq d,$$

$$\Psi(x) \leq A, \text{ for } x \leq -d, \quad \Psi(x) \geq B, \text{ for } x \geq d, \quad \text{with } B - A \geq \mathrm{osc}(E).$$

The most classical results on the periodically forced Liénard equation can be found in [167, 170]. These books cover comprehensively the volume of work on this topic up to the 60s. More important progress in this area was achieved in the early 70s with the works of Graef [93] and Mawhin [126, 129]. For more recent contributions see [138, 152, 153, 157–159, 186, 187] and the references therein.

Example 2.4 Another example of application of Theorem 2.2 may be given for the scalar equations. Namely, we have the following classical result which, otherwise, is usually also proved via the method of lower and upper solutions.

Proposition 2.4 *Suppose* $m = 1$ *and let* $A, B \in \mathbb{R}$ *such that*

$$f(t, A) \geq 0 \geq f(t, B), \quad \forall t \in [0, T]. \tag{2.17}$$

Then (2.6) *has at least one solution* $\tilde{x}$ *with* $\min\{A, B\} \leq \tilde{x}(t) \leq \max\{A, B\}, \forall t \in \mathbb{R}$.

Proof Clearly the result is trivial if $A = B$. So, assume $A < B$ and let $p := \frac{A+B}{2}$. Define $V(x) := (x - p)^2$ and take $c = (\frac{B-A}{2})^2$. It is immediately seen that $V^{-1}(c) = \{A, B\}$ and (2.17) is equivalent to (V_2) of Theorem 2.2. Then the result easily follows. In case that $A > B$, we just apply Remark 2.3 since now (V_3) holds. □

If f splits as $f(t, x) = -g(x) + e(t)$, so that (2.5) writes as

$$\dot{x} + g(x) = e(t), \tag{2.18}$$

with $[g, e] : \mathbb{R} \to \mathbb{R}$ continuous and $e(t + T) = e(t)$, for all t, then Proposition 2.4 applies provided that $\mathcal{R}(g) \supset \mathcal{R}(e)$. More precisely, given any pair of points A, B such that $[g(A), g(B)] \supset \mathcal{R}(e)$, we have a T-periodic solution of (2.18) "between" A and B. If, furthermore, $\bar{e} = 0$, then we can write (2.18) in the equivalent form $\dot{u} + g(u + E(t) - k) = 0$, via the change of variable $x = u + E(t) - k$, with $E(t) = \int_0^t e(s)\, ds$ and $k \in \mathbb{R}$ an arbitrary but fixed constant. In this case, applying Proposition 2.4 to the latter equation, we obtain the solvability of the T-periodic problem to (2.18) provided that we find two intervals J_1, J_2 of lengths greater than or equal to $\mathrm{osc}(E)$, such that

$$g(x) \leq 0, \ \forall x \in J_1 \quad \text{and} \quad g(x) \geq 0, \ \forall x \in J_2.$$

Observe that since in this situation, we consider an equation in which $E(t)$ appears, then $e \in L^1_{\mathrm{loc}}$ is sufficient. Finally, we note that by combining both approaches some "mixing" hypotheses can be introduced too, as, for instance if $\bar{e} = 0$ and

$$g(A) \leq \inf e(t) \leq 0 \leq g(x), \ \forall x \in [c, d], \quad d - c \geq \mathrm{osc}(E),$$

or analogous conditions that can be obtained reversing all the inequalities. The location of the solutions can be established from the informations contained in the assumptions.

We conclude this section with an example showing, for a simple model, some advantages that the study of the Poincaré map may have with respect to other methods based on more abstract functional-analytic approaches, in view of obtaining more detailed information about the asymptotic behaviour of the solutions of (2.5).

Example 2.5 Suppose $m = 1$ and assume the forward uniqueness for the solutions of the Cauchy problems associated to (2.5) (a condition which is always satisfied if f is nonincreasing or locally Lipschitzian in the x-variable). Then we have

Proposition 2.5 *Let $\alpha, \beta : \mathbb{R} \to \mathbb{R}$ be continuous and T-periodic functions, with $\int_0^T \alpha > 0 > \int_0^T \beta$, such that*

$$f(t, x) \geq \alpha(t) \quad \textit{for } x \leq -d, \qquad f(t, x) \leq \beta(t) \quad \textit{for } x \geq d,$$

for all $t \in [0, T]$. Then (2.6) has a minimal solution x_* and a maximal solution x^*, such that for each T-periodic solution $\tilde{x}$ we have $x_*(t) \leq \tilde{x}(t) \leq x^*(t)$, $\forall t$. Moreover, for any $s \in \mathbb{R}$ and $z \in \mathbb{R}$, there is a solution $\tilde{x}$ of (2.6) such that $\lim_{t\to+\infty} |x(t; s, z) - \tilde{x}(t)| = 0$.

Proof By the assumption, it is easy to find a function $q(t)$ such that

$$\frac{d}{dt}(x(t)^2 + 1) = 2f(t, x(t))x(t) \leq 2q(t)x(t) \leq q(t)(x(t)^2 + 1).$$

Thus, standard results on differential inequalities ensure the global existence in the future of the solutions to the Cauchy problems, via the fundamental theory of ODEs. Let $K := \max\{|\alpha|_1, |\beta|_1\}$, and set $r := d + K + 1$. Consider the solution $x(t) = x(t; z)$ of (2.5) for $z \geq r$.

We claim that $x(T; z) < z$. To see this, we observe that the result is obvious if $x(T) \leq d$. So, assume that $x(T) > d$ and examine the following two possibilities: either $x(t) > d$, for each $t \in [0, T]$. In this case, $x'(t) \leq \beta(t)$ and therefore $x(T) - z \leq \int_0^T \beta(t)\,dt < 0$. Or, there is a last $t \in (0, T)$ such that $x(t) = d$. In this case, we have $x(T) - d = \int_t^T x'(s)\,ds \leq \int_0^T |\beta(s)|\,ds = |\beta|_1 \leq K$, and the claim is proved too. Similarly, one can see that $x(T; z) > z$ if $z \leq -r$. Then $\Phi([-r, r]) \subset (-r, r)$, and all the possible fixed points of Φ are contained in $(-r, r)$. By Corollary 2.1, it follows that (2.6) has at least one solution $\tilde{x}$. If $\tilde{x}(t) \geq d$, for each t, then $\tilde{x}'(t) \leq \beta(t)$, and hence $0 = \tilde{x}(T) - \tilde{x}(0) \leq \int_0^T \beta < 0$, which is impossible. Similarly, one can check that $\max \tilde{x} > -d$. By these estimates, arguing as above, it is possible to see that $-r < \min \tilde{x} \leq \max \tilde{x} < r$.

Define z_* and z^* to be, respectively, the infimum and the supremum of the z's that are initial points (at $t = 0$) of T-periodic solutions. Then, clearly, $-r < z_* \leq z^* < r$ and $x_* := x(\cdot; z_*)$ and $x^* := x(\cdot; z^*)$ are the minimal and maximal solutions of (2.6) we are looking for. By the forward uniqueness, it follows that $x_*(t) \leq \tilde{x}(t) \leq x^*(t)$, for any $t \in \mathbb{R}$ and any $\tilde{x}$ solution of (2.6).

Finally, let $(s, z) \in \mathbb{R}^2$ be arbitrary and consider the solution $x(\cdot; s, z)$. Arguing as in the proof of Massera theorem, [125, 194] we set $z_0 := x(jT; s, z)$ where $j \in \mathbb{N}$ satisfies $jT \geq s$ and $z_k := x(kT; z_0)$. Note that $x(\tau; z_0) = x(jT + \tau; s, z)$, for all $\tau \geq 0$, and therefore, $x(t; s, z) = x(t - jT; z_0)$, for all $t \geq jT$. The forward uniqueness of the solutions it follows that either $(z_k)_k$ is eventually constant, or it is strictly monotone and bounded. In the former case, $x(\cdot + kT; z_0) = x(\cdot; z_k) = \tilde{x}(\cdot)$ is a T-periodic solution of (2.5) for some (large) k. In the latter situation, $z_k \to \tilde{z}$ and $x(\cdot + kT; z_0) \to x(\cdot; \tilde{z}) = \tilde{x}(\cdot)$, uniformly in $[0, T]$, with $\tilde{x}(\cdot)$ a T-periodic solution of (1). Therefore, in any case $x(t; s, z) \to \tilde{x}(t - jT) = \tilde{x}(t)$, as $t \to +\infty$ and we conclude with our thesis (see the proof of Massera theorem [125] for the missing details). □

Proposition 2.5 is true also if f satisfies the Carathéodory conditions. In this case, the inequalities assumed in the above hypothesis read for a.e. $t \in [0, T]$. For related results, see [43, 133, 185, 196] as well as [44] and the references therein.

It is easy to see that (2.6) has a *unique* solution if $f(t, \cdot)$ is decreasing (strictly). Clearly, in this case, $x_*(\cdot) = x^*(\cdot)$. There are however situations in which the uniqueness of the T-periodic solution can be achieved under some less restrictive conditions. An example is the following which is motivated by a typical equation for the dynamics of a single species population.

Consider the Kolmogorov equation with periodic coefficients

$$(K) \qquad \dot{u} = uh(t, u),$$

where $h : \mathbb{R} \times \mathbb{R}_+ \to \mathbb{R}$ is continuous and T-periodic in the first variable.

Proposition 2.6 *Assume* $\int_0^T h(t, 0)\, dt > 0 > \int_0^T h(t, M)\, dt$ *for some* $M > 0$. *Suppose also that* $h(t, \cdot)$ *is monotone nonincreasing for all* $t \in [0, T]$ *and let* $\hat{t} \in [0, T]$ *be such that* $h(\hat{t}, \cdot)$ *is (strictly) decreasing. Then equation* (K) *has a unique positive* T*-periodic solution* $\tilde{u}$, *such that* $\lim_{t\to+\infty} \big(u(t) - \tilde{u}(t)\big) = 0$, *for any solution* $u(\cdot)$ *with* $u(t_0) > 0$.

Proof First of all, observe that the set $\mathbb{R}^+$ is positively invariant, so that $u(t) > 0$, for all $t \geq t_0$ if $u(t_0) > 0$. Hence, if we restrict ourselves to the search of only positive solutions, then we can write $u(t) := \exp x(t)$ and pass to the equivalent equation $\dot{x} = g(t, x) := h(t, \exp x)$. For this equation, we have forward uniqueness, as $g(t, \cdot)$ is nonincreasing. Moreover, it is straightforward to apply Proposition 2.5, so that we find the minimal and the maximal solutions x_* and x^*. To conclude the proof, it is sufficient to check that $x_*(t) = x^*(t)$, for all t. To this end, take the difference $y(t) := x^*(t) - x_*(t)$ and have $\dot{y}(t) = g(t, x^*(t)) - g(t, x_*(t)) \leq 0$, so that $y(t) = \text{constant } = k \geq 0$, as $y(\cdot)$ is T-periodic. Now, we have $h(t, \exp x_*(t)) = h(t, \exp x^*(t))$, for all $t \in [0, T]$ and the assumption on $\hat{t}$ implies $x_*(\hat{t}) = x^*(\hat{t})$ and thus $x_* = x^*$. □

A simple example to which Proposition 2.6 applies is the logistic equation $u' = u\big(a(t) - b(t)u\big)$, where $a, b : \mathbb{R} \to \mathbb{R}$ are continuous and T-periodic with mean value $\bar{a} > 0$ and $b(t) \geq 0$ for all t with $\bar{b} > 0$. See [6, 8, 37, 40, 51, 182, 184, 196, 199] for some "ecological" models where similar results are applied. Actually, in the special case of the logistic equation with a and b satisfying the above conditions, it is possible to obtain explicitly the positive solution $\tilde{u}$, by a direct integration on the linear equation $\dot{y} + a(t)y = b(t)$, where $y := 1/u$. Indeed, we find

$$\tilde{u}(t) = \frac{\exp \int_0^t a(s)\, ds}{\tilde{u}(0)^{-1} + \int_0^t \big(b(s) \exp \int_0^s a(\xi)\, d\xi\big)\, ds},$$

where $\tilde{u}(0) = \big((\exp \int_0^T a(t)\, dt) - 1\big) / \int_0^T \big(b(t) \exp \int_0^t a(s)\, ds\big)\, dt$.

Remark 2.5 *On the existence of bounded solutions.* In some examples, the methods employed for the search of periodic solutions (especially those based on the use of flow—invariant sets) can be adapted to ensure the existence of bounded solutions for system (2.5). With this respect, suppose that $f : \mathbb{R} \times \Omega \to \mathbb{R}^m$ is a continuous function with $\Omega \subset \mathbb{R}^m$ an open set and let $\emptyset \neq M \subset \Omega$ be a compact set such that for each $n \in \mathbb{N}$ there is at least one solution $x_n(\cdot)$ of (2.5) defined on $[-n, n]$ and such that $x_n(t) \in M$, for all $t \in [-n, n]$. Then, via a diagonal argument, it is possible to show that x_n has a subsequence converging uniformly on compact intervals to a solution $\tilde{x}(\cdot)$ of (2.5) with $\tilde{x}$ defined on $\mathbb{R}$ and such that $\tilde{x}(t) \in M$ for all $t \in \mathbb{R}$ (see also [1, proof of Lemma 1] and [106, Lemma 8.1]). In some cases, it will be possible to obtain $x_n(\cdot)$ as a solution of (2.5) satisfying the periodic boundary condition $x_n(-n) = x_n(n)$.

There has been a great deal of work in the area of the theory of existence and multiplicity of periodic solutions for first-order scalar differential equations. See [36, 150, 151, 183] and the references therein for recent research in this area.

2.4 Application of the Poincaré-Bohl Theorem

In this section, we confine ourselves to the consideration of *planar systems* of the form

$$\begin{cases} \dot{u} = & U(t, u, y) \\ \dot{y} = & Y(t, u, y), \end{cases} \tag{2.19}$$

so that, with respect to (2.5) we have $F(t, x) = \big(U(t, x), Y(t, x)\big)$, $x = (u, y)$. We assume the usual continuity and periodicity conditions on F, or, alternatively, only $F : [0, T] \times \mathbb{R}^2 \to \mathbb{R}^2$, continuous, when the periodicity of $F(\cdot, x)$ is not needed. In the sequel, we present a transformation which makes use of a system of modified polar-like coordinates and turns out to have some applications for the periodic problem associated to system (2.19). Namely, we estimate the number of the rotations around the origin for the solutions of (2.19) using a comparison with suitable quadratic forms (see also [11, 31, 61, 62, 73, 135] for related results).

Let $V : \mathbb{R}^2 \to \mathbb{R}$ be a continuous and positively homogeneous function of degree two, with $V \in \mathcal{C}^1(\mathbb{R}^2)$ and $V(z) > 0$ for $z \neq 0$. In the sequel we say that any function V satisfying all these conditions is *positive definite*. Note that $\lim_{||z|| \to +\infty} V(z) = +\infty$ and $\nabla V(z) \neq 0$ for $z \neq 0$ follow from the above assumptions. We can project any $z \neq 0$ onto the set $V^{-1}(1)$, so that

$$z = V^{1/2}(z)\nu(z), \quad \text{with } \nu(z) := z / V^{1/2}(z) \in V^{-1}(1).$$

Observe that $V^{-1}(1)$ is homeomorphic to the unit circle $S^1 := \{z \in \mathbb{R}^2 : ||z|| = 1\}$, and therefore we can define a covering projection $\Pi : \mathbb{R} \to V^{-1}(1)$ and represent points of $V^{-1}(1)$ using angular coordinates. A suitable choice of these coordinates can be given as follows.

Consider the planar autonomous conservative system in $\mathbb{R}^2$,

$$\begin{cases} \frac{d\alpha}{d\theta} &= -\frac{1}{2}\frac{\partial V}{\partial y}(\alpha, \beta) \\ \frac{d\beta}{d\theta} &= \frac{1}{2}\frac{\partial V}{\partial u}(\alpha, \beta), \end{cases} \tag{2.20}$$

and let $\Pi : \theta \to (\alpha(\theta), \beta(\theta))$ be the solution of (2.20) such that $\Pi(0) = V^{-1}(1) \cap (\mathbb{R}^+ \times \{0\})$. It is straightforward to see that $\Pi : \mathbb{R} \to V^{-1}(1)$ is continuously differentiable and periodic with a minimal period $\sigma > 0$.

By the above positions, we have that $(V^{1/2}, \theta)$ plays the role of polar coordinates in $\mathbb{R}^2 \setminus \{0\}$, namely, we can write $u = V^{1/2}(u, y)\alpha(\theta)$, $y = V^{1/2}(u, y)\beta(\theta)$, with α and β replacing the trigonometric functions "cos" and "sin", respectively, so that we have $\nu(u, y) = (\alpha(\theta), \beta(\theta))$. Observe that $-\beta(\theta)\alpha'(\theta) + \alpha(\theta)\beta'(\theta) = (\alpha/2)\frac{\partial V}{\partial u}(\alpha, \beta) + (\beta/2)\frac{\partial V}{\partial y}(\alpha, \beta) = \big(\nabla V(\alpha, \beta)|(\alpha, \beta)\big)/2 = V(\alpha, \beta) = 1$. Suppose now that $x(t) = (u(t), y(t))$ is a solution of (2.19) defined on $[0, T]$ and such that $x(t) \neq 0$, for all $t \in [0, T]$. Recall that, if global existence is assumed, by Lemma 2.2 we can find a constant $R_0 := \eta(0)$ such that $x(t) \neq 0$, $\forall t \in [0, T]$, whenever $||x(0)|| \geq R_0$. For $x(t)$ as above, we can write $u(t) = v(t)\alpha(\theta(t))$, $y(t) = v(t)\beta(\theta(t))$, with $v(t) = V^{1/2}(x(t)) = V^{1/2}(u(t), y(t))$. Note that $t \mapsto \theta(t)$ is a lifting of the map $t \mapsto \nu(u(t), y(t)) \in V^{-1}(1)$, to the covering space $\mathbb{R}$ of $V^{-1}(1)$. Differentiating with respect to $t \in [0, T]$, we have $\dot{u} = \dot{v}\alpha(\theta) + v\frac{d\alpha}{d\theta}(\theta)\dot{\theta}$, $\dot{y} = \dot{v}\beta(\theta) + v\frac{d\beta}{d\theta}(\theta)\dot{\theta}$, hence $\dot{u}(t)y(t) - \dot{y}(t)u(t) = v(t)^2\dot{\theta}(t)\big(\beta(\theta(t))\dot{\alpha}(\theta(t)) - \alpha(\theta(t))\dot{\beta}(\theta(t))\big) = -v(t)^2\dot{\theta}(t)$. Recalling that (u, y) is a solution of (2.19) we obtain, for any $\tau : 0 < \tau \leq T$,

$$\int_0^\tau \frac{U(t, u(t), y(t))y(t) - Y(t, u(t), y(t))u(t)}{V(u(t), y(t))}\, dt = \theta(0) - \theta(\tau). \tag{2.21}$$

Suppose also that for some $\tau \in]0, T]$ and $\mu > 0$,

$$x(\tau) = \mu x(0). \tag{2.22}$$

Then we have $x(\tau)/V^{1/2}(x(\tau)) = \mu x(0)/V^{1/2}(\mu x(0)) = x(0)/V^{1/2}(x(0))$, as $V^{1/2}$ is positively homogeneous of degree one. Hence,

$$\big(\alpha(\theta(\tau)), \beta(\theta(\tau))\big) = \big(\alpha(\theta(0)), \beta(\theta(0))\big)$$

and therefore,

$$\exists\, n \in \mathbb{Z}: \quad \theta(\tau) - \theta(0) = n\sigma.$$

Coming back to (2.21) we can now write

$$\int_0^{\tau} (U(t,u,y)y - Y(t,u,y)u)/V(u,y)\,dt = -n\sigma.$$

Green's theorem implies that

$$\begin{aligned}
\sigma = \int_0^{\sigma} 1\,d\theta &= \int_0^{\sigma} (\alpha(\theta)\dot{\beta}(\theta) - \beta(\theta)\dot{\alpha}(\theta))\,d\theta \\
&= \oint_{\partial[V\le 1]^+} - x_2 dx_1 + x_1 dx_2 \\
&= \int\int_{[V\le 1]} 2\,dx_1 dx_2 = 2A(V,1)
\end{aligned}$$

where we have set $A(V,1) := \text{Area}\,([V \le 1])$. Thus, we have proved the following formula:

$$\frac{1}{2A(V,1)} \int_0^{\tau} \frac{U(t,u(t),y(t))y(t) - Y(t,u(t),y(t))u(t)}{V(u(t),y(t))}\,dt = -n \in \mathbb{Z}. \tag{2.23}$$

It can be of interest to remark that the area $A(V,1)$ can be computed using polar coordinates as follows. For $V(x_1,x_2) = 1$, set $x_1 = \rho\cos\xi$, $x_2 = \rho\sin\xi$, so that $\rho^2 = 1/V(\cos\xi,\sin\xi)$. Then from $A(V,1) = \frac{1}{2}\int_0^{2\pi} \rho^2(\xi)\,d\xi$, we have

$$\sigma = 2A(V,1) = \int_0^{2\pi} \frac{1}{V(\cos\xi,\sin\xi)}\,d\xi. \tag{2.24}$$

Conversely, assume that (2.23) holds for some $\tau \in]0,T]$. By (2.21) and the first identity in (2.24) we find $n\sigma = \theta(0) - \theta(\tau)$ for some $n \in \mathbb{Z}$. Hence, by the σ-periodicity of the functions α and β, we have

$$\begin{aligned}
x(\tau) &= \nu(x(\tau))V^{1/2}(x(\tau)) \\
&= (\alpha(\theta(\tau)), \beta(\theta(\tau)))V^{1/2}(x(\tau)) \\
&= x(0)V^{1/2}(x(\tau))V^{-1/2}(x(0)),
\end{aligned}$$

that is, (2.22) is satisfied for some $\mu > 0$.

At last, suppose that $W : \mathbb{R}^2 \to \mathbb{R}$ is another *positive definite* function, i.e. W is continuous and positively homogeneous of degree two, with $W \in \mathcal{C}^1(\mathbb{R}^2)$ and $W(z) > 0$ for $z \neq 0$. We claim that under the previous assumptions,

$$\frac{1}{2A(W,1)} \int_0^\tau \frac{U(t,u(t),y(t))y(t) - Y(t,u(t),y(t))u(t)}{W(u(t),y(t))}\,dt$$
$$= \frac{1}{2A(V,1)} \int_0^\tau \frac{U(t,u(t),y(t))y(t) - Y(t,u(t),y(t))u(t)}{V(u(t),y(t))}\,dt.$$

Indeed, it is sufficient to observe that the map

$$\lambda \to \frac{1}{2A(V_\lambda,1)} \int_0^\tau \frac{U(t,u(t),y(t))y(t) - Y(t,u(t),y(t))u(t)}{V_\lambda(u(t),y(t))}\,dt,$$

with $V_\lambda(x) := (1-\lambda)V(x) + \lambda W(x)$, for $\lambda \in [0,1]$ is continuous and therefore it must be constant since it takes only integer values.

Thus we are led to define a rotation number as follows. For $x(t) = (u(t), y(t))$ such that $x(t) \neq 0$, $\forall t \in [0,T]$ and V as above, we set, for any $s \in [0,T]$,

$$\operatorname{rot}_V(s; x(\cdot), F) := \frac{1}{2A(V,1)} \int_0^s \frac{U(t,u,y)y - Y(t,u,y)u}{V(u,y)}\,dt. \tag{2.25}$$

We'll omit to indicate the vector field F in the definition of the rotation, if no misunderstanding is possible. When $x(\cdot) = x(\cdot; z)$, uniqueness of the solutions being assumed, we'll also set $\operatorname{rot}_V(s; z) := \operatorname{rot}_V(s; x(\cdot; z))$. Then the following result is proved.

Lemma 2.3 *If $x(t) \neq 0$, for all $t \in [0,T]$, then $x(\tau) = \mu x(0)$, for some $\tau \in]0,T]$ and $\mu > 0$, if and only if there is $k \in \mathbb{Z}$ such that $\operatorname{rot}_V(\tau; x) = k$, and the number k is independent upon the particular positive definite function V which has been chosen in formula* (2.25).

Remark 2.6 Formulas (2.23) and (2.24) were previously obtained with a different approach in [31, 135] to estimate rotation numbers of periodic solutions. The choice $V(u,y) = u^2 + y^2$ yields the usual formula for the rotation in standard polar coordinates.

The case $V(u,y) = u^2 + (y/\kappa)^2$, for $\kappa > 0$, gives

$$\operatorname{rot}_\kappa(s; x(\cdot)) = \frac{\kappa}{2\pi} \int_0^s \frac{U(t,u(t),y(t))y(t) - Y(t,u(t),y(t))u(t)}{\kappa^2 u(t)^2 + y(t)^2}\,dt.$$

In this situation, Lemma 2.3 ensures that $\operatorname{rot}_\kappa(\tau; x(\cdot))$ is constant with respect to $\kappa > 0$ when $x(\tau) = \mu x(0)$, for some $\mu > 0$ (a fact already observed by Fabry in [61]) Applications of this property to various aspects of problem (2.6) for planar systems can be found in [62, 64]. Consider now the case when $V(u, y) = \Phi(u)^2 + y^2$, where $\Phi(u) = k_1 u^+ - k_2 u^-$, with $u^+ = \max\{u, 0\}$, $u^- = \max\{-u, 0\}$ and k_1, k_2 positive constants. In this situation, we find

$$
\begin{aligned}
\mathrm{rot}_V(s; x(\cdot)) &= \mathrm{rot}_{(k_1,k_2)}(s; x(\cdot)) \\
&= \frac{1}{\pi k_1^{-1} + \pi k_2^{-1}} \int_0^s \frac{U(t, u(t), y(t))y(t) - Y(t, u(t), y(t))u(t)}{\Phi(u(t))^2 + y(t)^2} dt \\
&= \frac{k_1 k_2}{\pi(k_1 + k_2)} \Big(\int_{\{t:u\geq 0\}} (\ldots)\, dt + \int_{\{t:u<0\}} (\ldots)\, dt \Big) \\
&= \frac{k_2}{\pi(k_1 + k_2)} \int_{\{t:u\geq 0\}} \frac{U\Phi'(u)y - Y\Phi(u)}{\Phi(u)^2 + y^2} dt + \\
&\quad + \frac{k_1}{\pi(k_1 + k_2)} \int_{\{t:u<0\}} \frac{U\Phi'(u)y - Y\Phi(u)}{\Phi(u)^2 + y^2} dt,
\end{aligned}
$$

where Φ' stands for the right derivative of Φ. Again, Lemma 2.3 ensures that $\mathrm{rot}_{(k_1,k_2)}(\tau; x(\cdot))$ is constant with respect to the choice of the pair $(k_1, k_2) \in (\mathbb{R}^+)^2$, when $x(\tau) = \mu x(0)$ with $\mu > 0$. These properties of $\mathrm{rot}_{(k_1,k_2)}$ have been used in [165] with an independent proof.

From Lemma 2.3 it is straightforward to prove the next theorem, assuming global existence in the future (or just in $[0, T]$) for the solutions of (2.17)

Theorem 2.3 *Suppose there is an open bounded set $\Omega \subset \mathbb{R}^2$, with $0 \in \Omega$, such that for any solution $x(t) = (u(t), y(t))$ of* (2.19) *with $x(0) \in \partial\Omega$, it follows that $x(t) \neq 0$, for all $t \in [0, T]$ and there is a positive definite function V such that* $\mathrm{rot}_V(T; x(\cdot)) \notin \mathbb{Z}$. *Then Eq.* (2.19) *has at least one T-periodic solution $\tilde{x}$ such that $\tilde{x}(0) \in \overline{\Omega}$.*

Proof All the solutions we consider in this proof are assumed to be defined on $[0, T]$. Let $\varepsilon > 0$ be such that $2\varepsilon \leq ||x(t)||$ for all $t \in [0, T]$ and each $x = (u, y)$ solution of (2.19) with $x(0) \in \partial\Omega$ and also $||x(t)|| \leq 1/2\varepsilon$ for all $t \in [0, T]$ and each x solution of (2.19) with $x(0) \in \overline{\Omega}$. (the existence of such an ε follows from a compactness argument). By Lemma 2.3 we can fix, once for all, a positive definite function W and observe that $\mathrm{rot}_W(T; x(\cdot), F) \notin \mathbb{Z}$ for any solution $x(\cdot)$ of (2.19) such that $x(0) \in \partial\Omega$.

Let $(f_n)_n = (U_n, Y_n)_n$ be a sequence of $\mathcal{C}^1$-vector fields such that $f_n \to F$ uniformly on compact sets. By a standard result [97, Th.3.2] we know that for n sufficiently large, say $n \geq n_1$, any solution $x(\cdot)$ to $\dot{x} = f_n(t, x)$, with $x(0) = z \in \overline{\Omega}$, is defined in $[0, T]$, uniquely determined by its initial point z and $\varepsilon \leq ||x(t)||$ for all $t \in [0, T]$ if $z \in \partial\Omega$, as well as $||x(t)|| \leq 1/\varepsilon$ for all $t \in [0, T]$ if $z \in \overline{\Omega}$. Moreover, since $(x, F) \mapsto \mathrm{rot}_W(T; x(\cdot), F)$ is continuous with respect to the compact-open topology, we can also assume that $\mathrm{rot}_W(T; z, f_n) \notin \mathbb{Z}$, for all $z \in \partial\Omega$ and $n \geq n_2 \geq n_1$. Hence, when $n \geq n_2$, Lemma 2.3 implies that $x(T; z) \neq \mu z, \forall z \in \partial\Omega, \forall \mu > 0$. Then, Theorem 2.1 applies and the existence of at least one T-periodic solution $\tilde{x}_n$ to $\dot{x} = f_n(t, x)$, with $\tilde{x}_n(0) \in \overline{\Omega}$ is guaranteed, for any $n \geq n_2$. Since $||\tilde{x}_n(t)|| \leq 1/\varepsilon$,

$\forall t \in [0, T]$, arguing like in Remark 2.2 and in the proof of Theorem 2.2, we obtain the conclusion. □

Example 2.6 Let $\psi : \mathbb{R} \to \mathbb{R}$ be an increasing homeomorphism with $\psi(0) = 0$ and let $g : \mathbb{R} \to \mathbb{R}$ and $e : \mathbb{R}^3 \to \mathbb{R}$ be continuous functions with e, T-periodic in the first variable and bounded on all its domains. Some physical examples (see [19]) suggest to consider equations of the form

$$(\mathcal{P}) \qquad \frac{d}{dt}(\psi(\dot{u})) + g(u) = e(t, u, \dot{u}).$$

By a solution $u(\cdot)$ to $(\mathcal{P})$ we mean a continuously differentiable function with $\psi(u')$ differentiable and satisfying $(\mathcal{P})$ for all $t \in \mathbb{R}$ (as in [92], weaker formulations are also possible). To be more specific, we assume linear growth for ψ and g, that is $0 < \psi_{\pm} \leq \liminf_{s\to\pm\infty} \psi(s)/s \leq \limsup_{s\to\pm\infty} \psi(s)/s \leq \psi^{\pm} < +\infty$, $0 < g_{\pm} \leq \liminf_{s\to\pm\infty} g(s)/s \leq \limsup_{s\to\pm\infty} g(s)/s \leq g^{\pm} < +\infty$. Then we have the following result.

Proposition 2.7 *Suppose that, for each $n \in \mathbb{N}$, $\frac{2T}{\pi n} \notin [\Lambda_*, \Lambda^*]$, where*

$$\Lambda_* = (\psi_+/g^+)^{\frac{1}{2}} + (\psi_+/g^-)^{\frac{1}{2}} + (\psi_-/g^+)^{\frac{1}{2}} + (\psi_-/g^-)^{\frac{1}{2}}$$

and

$$\Lambda^* = (\psi^+/g_+)^{\frac{1}{2}} + (\psi^+/g_-)^{\frac{1}{2}} + (\psi^-/g_+)^{\frac{1}{2}} + (\psi^-/g_-)^{\frac{1}{2}}.$$

Then equation $(\mathcal{P})$ has at least one T-periodic solution.

Proof First of all, we write $(\mathcal{P})$ as an equivalent system of the form (2.17) by means of the position $\psi(\dot{u}) = y$, in order to have

$$\begin{cases} \dot{u} = & \psi^{-1}(y) \\ \dot{y} = & -g(u) + e(t, u, \psi^{-1}(y)) \end{cases} \tag{2.26}$$

Note that since ψ^{-1} and g have linear growth and e is bounded, then the global existence (backward and forward) of the solutions for the Cauchy problems associated to (2.26) is guaranteed. Now we take some positive constants ε, M, with $0 < a_{\pm} = \psi_{\pm} - \varepsilon$, $\psi^{\pm} + \varepsilon = a^{\pm}$, $0 < b_{\pm} = g_{\pm} - \varepsilon$, $b^{\pm} = g^{\pm} + \varepsilon$, such that,

$$(1/a^+)(s^+)^2 + (1/a^-)(s^-)^2 - M \leq \psi^{-1}(s)s \leq (1/a_+)(s^+)^2 + (1/a_-)(s^-)^2 + M,$$

$$b_+(s^+)^2 + b_-(s^-)^2 - M \leq g(s)s \leq b^+(s^+)^2 + b^+(s^+)^2 + M,$$

holds for all $s \in \mathbb{R}$, where we have set, for the free variable s, $s^+ := \max\{s, 0\}$, $s^- := \max\{-s, 0\}$. By means of the constants we have just introduced, we define the two positive definite functions

$$V(u, y) := b_+(u^+)^2 + b_-(u^-)^2 + (1/a^+)(y^+)^2 + (1/a^-)(y^-)^2,$$

$$W(u, y) := b^+(u^+)^2 + b^-(u^-)^2 + (1/a_+)(y^+)^2 + (1/a_-)(y^-)^2$$

and consider the corresponding rotations $r_-(x) := \mathrm{rot}_V(T; x(\cdot))$, $r_+(x) := \mathrm{rot}_W(T; x(\cdot))$, for a solution $x(\cdot) = (u(\cdot), y(\cdot))$ of (2.26) such that $x(t) \neq 0$, $\forall t \in [0, T]$.

Let $R > 0$ be a given constant and define $\Omega := B(\eta(R))$, where the function η comes from Lemma 2.2. By Lemma 2.2 and Remark 2.1 we know that $W(x(t)) \geq V(x(t)) \geq c_R := \min_{||z||=R} V(z)$, for all $t \in [0, T]$ and any solution $x(\cdot)$ such that $x(0) \in \partial\Omega$. Note also that $c_R \to +\infty$ as $R \to +\infty$. Then, we can conclude that for any x as above,

$$\begin{aligned} r_-(x) &= (2A(V, 1))^{-1} \int_0^T \frac{\psi^{-1}(y)y + g(u)u - e(t, u, \psi^{-1}(y))u}{V(u, y)}\, dt \\ &\geq T(2A(V, 1))^{-1} - (K/\sqrt{c_R}) \end{aligned}$$

and, by a similar computation,

$$r_+(x) \leq T(2A(W, 1))^{-1} + (K/\sqrt{c_R}),$$

where $K > 0$ is a suitable constant depending on M and the bound for e.

Computing the two areas, we have

$$2A(V, 1) = \frac{\pi}{2}\left(\left(\frac{a^+}{b_+}\right)^{\frac{1}{2}} + \left(\frac{a^+}{b_-}\right)^{\frac{1}{2}} + \left(\frac{a^-}{b_+}\right)^{\frac{1}{2}} + \left(\frac{a^-}{b_-}\right)^{\frac{1}{2}}\right),$$

$$2A(W, 1) = \frac{\pi}{2}\left(\left(\frac{a_+}{b^+}\right)^{\frac{1}{2}} + \left(\frac{a_+}{b^-}\right)^{\frac{1}{2}} + \left(\frac{a_-}{b^+}\right)^{\frac{1}{2}} + \left(\frac{a_-}{b^-}\right)^{\frac{1}{2}}\right),$$

from which we immediately find the conclusion, via Theorem 3, provided that, at the beginning of the proof, ε was chosen sufficiently small and, subsequently, we take R large enough. □

Remark 2.7 In the case when $\psi(s) = s$, so that equation $(\mathcal{P})$ assumes the form $(\mathcal{D})$ and $g_\pm = g_*$, $g^\pm = g^*$, we obtain a classical theorem by Mawhin [130], according to which $(\mathcal{D})$ has a T-periodic solution if $[g_*, g^*]$ has empty intersection with the spectrum $\{\lambda_j = (2\pi j/T)^2 : j = 0, 1, 2, \dots\}$. The unsymmetric version of Proposition 6, again for $\psi(s) = s$, (i.e. if $g_- \neq g^+$ or $g^- \neq g^+$) covers the "nonresonance" with respect to the periodic Fučik spectrum (see [42, 73, 90] for more details, references and related results).

In this section, we have furnished a typical example of the application of the Poincaré-Bohl Theorem (see Theorem 2.1), by evaluating the angular rotations of the solutions around the origin. Such rotation number techniques although quite classical [11] are very effective and applied in connection with other fixed points as well, as we will see in the next section as well. For recent works applying such methods see for example [63, 79] and the references therein.

2.5 Topological Degree

In this section, we introduce some basic definitions and results concerning the topological degree in finite and infinite dimensional normed spaces. Applications are given to the solvability of some nonlinear boundary value problems for ordinary differential equations. In many applications concerning the solvability of boundary value problems for ordinary and partial differential equations and systems, one is led to consider some equations of the form

$$x = \phi(x), \tag{2.27}$$

or

$$f(x) = p, \tag{2.28}$$

where, $\phi : X \supset \text{dom}\,\phi \to X$ (or, respectively, $f : X \supset \text{dom}\, f \to X$) is a suitable (e.g. continuous, compact, etc.) operator defined on some domain of a topological space X. When X has also a linear structure, for instance, when X is a real or complex vector space, problem (2.27), which is the problem of the search for the fixed points of ϕ and problem (2.28), with concerns the search of the zeros of $f(\cdot) - p$, are, in some sense, equivalent, as any equation like (2.27) can be written in the form $f(x) := p + \alpha(x)(x - \phi(x)) = p$, with $\alpha : X \to \mathbb{R} \setminus \{0\}$ and conversely, any equation like (2.28) fits into the form $x = \phi(x) := x + \alpha(x)(p - f(x))$. However, for some problems and domains (like, e.g. if X is a manifold), the approach of solving an equation via the search of the zeros of a map and that of looking for the fixed points of an operator, are, in general, not equivalent. Moreover, the regularity conditions required for ϕ in (2.27) will be not exactly the same as those assumed for f in (2.28). Therefore, some different (although similar) tools have been introduced. Here we draw the reader's attention to the *Fixed point index*, addressed to problem (2.27) and to the *Topological degree*, dealing with Eq. (2.28).

Consider first Eq. (2.28), where $f : X \supset \mathcal{D} \to X$, is a continuous map and let $p \in X$ be a given point. We are interested in discussing the following problems:

- existence of solutions for Eq. (2.28);
- multiplicity results for (2.28);
- structure of the solution set when f or p depends on a parameter.

The aim of this section is to give a brief introduction to the main properties of the topological degree and to outline some applications of this technique to nonlinear boundary value problems for ordinary differential equations. The use of degree theory to nonlinear problems for ordinary, integral and partial differential equations has become a standard tool.

2.6 Axioms for the Brouwer Degree

In this section, we introduce in an axiomatic way the topological degree for continuous maps on bounded domains of Euclidean spaces. This is the so-called *Brouwer degree* from L.E.J. Brouwer (1912). We refer to [177] for a history of the development of the concept of degree starting with the *winding number* (which is an equivalent formulation of the Brouwer degree in $\mathbb{R}^2$) and with the works of Gauss, Cauchy, Kronecker, Hadamard, and Poincaré.

Let $f : \overline{\Omega} \to \mathbb{R}^n$ be a continuous function, where $\Omega \subset \mathbb{R}^n$ is an open bounded set and let $p \in \mathbb{R}^n$ be a given point. We say that the triplet (f, Ω, p) is *admissible* provided that $f(x) \neq p$, for all $x \in \partial\Omega$. To any admissible triplet (f, Ω, p) we can associate an integer $d(f, \Omega, p)$, the "Brouwer degree" which satisfies the following properties (axioms of the degree, see, e.g. [46, 120]).

(d_1) `Additivity/Excision:` *Let* Ω_1, Ω_2 *be open subsets of* Ω *with* $\Omega_1 \cap \Omega_2 = \emptyset$ *and assume* $f(x) \neq p$, $\forall x \in \overline{\Omega} \setminus (\Omega_1 \cup \Omega_2)$. *Then*

$$d(f, \Omega, p) = d(f, \Omega_1, p) + d(f, \Omega_2, p).$$

(d_2) `Invariance under homotopy:` *Let* $h = h_\lambda(x) : \overline{\Omega} \times [0, 1] \to \mathbb{R}^n$ *and* $p = p_\lambda : [0, 1] \to \mathbb{R}^n$ *be continuous functions such that* $h_\lambda(x) \neq p_\lambda$, $\forall x \in \partial\Omega$ *and* $\lambda \in [0, 1]$. *Then* $d(h_\lambda, \Omega, p_\lambda) = \mathrm{constantwithrespectto}\lambda$.

(d_3) `Normalization:` $d(I, \Omega, p) = 1$, *if* $p \in \Omega$, *where* $I(x) = x$ *is the identity in* $\mathbb{R}^n$.

An apparently weaker form of the normalization axiom is the following: $d(I, \Omega, 0) = 1$, if $0 \in \Omega$. It is not difficult to see that under (d_1) and (d_2) it follows that this latter condition is equivalent to (d_3).

From (d_1), using additivity (with $\Omega = \Omega_1 = \Omega_2 = \emptyset$), it is easy to obtain $d(f, \emptyset, p) = 0$ and that $d(f, \Omega, p) = d(f, \Omega_0, p)$ for any open set $\Omega_0 \subset \Omega$ such that $f(x) \neq p$ for all $x \in \overline{\Omega} \setminus \Omega_0$ (this is the "excision property"). Then, by excision, we find $d(f, \Omega, 0) = 0$ if $f(x) \neq 0$, for all $x \in \overline{\Omega}$ and then can state the following property (usually referred to as Kronecker).

(d_0) `Existenceofsolutions:` *The equation* $f(x) = p$ *has at least one solution in* Ω *if* $d(f, \Omega, p) \neq 0$.

From the homotopic invariance property it can be seen that

$$d(f, \Omega, p) = d(f - p, \Omega, 0),$$

and, for any $\varepsilon > 0$ sufficiently small (say $0 < \varepsilon \leq \varepsilon_0$, with ε_0 depending on $\min\{||f(x) - p|| : x \in \partial\Omega\}$),

$$d(f, \Omega, p) = d(g, \Omega, q), \text{ if } ||f - g|| \leq \varepsilon \text{ on } \partial\Omega \text{ and } ||p - q|| \leq \varepsilon.$$

In particular, one has the *boundary value dependence* property, i.e.

$$d(f_1, \Omega, p) = d(f_2, \Omega, p), \quad \text{when } f_1(x) = f_2(x), \ \forall\, x \in \partial\Omega.$$

Another related property which is easy to prove from the invariance under homotopy is that $d(f, \Omega, p_1) = d(f, \Omega, p_2)$ when p_1 and p_2 belong to the same connected component of $\mathbb{R}^n \setminus f(\partial\Omega)$.

Perhaps the simplest choice of a continuous homotopy between two maps f_1 and f_2 is the convex one, given by $h_\lambda(x) = (1 - \lambda) f_1(x) + \lambda f_2(x)$, with $\lambda \in [0, 1]$. In this case, the homotopy is admissible provided that the segment $[f_1(x), f_2(x)]$ does not contain p. This yields the following result, namely, the
`Poincaré-Bohl theorem:` $d(f_1, \Omega, p) = d(f_2, \Omega, p)$, *if* $p \notin [f_1(x), f_2(x)]$, *for all* $x \in \partial\Omega$.

Properties (d_1), (d_2), (d_3) define *axiomatically* the Brouwer degree (see [4, 46, 120, 198] and the references therein). On the other hand, if we wish to *construct* the degree, then, using Sard's lemma we can start from a generic situation, assuming $f \in C^1(\Omega)$ and p a regular value of f, that is, the (possibly empty) set $f^{-1}(p)$ contains only points z where the Jacobian matrix $f'(z)$ is nonsingular. Then, using a compactness argument and the local inversion theorem, it is possible to see that $f^{-1}(p)$ is finite and thus we can *define*,

$$(D_0) \qquad \deg(f, \Omega, p) = \sum_{z \in f^{-1}(p)} \operatorname{sgn} \det f'(z).$$

Note that, with this approach, (d_1), (d_2), (d_3) are theorems that can be derived by (D_0). Conversely, if we assume (d_1), (d_2), (d_3) as axioms then we can prove that $\deg(M, \Omega, 0) = \operatorname{sgn} \det M$, for any nonsingular $n \times n$ matrix M and for $0 \in \Omega$. From that, we can derive (D_0) as a formula to compute the degree. In the sequel, in order to avoid further explanations, when we write $\deg(f, \Omega, p)$, we implicitly assume that the triplet we are considering is admissible.

Other useful properties of the degree are

$$\deg(Mf, \Omega, 0) = (\operatorname{sgn} \det M) \deg(f, \Omega, 0),$$

for any nonsingular $n \times n$ matrix M (a very special form of the Leray product formula which in this case can be deduced directly from (D_0)), and

$$\deg(f_1 \times f_2, \Omega_1 \times \Omega_2, (p_1, p_2)) = \deg(f_1, \Omega_1, p_1) \times \deg(f_2, \Omega_2, p_2),$$

where $f_i : \overline{\Omega}_i \to \mathbb{R}^{n_i}$, $p_i \in \mathbb{R}^{n_i}$, $\Omega_i \subset \mathbb{R}^{n_i}$.

Suppose now that we are interested in solving the fixed point equation

$$x = \phi(x), \tag{2.29}$$

where $\phi : \mathbb{R}^n \to \mathbb{R}^n$ is a continuous map. A *Continuation method* for the solvability of (2.29) makes use of the homotopic invariance of the degree as follows: first we "homotopize", that is we continuously embed problem (2.29) into a one - parameter family of problems of the form

$$h_\lambda(x) = 0 \tag{1_λ}$$

with

$$f_1(x) = I - \phi(x), \quad \forall x \in \mathbb{R}^n. \tag{2.30}$$

Then we look for an open bounded set Ω such that the homotopy $(x, \lambda) \to h_\lambda(x)$ is admissible with respect to Ω and $p = 0$, according to (d_2). Finally, we ask that the problem we have reached for $\lambda = 0$ is "simpler" than the original one, in the sense that we can prove that its degree is nonzero. Describing this procedure in a formal statement, we obtain:

Lemma 2.4 *Let $(x, \lambda) \to h_\lambda(x)$ be a continuous function from $\mathbb{R}^n \times [0, 1]$ to $\mathbb{R}^n$ satisfying (2.30) and assume that there is an open bounded set $\Omega \subset \mathbb{R}^n$, such that the following properties hold:*

(a) `admissibility:`

$$h_\lambda(x) \neq 0, \quad \forall x \in \partial\Omega, \ \forall \lambda \in]0, 1[;$$

(b) `degree condition:`

$$\deg(h_0, \Omega, 0) \neq 0.$$

Then ϕ has at least one fixed point $\tilde{x}$ with $\tilde{x} \in \overline{\Omega}$.

Proof First observe that if ϕ has fixed points belonging to $\partial\Omega$, then we have the result. Hence, we can assume that (2.29) has no solutions on $\partial\Omega$. On the other hand, in (b) it is implicitly assumed that the triplet $(h_0, \Omega, 0)$ is admissible (see the corresponding remark above), so that we can restate more precisely (a) in the form: $h(x, \lambda) \neq 0$, $\forall x \in \partial\Omega$, $\forall \lambda \in [0, 1]$. Now, we recall (d_2) and (2.30), and using (b) we find $\deg(I - \phi, \Omega, 0) = \deg(h_1, \Omega, 0) = \deg(h_0, \Omega, 0) \neq 0$. Then, (d_0) yields the existence of a zero for $I - \phi$ belonging to the set Ω. In any case, we find a fixed point of ϕ in $\overline{\Omega}$ and hence the result is proved. □

Choosing the convex homotopy $h_\lambda := I - \lambda\phi$, $\lambda \in [0, 1]$, we have $h_0 = I$ and thus, in view of the normalization property of the degree, (b) is satisfied if and only if $0 \in \Omega$. Then we can state the following result (a consequence of the `Poincaré-Bohl` `theorem`, as seen earlier in Theorem 2.1, which is also known as the `Omitted rays theorem`).

Theorem 2.4 *Let $\Omega \subset \mathbb{R}^n$ be an open bounded set with $0 \in \Omega$, such that*

$$\phi(x) \neq \mu x, \quad \forall x \in \partial\Omega, \ \forall \mu > 1. \tag{2.31}$$

Then ϕ has at least one fixed point in $\overline{\Omega}$.

Proof Put $\lambda = 1/\mu$ and observe that (2.31) makes (a) fulfilled by the choice of the homotopy $h_\lambda = I - \lambda\phi$ described above. □

We notice that for the validity of Theorem 2.4 it is sufficient to assume ϕ defined in $\overline{\Omega}$ Indeed, we can always enter in the setting of Lemma 2.4, with $\phi \in \mathcal{C}(\mathbb{R}^n)$, using the Tietze extension theorem (which is stated below). It is also easy to see that, instead of $0 \in \Omega$ and (2.31), we can suppose that there is a point $z_0 \in \Omega$ such that $\phi(x) \neq z_0 + \mu(x - z_0)$, $\forall x \in \partial\Omega$, $\forall \mu > 1$. An obvious consequence of Theorem 2.4, which generalizes the Brouwer fixed point theorem, is the following variant of Rothe's theorem (see [46, Ex. 5, p. 33]) or Birkhoff-Kellogg theorem [17, Theorem I] (Fig. 2.1).

Corollary 2.2 *Let $\Omega \subset \mathbb{R}^n$ be an open bounded set such that $\overline{\Omega}$ is strictly star shaped and let $\phi \in \mathcal{C}(\overline{\Omega})$ be such that $\phi(\partial\Omega) \subset \overline{\Omega}$. Then ϕ has at least one fixed point in $\overline{\Omega}$.*

Proof Let $\phi_0(x) := \phi(x + p) - p$, where $\overline{\Omega}$ is strictly star shaped with respect to $p \in \Omega$, and set $\Omega_0 := \Omega - p$. Then, as $\phi_0(\partial\Omega_0) \subset \overline{\Omega}_0$ and $0 \in \Omega_0$, we can apply Theorem 2.4 to get a fixed point $w \in \overline{\Omega}_0$, for ϕ_0. Clearly, $\tilde{z} = w + p$ is a fixed point for ϕ in $\overline{\Omega}$. □

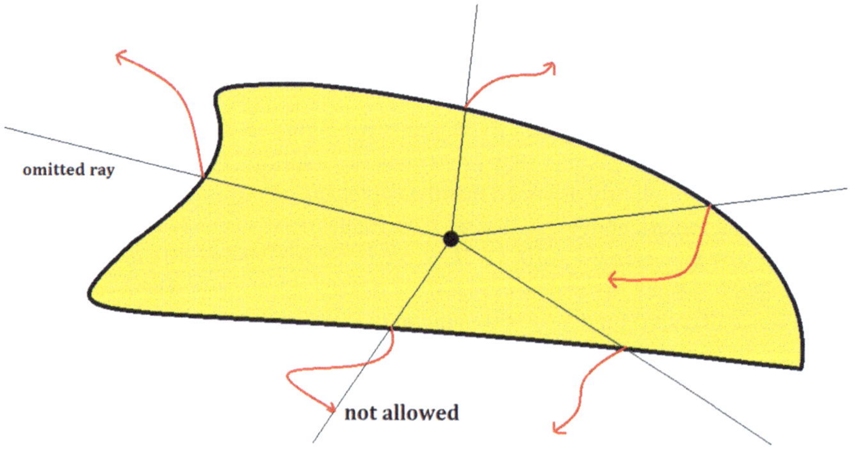

Fig. 2.1 A pictorial explanation of Theorem 2.4

From Corollary 2.2 one easily gets the Brouwer fixed point theorem ensuring the existence of a fixed point for a continuous function ϕ, mapping the closed ball $B[z_0, R] := \{z \in \mathbb{R}^n : ||z - z_0|| \leq R\}$, into itself. Alternatively, we can use directly the Poincaré-Bohl theorem, observing that, by hypothesis, $||\phi(x) - z_0|| \leq ||x - z_0||$ for all $x \in \partial B(z_0, R)$ and thus $\phi(x) \neq z_0 + \mu(x - z_0)$ for all $x \in \partial B(z_0, R)$ and $\mu > 1$.

Next, using the fact that the fixed point property is preserved under retractions, one can immediately obtain that if $\phi : K \to K$ is continuous, where $K \subset \mathbb{R}^n$ is a compact convex set, then ϕ has at least one fixed point in K (to prove this, just take a large ball $B[0, R] \supset K$ and a continuous retraction $r : B[0, R] \to K$. The continuous map $r \circ \phi : B[0, R] \to B[0, R]$ has a fixed point which indeed belongs to K and is a fixed point of ϕ). Finally, using the fact that the fixed point property is preserved under homeomorphisms, one can prove that *any continuous map $\phi : D \to D$ has a fixed point when D is any domain which is homeomorphic to a compact convex set of an Euclidean space.*

2.7 Some Remarks About the Uniqueness and the Existence of the Brouwer Degree

We have considered a degree function $d : (f, \Omega, p) \mapsto d(f, \Omega, p) \in \mathbb{Z}$, which associates an integer to any admissible triplet (f, Ω, p) in such a manner to have fulfilled the three main properties (d_1), (d_2), (d_3). The aim of this section is to sketch how a degree can be defined and that there is "essentially" a unique way how to define it. The idea of the proof is to define the degree for linear maps and starting from (D_0), then go up to the general case, obtaining (d_1), (d_2), (d_3) as theorems from the basic definition (this gives the existence of a degree function). Conversely, to assume that a degree function d satisfying (d_1), (d_2), (d_3) does exist for continuous maps and then to go down to the particular case of the linear maps and show that, necessarily, in this case, the degree on an open bounded set containing the origin is the sign of the determinant (this gives the uniqueness of a degree function). We present first an approach for the uniqueness (moving from the "complex" down to the "simple" case). This will motivate the formula (D_0) to define the degree in the generic case which is also the first step to "construct" it. The uniqueness of the Brouwer degree was obtained by Führer [87] in 1972 (see also [198, Ch. 12]). The corresponding uniqueness result for the Leray-Schauder degree was obtained in 1973 by Amann and Weiss [4] (for more details, see, for instance, [46, 120, 198]).

Assume that a degree function d is defined on the admissible triplets (f, Ω, p). As a simple exercise, we show how to compute it in the case when $n = 1$ and $\Omega =]a, b[\subset \mathbb{R}$.

Example 2.7 Let $f : [a, b] \to \mathbb{R}$ be a continuous function and assume that $f(a)f(b) \neq 0$. We wish to compute the degree $d(f,]a, b[, 0)$, using the axioms. If $f(a)f(b) > 0$, we define the map $g(x) := f(a) + \big((f(b) - f(a))/(b - a)\big)(x -$

a). By the boundary dependence of the degree, we have that $d(f,]a, b[, 0) = d(g,]a, b[, 0)$ and, since $g(x) \neq 0$, for all $x \in [a, b]$, we conclude (from (d_0)), that $d(g,]a, b[, 0) = 0$.
Suppose now that $f(a)f(b) < 0$ and assume $f(a) < 0$. Let $c := (a + b)/2$ and define $h(x) := x - c$. From the Poincaré - Bohl theorem it is straightforward to see that $d(f,]a, b[, 0) = d(h,]a, b[, 0)$. On the other hand, $d(h,]a, b[, 0) = d(I,]a, b[, c) = 1$ (by the normalization property).
Finally, we consider the case when $f(a)f(b) < 0$ with $f(a) > 0$. We take $r > b$ and choose any continuous extension $\tilde{f}$ of f to $[a, r]$ such that $\tilde{f}(r) > 0$. By the previous two cases and applying the additivity property of the degree, we find $0 = d(\tilde{f},]a, r[, 0) = d(\tilde{f},]a, b[, 0) + d(\tilde{f},]b, r[, 0) = d(f,]a, b[, 0) + 1$ and so, $d(f,]a, b[, 0) = -1$.

As a next step, we consider now the degree for linear maps. Let $L : \mathbb{R}^n \to \mathbb{R}^n$ be a linear map. We denote in the sequel by the same symbol L the $n \times n$ square matrix associated to L with respect to the standard orthonormal base of $\mathbb{R}^n$. We also denote by $\mathcal{L}(\mathbb{R}^n)$ the set of linear maps of $\mathbb{R}^n$ endowed with the operator norm and by $GL(\mathbb{R}^n) \subset \mathcal{L}(\mathbb{R}^n)$ the set of linear invertible maps of $\mathbb{R}^n$. Let $\Omega \subset \mathbb{R}^n$ be an open bounded set with $0 \in \Omega$. It is clear that the degree $d(L, \Omega, 0)$ is defined (in the sense that the triplet $(L, \Omega, 0)$ is admissible) if and only if $\ker L = \{0\}$, that is L is nonsingular or, equivalently (in the matrix language), $\det L \neq 0$. In the same manner, one can easily check that a (continuous) homotopy of linear maps $[0, 1] \ni \lambda \mapsto L_\lambda \in \mathcal{L}(\mathbb{R}^n)$ is admissible with respect to Ω if and only if $\det L_\lambda \neq 0$ and $\operatorname{sgn} \det L_\lambda =$ constant w.r. to $\lambda \in [0, 1]$ (note that $\det L$ depends continuously on L).

Now, suppose that L_1 and L_2 are two invertible linear maps and there is an admissible homotopy joining them. Then, necessarily, we have that L_1 and L_2 belong to the same connected component of $GL(\mathbb{R}^n)$ and $\operatorname{sgn} \det L_1 = \operatorname{sgn} \det L_2$. Conversely, assume that L_1 and L_2 belong to the same connected component of $GL(\mathbb{R}^n)$. In this case, there is a continuous nonsingular path $\lambda \mapsto L_\lambda$ joining L_1 to L_2 and therefore, by the homotopic invariance of the degree, $d(L_1, \Omega, 0) = d(L_2, \Omega, 0)$ (and also L_1 and L_2 have the determinant with the same sign).

Finally, we note that, by the excision property, $d(L, \Omega_1, 0) = d(L, \Omega_2, 0)$, for any Ω_1 and Ω_2 open and bounded sets of $\mathbb{R}^n$ with $0 \in \Omega_1 \cap \Omega_2$, provided that $L \in GL(\mathbb{R}^n)$.

We show now how to compute $d(L, \Omega, 0)$ in the case of diagonal matrices.

Example 2.8 Let $L = \operatorname{diag}(\delta_i)$, for $i = 1, \dots, n$, with $\delta_i \neq 0$, for all $i = 1, \dots, n$ and let $0 \in \Omega$. Without restriction, we can assume $\Omega =]-a, a[^n$, with $a > 0$. We also define $I_1 := \operatorname{diag}(\delta_i)$, where $\delta_1 = -1$ and $\delta_j = 1$, for $j = 2, \dots, n$.

We already know that $d(I, \Omega, 0) = 1$ (the normalization property), and, working on the first component as in the third case of Example 2.7 and keeping fixed the other components, it is easy to see that $d(I_1, \Omega, 0) = -1$.

Consider, for a moment, the matrix $-I$ in a two-dimensional space. It is standard to find a (continuous) nonsingular path in $GL(\mathbb{R}^2)$ joining $-I$ with the rotation matrix

$J = \begin{bmatrix} 0 & -1 \\ 1 & 0 \end{bmatrix}$ (the convex homotopy works well). In the same manner, from J, then one can move to I. Thus, we find that $-I$ and I belong to the same connected component of $GL(\mathbb{R}^2)$.

The same construction (repeated for any 2×2 sub-block like $-I$), yields that each $n \times n$ diagonal matrix having only -1 or $+1$ entries on the diagonal belong to the same connected component of I or to the same connected component of I_1 according to the fact that the number of negative entries is even or odd. Then, from the above remark, we know that the degree is 1 or -1, accordingly.

Finally, with a minimal effort, one can see that for the diagonal matrix L considered at the beginning of this example, one has

$$d(L, \Omega, 0) = \text{sgn det } L = \text{sgn} \prod_{i=1}^{n} \delta_i = (-1)^k,$$

where k denotes the number of the negative eigenvalues of L counting their algebraic multiplicity.

Up to now, we have seen that nonsingular matrices have a determinant with different sign belong to different connected components of $GL(\mathbb{R}^n)$ and have different degrees with respect to Ω and $p = 0$. We also know that I and I_1 (the $n \times n$ diagonal matrix having -1 entry at the first place and 1 entries at all the other positions) belong to different connected components of $GL(\mathbb{R}^n)$ and that $d(I, \Omega, 0) = 1 = -d(I_1, \Omega, 0)$. Now we can conclude the discussion for the linear case, taking advantage of the following result from linear algebra: *The set $GL(\mathbb{R}^n)$ has exactly two connected components which are labelled by the sign of the determinant.* (For a proof, see [67, Prop. 1.1], [198, Prob. 12.4]).

From this result, we immediately find that

$$\begin{aligned} &d(L, \Omega, 0) = \text{sgn det } L, \quad \forall\, L \in GL(\mathbb{R}^n), \\ &\forall\, \Omega, \text{ open and bounded, with } 0 \in \Omega \subset \mathbb{R}^n. \end{aligned} \tag{2.32}$$

(For a direct proof of (2.32) which with minor modifications can work also as a proof of also a proof of the above-recalled result on the two components of $GL(\mathbb{R}^n)$, see, for instance, [46, Prop.1.5]).

Before proceeding further, we re-write (2.32) in a slightly different form which fits better in view of an extension to infinite dimensional spaces.

Let L (linear, nonsingular) as above and let λ be a (real) eigenvalue of L. As well known from linear algebra, the algebraic multiplicity α of λ (that is the multiplicity of λ as a root of the characteristic polynomial $\det(\lambda I - L) = 0$), coincides with the dimension of the *generalized eigenspace* $N(\lambda)$ which is defined as $\ker\,(\lambda I - L)^j$, where $j \geq 1$ is the smallest positive integer such that $\ker\,(\lambda I - L)^j = \ker\,(\lambda I - L)^{j+1}$.

Now, it is clear that

$$\text{sgn det } L = (-1)^{\alpha_1 + \cdots + \alpha_m},$$

where $\alpha_i = \dim N(\lambda_i)$ is the *algebraic multiplicity* of λ_i and $\lambda_1, \ldots, \lambda_m$ are all the negative eigenvalues of L. Thus we can write

$$d(L, \Omega, 0) = (-1)^k, \quad \forall L \in GL(\mathbb{R}^n), \\ \forall \Omega, \text{ open and bounded, with } 0 \in \Omega \subset \mathbb{R}^n. \tag{2.33}$$

where k is the sum of the algebraic multiplicities of the (real) negative eigenvalues of L.

In the sequel, it will be also interesting to compute the degree for $L = I - M$, with $I - M$ nonsingular (that is M has not $\lambda = 1$ as an eigenvalue). In this case, it is obvious that, for Ω as above, one has

$$d(I - M, \Omega, 0) = (-1)^m, \tag{2.34}$$

where m is the sum of the algebraic multiplicities of the (real) eigenvalues μ of M with $\mu > 1$.

Now we come back to the general situation and find a "formula" to compute the degree $d(f, \Omega, p)$, for an admissible triplet (f, Ω, p). The idea is to approximate f with smooth maps, p with regular values in order to reduce our problem to finite number of linear approximations. Then, we can use the formula of the degree for linear invertible maps and conclude our proof.

STEP 1. *Extension.*

Recall the `Tietze extension theorem`: *Given any closed subset C of a metric space X and a continuous map $h : C \to \mathbb{R}$, there is a continuous extension $\tilde{h} : X \to \mathbb{R}$ of h (i.e. $\tilde{h}|_C = h$) such that $\mathcal{R}(\tilde{h}) \subset \mathrm{co}\mathcal{R}(h)$.* (For a proof, see [58], [169]).

Using Tietze's theorem on any component $f_i : \mathbb{R}^n \supset \overline{\Omega} \to \mathbb{R}$, $(i = 1, \ldots, n)$ of f, we can define a continuous map $\tilde{f} = \mathrm{col}(\tilde{f}_1, \ldots, \tilde{f}_n) : \mathbb{R}^n \to \mathbb{R}^n$ with $\tilde{f}|_{\overline{\Omega}} = f$. (This step may be avoided through some other approaches).

STEP 2. *Smoothness.*

For any $\epsilon > 0$, we can find a C^∞- function $g = g_\epsilon : \mathbb{R}^n \to \mathbb{R}^n$ such that $||\tilde{f}(x) - g(x)|| = ||f(x) - g(x)|| \leq \epsilon$, for all $x \in \partial\Omega$.

Indeed, following (see [46]), we can take a family of mollifiers $\varphi_\alpha : \mathbb{R}^n \to \mathbb{R}$, for $\alpha > 0$, defined by $\varphi_\alpha(x) = \alpha^{-n}\varphi_1(x/\alpha)$ $\varphi_1(x) = c\,\psi(||x||)$, where $\psi : \mathbb{R}^+ \to \mathbb{R}^+ := [0, +\infty)$ is a C^∞-function positive and strictly decreasing on $[0, 1]$, with $\psi(s) = 0$ for $s \geq 1$, and $\psi'(0) = 0$. Moreover, $c > 0$ is such that $\int_{\mathbb{R}^n} \varphi_1(x)\, dx = 1$. Then, it follows that the sequence $(g_k)_k$ of C^∞-functions defined by $g_k(x) := \int_{\mathbb{R}^n} \tilde{f}(u)\varphi_{1/k}(u - x)\, du$, converges uniformly to $\tilde{f}$ on compact sets in $\mathbb{R}^n$.

Alternatively, we could apply the following form of the Weierstrass approximation theorem: *given any continuous function $h : \mathbb{R}^n \to \mathbb{R}^n$, there is a sequence $(g_k)_k$ of smooth functions $g_k : \mathbb{R}^n \to \mathbb{R}^n$, such that $g_k \to h$, uniformly on each compact subset of $\mathbb{R}^n$.* (For a proof, see [169]).

In any case, as a consequence of the above result and the stability of the degree with respect to small perturbations, we can conclude that

$$d(f, \Omega, p) = d(g, \Omega, p), \ g \in C^{\infty}, \ (\mathbb{R}^n), \ \text{with } ||g - f|| \leq \epsilon \text{ on } \partial\Omega,$$

where $\epsilon > 0$, sufficiently small.

STEP 3. *Regular values.*
Consider for a moment the following situation. Let $h : \mathbb{R}^n \supset \mathcal{O} \to \mathbb{R}^m$, with $\mathcal{O}$ an open set and h of class C^r with $1 \leq r \leq \infty$. We focus our attention to the case $n \geq m$. If $z \in \mathcal{O}$ is such that the differential $dh(z) : \mathbb{R}^n \to \mathbb{R}^n$ is not surjective, that is the Jacobian matrix $h'(z)$ has rank strictly less than m, we say that z is a *critical point* of h. We denote by C_h the set of critical points of h and we say that $q \in \mathbb{R}^n$ is a *regular value* of h if $q \in \mathbb{R}^n \setminus h(C_h)$. Then we recall that, according to the `Morse - Sard theorem`, *If* $r > \max\{0, n - m\}$, *then* $h(C_h)$ *has* m*-dimensional Lebesgue measure equal to zero, that is,* $\mu_m(h(C_h)) = 0$. (For a proof, see [101]).

It is interesting to remark that, in general, the smoothness assumption on h cannot be relaxed by lowering the value r. Indeed, as shown by Whitney in [191] (see also [101]), there is a C^1 map $h : \mathbb{R}^2 \to \mathbb{R}$ such that $C_h = \{(x, y) \in \mathbb{R}^2 : (\partial f/\partial x)(x, y) = (\partial f/\partial x)(x, y) = 0\}$ contains a continuum S where h is not constant and thus $h(S)$ contains an interval.

Now, when $m = n$, z is a critical point of h if and only if $\det(h'(z)) = 0$, and thus the above result ensures that for a C^1-map $h : \mathbb{R}^n \supset \mathcal{O} \to \mathbb{R}^n$, the set $h(C_h)$ of critical values has zero n-dimensional measure and this in turn implies that the set of the regular values of h is dense.

Perhaps it can be of some interest to state another version of these results, quoted as `Sard lemma`: *For any* C^1 *map* $h : \mathbb{R}^n \supset \mathcal{O} \to \mathbb{R}^n$ *and any measurable set* $S \subset \mathcal{O}$, *one has* $\mu_n(h(S)) \leq \int_S |\det h'(x)| \, dx$. (For a proof, see [175]). In this case, it is clear that $\mu_n(h(C_h)) = 0$ as the integrand in the above formula vanishes on C_h.

By Sard's result, we have that if g is a smooth approximation of f (as in Step 2), then in any neigbourhood of the point p, we can find some q such that q is a regular value of g. Hence, thanks to the constancy of the degree with respect to small perturbations of the "target" point, we conclude that

$$\begin{aligned} &d(f, \Omega, p) = d(g, \Omega, q) \text{ with } g \in C^{\infty}(\mathbb{R}^n), \ ||g - f|| \leq \epsilon \text{ on } \partial\Omega, \\ &\text{and } q \in \mathbb{R}^n, \ ||q - p|| \leq \epsilon, \ \text{with } q \text{ a regular value of } g, \end{aligned} \tag{2.35}$$

where $\varepsilon > 0$ is sufficiently small.

STEP 4. *Linearization.*
By the previous construction, we are led to compute $d(g, \Omega, q)$ for g smooth (C^1 is enough) and q a regular value of g.

Consider the compact set $g^{-1}(p) \subset \Omega$ (where, from now on, we consider the restriction of g to $\overline{\Omega}$). For any $z \in g^{-1}(p)$ we have that $g'(z)$ is nonsingular and hence the local inversion theorem implies that there is an open set U with $z \in U \subset \Omega$

and an open set V with $q \in V \subset \mathbb{R}^n$ such that $g : U \to V$ is a homeomorphism. In particular, $U \cap g^{-1}(q) = \{z\}$ and thus $g^{-1}(q)$ is a compact set with the discrete topology. Then, necessarily, $g^{-1}(q)$ is finite (possibly empty).

Assume $\emptyset \neq g^{-1}(q) = \{z_1, \dots, z_k\}$. Then, by the additivity excision property of the degree, we find

$$d(g, \Omega, q) = \sum_{i=1}^{k} d(g, B(z_i, r_i), q),$$

where $B(z_i, r_i)$ denotes the open ball of centre z_i and radius $r_i >$ and we take the r_i's sufficiently small as to have pairwise disjoint balls.

Now we compute $d(g, B(z, r), q)$ with $g(x) = q$ with $x \in B(z, r)$ if and only if $x = z$ (this will make the next computations valid for an arbitrary of the z_i's).

As g is of class C^1, we have that $g(x) - g(z) = g(x) - q = g'(z)(x - z) + h(x)$ with $||h(x)||/||x - z|| \to 0$ as $x \to z$. This, by the fact that $g'(z)$ is nonsingular and by the stability of the degree with respect to small perturbations, allows us to conclude that

$$d(g, B(z, r), q) = d(g - q, B(z, \delta), 0) = d(g'(z)(\cdot - z), B(z, \delta), 0),$$

with $0 < \delta \leq r$ a sufficiently small constant.

Finally, let us take $R > ||z|| + \delta$ and observe that

$$d(g'(z)(\cdot - z), B(z, \delta), 0) = d(g'(z)(\cdot - z), B(0, R), 0) = d(g'(z), B(0, R), 0)$$

(the last equality follows from the admissible homotopy $\lambda \mapsto g'(z)(x - \lambda z)$).

Now, we are in position to apply the formula for the computation of the degree for linear maps (2.32), or (2.33) and get

$$d(g, B(z, r), q) = \text{sgn}\ \det g'(z).$$

Thus, doing the same for any of the z_i's we find that

$$d(g, \Omega, q) = \sum_{z \in g^{-1}(q)} \text{sgn}\ \det g'(z). \tag{2.36}$$

With the agreement that the summation on an empty set of terms is 0, we have the validity of (2.36) in the case when $g(x) \neq q$, for all $x \in \overline{\Omega}$, as well.

From (2.35) and (2.36) we ensure the validity of (D_0) announced before and this proves the "uniqueness" of the degree.

Conversely, to construct the degree, we can define $d(f, \Omega, p)$ by formula (D_0) when f is smooth (C^1) and p is a regular value of f. Coming back to Sard's lemma, we can pass to the situation when p is a possibly singular value for f. Then we can consider the case of f continuous, by an approximation of f with smooth functions. The steps to do this are clearly the "reverse" of those considered above, even if the technical details are not the same. Actually, to show from the definition (D_0) that a degree can be defined for a singular value p, we need to prove that $d(f, \Omega, q)$ will be the same for all regular values q sufficiently near to p and, similarly, if f is only continuous, we need to prove that $d(g, f, p)$ will be the same for all the smooth functions near f. For technical reasons, one has usually to pass through C^2 functions and apply some non-trivial results of differential and integral calculus, using also the Heinz formula

$$d(f, \Omega, p) = \int_\Omega \varphi_\alpha(f(x) - p) f'(x)\, dx,$$

which holds in the same situation of (D_0) for $\alpha > 0$ sufficiently small, where φ_α's are the mollifiers considered in Step 2 (we omit all these details and we refer to one of the texts quoted in the bibliography like [46, 120].)

At last, we remark that there are alternative approaches yielding to the construction of the Brouwer degree. The interested reader may consult [2, 39, 57, 59] for construction of the degree using concepts from algebraic topology, or [95, 101, 140] for presentations using arguments of algebraic topology. Indeed, it is not a surprise that more "topological" approaches can be found (in line with Brouwer's original construction), as (d_1), (d_2), *and* (d_3) make use only of concepts like open sets and continuous maps.

2.8 Remarks on the Existence of the Leray-Schauder Degree

A natural question to ask is if there exists a degree theory for mappings on infinite dimensional spaces. The answer to this question is not so straightforward as the following example, due to Kakutani, will show.

Consider the space of square summable sequences defined by

$$\ell^2 := \{x = (x_1, x_2, \cdots) | \sum_i x_i^2 < \infty\}.$$

The space $\ell^2 \cong \mathbb{R}^\infty$ has a natural norm $\| x \|_2^2 := \sum_i x_i^2$ and inner product defined by $\langle x, y \rangle := \sum_i x_i y_i$ and is a complete normed space, infact, a Hilbert space. Let $B = \{x \in \ell^2 | \, \| x \|_2^2 \leq 1\}$ be the closed unit ball and define a mapping f on B as follows:

$$f(x) = \left(\sqrt{1 - \| x \|_2^2},\ x_1, x_2, \cdots \right),$$

which is a continuous mapping from B to $\partial B =: S^\infty$. The mapping f has no fixed points in B. Indeed, for $x \in S^\infty$ it holds that $f(x) = (0, x_1, x_2, \cdots) \neq x$. On the other hand $f(B) \subset B$ and the mapping f satisfies the requirements of the Brouwer fixed point theorem, which therefore does not hold for continuous mappings on ℓ^2. If a degree theory for continuous maps on ℓ^2 exists so does the Brouwer fixed point theorem. This is already an indication that a degree theory for continuous mappings on infinite dimensional spaces cannot be defined with the usual properties.

The algebraic topology of Banach spaces, and its application to nonlinear equations, started with the work of Juliusz Schauder in the 5 years period 1927–1932 [171, 173]. Schauder identified an important class of nonlinear operators in a Banach space, the completely continuous[1] perturbations of identity, for which he could generalize two important results of Brouwer in finite-dimensional topology: the fixed point and invariance of domain theorems. Schauder applied the first extension called the Schauder fixed point theorem [173] to the existence of solutions of differential equations for which uniqueness does not necessarily hold. Schauder applied his invariance of domain theorem [172, 174] to exhibit nonlinear elliptic problems for which uniqueness implies existence.

Leray and Schauder extend the Brouwer degree to compact[2] perturbations of the identity in a Banach space X as follows:
If $U \subset X$ is an open bounded set, $f : \overline{U} \to X$ is compact, and $z \notin (I - f)(\partial U)$, the Leray-Schauder degree $deg_{LS}[I - f, U, z]$ of $I - f$ in U over z is constructed from the Brouwer degree by approximating the compact mapping f over $\overline{U}$ by continuous mappings f_ϵ (such that $f_\epsilon \to f$ uniformly on $\overline{U}$ for $\epsilon \to 0$) with range in a finite-dimensional subspace X_ϵ (containing z) of X, and showing that the Brouwer degrees $deg_B[(I - f_\epsilon)|_{X_\epsilon}, U \cap X_\epsilon, z]$ stabilize for sufficiently small positive ϵ to a common value defining $deg_{LS}[I - f, U, z]$. This topological degree "algebraically counts" the number of fixed points of $f(\cdot) + z$ in U, and, for f of class C^1, and $I - f'(a)$ invertible for each fixed point a of $f(\cdot) + z$ in U, Leray and Schauder show that

$$deg_{LS}[I - f, U, z] = \sum_{a \in (I-f)^{-1}(z)} (-1)^{\sigma_j(a)}$$

where $\sigma_{j(a)}$ is the sum of the algebraic multiplicities of the eigenvalues of $f'(a)$ contained in $]1, +\infty[$. The Leray-Schauder degree conserves the basic properties of the Brouwer degree, as listed in [111]. In what follows, when we consider the Leray-Schauder degree $deg_{LS}[I - f, U, z]$, we always implicitly assume that $f : \overline{U} \to X$ is a compact mapping, $U \subset X$ is an open bounded set and $x - f(x) \neq z \; \forall \, x \in \partial U$.

Theorem 2.5 *The Leray-Schauder degree has the following properties.*

1. (Additivity). If $U = U_1 \cup U_2$, where U_1 and U_2 are open and disjoint, and if

$$x - f(x) \neq z, \; \forall \, x \in \overline{U} \setminus (U_1 \cup U_2)$$

---

[1] $f : X \to X$ is completely continuous if f is compact on each bounded subsets of X.

[2] Let X be a metric space, Y, a subspace of X, then a map $f : Y \to X$ is compact on Y iff f is continuous and $\overline{f(Y)}$ is compact.

then

$$deg_{LS}[I - f, U, z] = deg_{LS}[I - f, U_1, z] + deg_{LS}[I - f, U_2, z].$$

2. *(Existence and normalization). If* $deg_{LS}[I - f, U, z] \neq 0$, *then* $z \in (I - f)(U)$. *Moreover,* $deg_{LS}[I, U, z] = 1$ *if* $z \in U$.
3. *(Homotopy invariance). Let* $\Omega \subset X \times [a, b]$ *be a bounded open set, and let* $F : \overline{\Omega} \to X$ *be compact. If* $x - F(x, \lambda) \neq z$ *for each* $(x, \lambda) \in \partial\Omega$, *then* $deg_{LS}[I - F(\cdot, \lambda), \Omega_\lambda, z]$ *is independent of* λ.[3]

An important consequence of the existence and homotopy invariance properties of Leray-Schauder's degree is the celebrated Leray-Schauder continuation theorem.

Theorem 2.6 *Let* $\Omega \subset X \times [a, b]$, *be a bounded open set, and let* $F : \overline{\Omega} \to X$ *be compact. We define* $\Sigma := \{(x, \lambda) \in \overline{\Omega} | x = F(x, \lambda)\}$ *and assume that the following conditions hold:*

(i) $\Sigma \cap \partial\Omega = \emptyset$ *(a priori estimate),*
(ii) $deg_{LS}[I - F(\cdot, 0), \Omega_0, 0] \neq 0$ *(degree condition),*

then Σ *contains a continuum* C *along which* λ *takes all values in* $[0, 1]$.
In other words, under the above assumptions, Σ *contains a compact connected subset* C *connecting* Σ_0 *to* $\Sigma_1 \subset \Omega_1$. *In particular, the equation* $x = F(x, 1)$ *has a solution in* Ω_1.

Leray and Schauder use instead of (ii) a slightly less general condition requiring in addition that Σ_0 is a finite nonempty set $\{a_1, a_2 \ldots a_\mu\}$ and observe furthermore that, by refining Assumption (ii), one can obtain a more precise conclusion, which already anticipates subsequent results in bifurcation theory and alternative theorems described later. Let a be an isolated fixed point of f define its local Leray-Schauder index by

$$ind_{LS}[I - f, a] := deg_{LS}[I - f, B(a, r), 0]$$

for $r > 0$ small. Notice that the right-hand side of the above identity is constant with respect to r for $r > 0$ small.

Theorem 2.7 *If* a_1 *is an isolated fixed point of* $F(\cdot, 0)$ *and if*

$$ind_{LS}[I - F(\cdot, 0), a_1] \neq 0,$$

then $(a_1, 0)$ *belongs to a continuum in* Σ *which contains one of the points* $(a_2, 0), \ldots (a_\mu, 0)$, *or to a continuum in* Σ *along which* λ *takes all the values in* $[0, 1]$.

Leray and Schauder emphasize the following important special case of condition (ii) whose statement does not involve explicitly the concept of degree:

[3] We denote by Ω_λ= section of Ω at $\lambda \in [a, b]$, namely, $\Omega_\lambda := \{x | (x, \lambda) \in \Omega\}$.

Corollary 2.3 *If condition (i) holds, and if*

$$(ii^*) \quad 0 \in \Omega_0, \quad F(\cdot, 0) \equiv 0,$$

then Σ contains a continuum $C \ni (0, 0)$ along which λ takes all values in $[0, 1]$.

Motivated by applications to nonlinear elliptic problems, Leray and Schauder have formulated continuation theorems for more general equations, which are reducible to some compact perturbations of identity. Essentially, they consider equations which can be written in the form

$$G(x, x, \lambda) = 0, \tag{2.37}$$

where X, X_0, Y are Banach spaces with X_0 compactly imbedded in X, $G : X \times X_0 \times [0, 1] \to Y$ is Fréchet differentiable and such that, near each solution (x^*, λ^*) of (2.37) equation in y

$$G(x, y, \lambda) = 0$$

has a unique solution $y = F(x, \lambda)$ close to x^*, which defines locally a completely continuous mapping F. Thus, near (x^*, x^*, λ^*), Eq. (2.37) is equivalent to the fixed point problem

$$x = F(x, \lambda)$$

Leray and Schauder assume that the equation $x = F(x, 0)$ has a finite number of solutions and the sum of their indices is not zero. Then they conclude that there exists in $X \times [0, 1]$ a continuum of solutions along which λ takes all the values of $[0, 1]$. For further details and useful information from a historical point of view, see [136].

2.9 The Classical Fixed Point Index

Suppose that X is a topological space, Ω is an open subset of X and $f : \Omega \to X$ is a continuous map such that $S := \{x \in \Omega | f(x) = x\}$ is compact (possibly empty). Under further assumptions on f and X, one wants to associate in this situation an integer $i_X(f, \Omega)$ (sometimes written $i(f, \Omega, X)$), called the fixed point index of $f : \Omega \to X$. Roughly speaking, $i_X(f, \Omega)$ will be an algebraic count of the number of fixed points of f in Ω. If X is a Banach space or normed linear space and $\overline{f(\Omega)}$ is compact, one wants $i_X(f, \Omega)$ to be related to the Leray-Schauder degree of $I - f$ in the natural way:
$i_X(f, \Omega) = deg(I - f, \Omega, 0)$ = the Leray-Schauder degree of $I - f$ on Ω with respect to 0. Thus the fixed point index generalizes the Leray-Schauder degree. However, it is also defined in situations where degree theory is not directly applicable, for example, when X is a closed convex set with an empty interior in a Banach space Y or X does not have a linear structure. The fixed point index is a powerful, though surprisingly little-known, tool for the study of nonlinear problems in analysis. Indeed the fixed point index is even useful in the study of linear positive operators.

The first difficulty in defining the fixed point index is that one must find a suitable class of spaces X. A reasonable class of spaces X in which to try to study the fixed point index is provided by the so-called metric *absolute neighbourhood retracts* or metric ANRs. A typical example is the following: Suppose that C_i, $1 \leq i \leq m$, are closed convex subsets of a locally convex topological vector space E and that $X = \cup_{i=1}^{m} C_i$ is metrizable. Then X is a metric ANR [149]. Classical results about the fixed point index can be found in [94, 149] and the references therein.

Fixed point index for ANRs

We give a short introduction on the fixed point index for ANRs following [149]. As mentioned above, we also refer to [94] and [136] and the references therein. Let us consider a fixed point index category of spaces, the ANRs.

Definition 2.2 Let X be a metric (metrizable) space. X is called an ANR iff the following property holds: Whenever we have a metric space M and a continuous map $f : M \supseteq C \to X$, with C closed, then f has a continuous extension to an open set $\mathcal{O}$ with $M \supseteq \mathcal{O} \supseteq C$.

To put it simply, this says that if there is a map $f : C \to X$ then there is an extension of f to an open neighbourhood of C, $\mathcal{O}$.

Definition 2.3 X is called an Absolute Retract (AR), if, in the same situation as above, $f : M \supseteq C \to X$, with C closed, then there exists a continuous extension to the whole set M.

Example 2.9 Any compact interval $[a, b]$ is an AR. By the Tietze extension theorem, if there is a map $f : M \supseteq C \to [a, b]$, then this map extends continuously to the whole space M. These extensions also hold good to the product of intervals. Indeed, let us consider $\mathbf{f} : C \to \prod_{i=1}^{N} [a_i, b_i]$, with the different components of the function as $\mathbf{f} = (f_1, f_2, \cdots f_n)$, then each component can be extended individually and hence the extension applies to the product space.

Thus an AR is equivalent to a closed set C admitting an extension to the whole space, whereas an ANR is equivalent to the case where an extension exists to a neighbourhood of the set C. A retraction may be explained as follows: If Z is a retract of Y, the $Z \subset Y$ and there exists a continuous map $r : Y \to Z$ such that $r|_z = id_z$. (id_z is the Identity map of Z). Assume that Z is a closed subset of Y, consider the following scheme $Y \supseteq Z \overset{id}{\to} Z$. If Z is AR then the identity can be continuously extended to Y and therefore Z is a retract of Y. This means that if Z is an AR, then it is a retract of any space in which Z can be embedded as a closed subspace. Similarly, if Z is an ANR then whenever Z can be embedded as a closed subspace of a space Y, then there is a retraction of a neighbourhood of Z onto Z.

Some typical examples of an AR are a convex subspace of a Banach space (Dugundji extension theorem) (See J. Dugundji [58]). Suppose that X is a convex subset of a normed space E, A is a closed subset of a metric space M, and $f : A \to X$ is a continuous map then f has a continuous extension $F : M \to X$.

More generally speaking, if X is a closed, convex subset of a metrizable locally convex topological vector space E (or lctvs E, for short), there exists a retraction $r : E \to X$. (Take $X = A$, $M = E$ and $f(x) = x$ for all $x \in A$). Some typical examples of ANRs are given by the union of finitely many closed convex subspaces of a normed space. Moreover for a metrizable space, the property of being an AR or ANRs is preserved by the homeomorphisms, by definition. Consider as an example in $\mathbb{R}^2$ the perimeter of a square, this is an ANR as it is the union of four convex sets. Also, the circumference of a circle is an ANR as the upper semi-circle and the lower semi-circle are homeomorphic to two line segments.

The advantage of the fixed point index as against the topological degree theory is that the topological degree is applicable to open subsets of a normed space. One needs a linear structure. On the other hand, the fixed point index is applicable to open subsets of ANRs, therefore this need not be a linear structure. For example, in $\mathbb{R}^n$ consider the surface of a sphere, which is an ANR, because if one considers the upper and lower hemispheres of the sphere, these are homeomorphic to a disc which is compact. Thus we can say that the sphere is homeomorphic to the union of finitely many closed convex sets. Thus we have a fixed point theory for maps defined on the open subsets of the sphere with values in the sphere.

Axioms of the fixed point index

Recall that for the Brouwer degree, an admissible triplet (f, Ω, p) consists of f a continuous map, Ω, an open set and p, a point, with the requirement that $f(x) \neq p \ \forall x \in \partial\Omega$. In the setting of the fixed point index, we consider now the admissible triplet with X, a metric space, which is an ANR, U an open subset of X and $\phi : U \to X$ a continuous map with the assumption that the set of fixed points denoted by $\mathrm{Fix}(\phi, U) = \{x \in U : \phi(x) = x\}$ is a compact subset of U and ϕ is a compact map restricted to a neighbourhood of $\mathrm{Fix}(\phi, U)$.

In the sequel, we shall focus our attention on the pair (ϕ, U) and ignore the ambient space X as it is apparent from the context.

- Consider a special case in which the above conditions are applicable. Assume $\phi : \overline{U} \to X$ is compact (that is, ϕ is continuous and $\overline{\phi(\overline{U})}$ is compact).
- $\phi(x) \neq x, \ \forall \ x \in \partial U$.

Then we claim that the fixed point set, $\mathrm{Fix}(\phi, U) \subset U$ is compact.

Consider a sequence $x_n \in \mathrm{Fix}(\phi, U)$ and we prove that this sequence has a convergent subsequence. We have $x_n = \phi(x_n)$ with $\phi(x_n) \in \overline{\phi(\overline{U})}$, which is compact. Therefore there exists a convergent subsequence $x_{k_n} \to x^*$, $x_n \in U$, $x_{k_n} \in U$, $\implies$

$x^* \in \overline{U}$. We have $x_n = \phi(x_n)$ and taking limits as $n \to \infty$, gives $x^* = \phi(x^*)$ since ϕ is a continuous map. This shows that x^* is a fixed point of ϕ, and by assumption we have that $x^* \notin \partial U$. Thus we have the following: U is an open set, $x^* \in \overline{U},\ x^* \notin \partial U$. Hence $x^* \in U$, thus we have proved that $x^* \in \text{Fix}(\phi, U)$. Any sequence of points in $\text{Fix}(\phi, U)$ has a convergent subsequence in $\text{Fix}(\phi, U)$. Hence $\text{Fix}(\phi, U)$ is compact. Of course, ϕ is compact in a neighbourhood of $\text{Fix}(\phi, U)$. As we assumed that ϕ is compact on $\overline{U}$ and therefore ϕ is compact in $U \implies \phi$ is compact in a neighbourhood of $\text{Fix}(\phi, U)$.

As mentioned before, the general theory can be developed for the following case:

- X is an ANR, U is open in X, $\phi : U \to X$ is continuous and such that $\text{Fix}(\phi, U)$ is compact and ϕ is compact in a neighbourhood of $\text{Fix}(\phi, U)$.

For the general theory see [94, 136, 149]. For the case under consideration here we restrict our attention to the following simplified scenario.

- Admissible triplet: A triplet of the form X, U, ϕ, where X is ANR, U is open in X (not necessarily bounded) and $\phi : \overline{U} \to X$ is continuous and $\overline{\phi(\overline{U})}$ is compact such that there are no fixed points on the boundary, i.e. $\phi(x) \neq x,\ \forall\, x \in \partial U$.
- Analogously, if we consider a homotopy, $h : \overline{U} \times [0, 1] \to X$ we assume that this is compact and $h(x, \lambda) \neq x,\ \forall\, x \in \partial U$ and $\forall\, \lambda \in [0, 1]$.

We now give the axioms for the fixed point index. To each admissible triplet (X, U, ϕ) we associate an integer, which will denote by $i_X(\phi, U) \in \mathbb{Z}$, called the fixed point index of ϕ in U relatively to X. This satisfies the following conditions:

- Additivity/Excision: If $\phi(x) \neq x\ \ \forall\ \ x \in \overline{U} \setminus (U_1 \cup U_2)$ where $U_1,\ U_2$ are open and disjoint subsets of U, then the $i_X(\phi, U) = i_X(\phi, U_1) + i_X(\phi, U_2)$.
- Invariance under homotopy: let $\psi : \overline{U} \times [0, 1] \to X$, be a compact map such that $\psi(x, \lambda) \neq x,\ \ \forall\ x \in \partial U,\ \forall\, \lambda \in [0, 1]$, then $i_X(\psi(\cdot, \lambda), U) = \text{constant}$ with $\lambda \in [0, 1]$.
- The normalization property: The general normalization property for the fixed point index relates to the fixed point index with an object introduced in algebraic topology, that is the Lefschetz number. But the normalization property may also be presented without invoking the Lefschetz number. We denote by "strong normalization property" as that which refers to the Lefschetz number. There are some weaker forms of the normalization property which can be expressed without an explicit reference to the Lefschetz number.
  "Weak normalization property": In the degree theory we have, $deg_{LS}(I, \Omega, p) = deg(I - p, \Omega, 0) = 1$ if $p \in \Omega$. If we take $\phi(x) = p = $ constant map, then we have $i_X(p, U) = 1$ if $p \in U$.
  There are other properties of the fixed point index, namely, the
- Product property: If X_1 and X_2 are ANR, U_1 and U_2 are open in X_1 and X_2, respectively, then we have, and $\phi_1 : \overline{U_1} \to X_1,\ \ \phi_2 : \overline{U_2} \to X_2$, such that the triplet $\{X_1, U_1, \phi_1\}$ and $\{X_2, U_2, \phi_2\}$ are admissible, then we have,
  $i_{(X_1 \times X_2)}(\phi_1 \times \phi_2, U_1 \times U_2) = i_{X_1}(\phi_1, U_1) \times i_{X_2}(\phi_2, U_2)$

- Commutativity property: This implies that fixed point index is invariant by homeomorphism, (that is homeomorphic spaces have the same fixed point index.) The precise details of this property are not given here because these properties are not used in the succeeding sections. However, we will apply an important result (which in the Leray-Schauder degree theory is a direct consequence of the construction of the degree from the finite-dimensional approximation), which is usually known as "contraction property" or "reduction property" and claims that if the range of ϕ is contained in a subspace Y of X, then

$$i_X(\phi, U) = i_Y(\phi|_Y, U \cap Y)$$

(see [149, p. 149] and also [49, Sect. 1.3.4] for further information.)

If we compare the degree theory with the fixed point index we can see that we have the following correspondence:
Take

$$\begin{aligned} X &= \text{a normed space} \\ U &= \Omega, \quad \text{which is open and bounded} \end{aligned}$$

then we have $deg_{LS}(I - \phi, U, 0) = i_X(\phi, U)$.

As mentioned earlier, there is a more general form of the normalization property which connects the fixed point index to the so-called Lefschetz number $\Lambda(\phi)$. (see [149]). This is defined using homology theory. We use a simpler form of the normalization property which can be useful to compute the fixed point index and which is as follows:
Assume that X is a compact ANR (This is a striking difference with the degree theory, X is a Banach space and not compact.). Let $U = X$, which makes U open and $\partial U = \emptyset$. By ∂U, we mean the boundary of U with respect to X. In this case we do not have fixed points as the boundary is empty. For example consider $X = S^2$ where $S^2 \subset \mathbb{R}^3 = \{(x, y, z) : x^2 + y^2 + z^2 = 1\}$ $\partial S^2 = \emptyset$ with respect to the space S^2. ($\partial S^2 = S^2$ in $\mathbb{R}^3$.) Consider the identity map in X, i_X If X is compact, the i_X is a compact map. Then, the normalization property maybe expressed by the formula

$$i_X(I_X, X) = \chi(X),$$

where $\chi(x)$ is the Euler-Poincaré characteristic of X. The characteristic of any space, $\chi(X) \in \mathbb{Z}$. This is a generalization of the Euler-Descartes formula for polyhedra, which is usually defined as (the number of faces $-$ the number of edges $+$ the number of vertices). For example for a cube this number is $6 - 12 + 8 = 2$. The properties of the Euler-Poincaré characteristic for compact ANRs are as follows:

- $\chi(\emptyset) = 0 \quad \chi(\{\cdot\}) = 1$
- $\chi(X_1 \cup X_2) = \chi(X_1) + \chi(X_2) - \chi(X_1 \cap X_2)$

- $X_1 \simeq X_2$ homotopically equivalent, then $\chi(X_1) = \chi(X_2)$
- $\chi(X_1 \times X_2) = \chi(X_1) \times \chi(X_2)$

Examples of homotopically equivalent spaces:

- A compact convex set in $\mathbb{R}^n$ or a compact star-shaped set. These sets are homotopic to a point. Indeed, X is homotopically equivalent to p where p is any point of the convex set or, respectively, the centre of the star-shaped set. If X is a compact, convex set, $X \simeq \{p\}$. The $\chi(X) = \chi(\{\cdot\}) = 1$. So the characteristic for both a compact convex set and a compact star-shaped set is 1.
- If X_1 and X_2 are homeomorphic, they have the same characteristic.
- For a disc, the characteristic $\chi = 1$. By the boundary of a disc we mean, $S^1 = \{(x, y) : x^2 + y^2 = 1\}$ which may be considered as the union of two semi-circles. The intersection of these semi-circles consists of two points. These semi-circles are homotopic to two line segments. Then $\chi(S^1) = \chi(\text{upper semi-circle}) + \chi(\text{lower semi-circle}) - \chi(\text{the two points of intersection}) = 0$
- The annulus is homotopic to a circle, so $\chi(\text{annulus}) = 0$.
- The torus is $T_2 = S^1 \times S^1$ so $\chi(T_2) = 0 \times 0 = 0$. Another example of a product space is that of the cylinder = $C_2 = S^1 \times [0, 1]$ and $\chi(C_2) = 0 \times 0 = 0$.
- Let us consider the example of the surface of a sphere S^2, it is the union of the upper hemisphere and the lower hemisphere, intersecting in a circle, then its characteristic is $\chi(S^2) = \chi(\text{upper hemisphere}) + \chi(\text{lower hemisphere}) - \chi(\text{the circle of intersection}) = 1 + 1 - 0 = 2$. The two hemispheres being homotopic to a disc have $\chi = 1$, and $\chi(\text{of the circumference of a circle}) = 0$.

Corollary 2.4 *Let X be a compact ANR, with $\chi(X) \neq 0$ and suppose that $\phi : X \to X$ is a continuous map, homotopic to the Identity map, then ϕ has fixed points.*

In fact, this implies that $\exists\, h : X \times [0, 1] \to X$ a continuous map such that $h(\cdot, 0) = I_x : h(\cdot, 1) = \phi$. (Of course there are no fixed points on $\partial X = \emptyset$.) Then by the homotopic invariance of the fixed point index, we have that $i_X(\phi, X) = i_X(I_x, X) \overset{\text{def}}{=} \chi(X) \neq 0$. Then ϕ has a fixed point.

As an application, consider ϕ to be the Poincaré map associated with a tangent vector field on the sphere S^2. Consider the differential equation where the vector field at any point is tangent to the surface of the sphere, then the solutions, from the theory of positively invariant sets, go from one point of the sphere to another point of the sphere. So $\phi : S^2 \to S^2$ is homotopic to the identity and hence, ϕ has fixed points. This is because the characteristic of a sphere $\neq 0$. This would not be true if ϕ is not homotopic to the identity. For example if we take $\phi(x) = -x$, this has no fixed points on the sphere. This is also the case if $\chi(X) = 0$. For example in S^1, if we take a rotation of $\pi/2$, there are no fixed points on S^1. But the rotation of $\pi/2$ is homotopic to the identity. This is because $\chi(S^1) = 0$. It is the same also in the case of the annulus A. Let us consider a continuous map $\phi : A \to A \subset \mathbb{R}^2$, on a compact annular domain. If ϕ is homotopic to the identity map, for example, if ϕ is a Poincaré map[4] because if we denote then, $i_A(\phi, A) = i_A(I_A, A) = \chi(A) = 0$. So if the index $= 0$, we are not guaranteed that there are fixed points. This fact becomes relevant to the section concerning the Poincaré-Birkhoff theory. The Poincaré-Birkhoff theory applies in case the fixed point index $= 0$ (because of the fact that the Euler-Poincaré characteristic of the annulus $= 0$). There is an analogy from the results of the degree.

Consider $\deg(f, U, 0)$, for $U =]a, b[$ with $f(a) > 0$ and $f(b) > 0$. We have already shown that the degree becomes $= 0$ and either $f(x) = 0$ has no solutions in U or, typically, the solutions occur in pairs. The same result is obtained from the Poincaré-Birkhoff theorem. In the Poincaré-Birkhoff theorem, the index $= 0$ and we obtain pairs of solutions. Indeed according to the Poincaré-Birkhoff theorem there are at least two fixed points. Usually, the solutions are such that one has a positive index $= 1$ and the other has a negative index $= -1$ and the total index $= 0$. This arises sometimes because there are two fixed points, a saddle point and a centre. Figure 2.2 shows the portrait of the pendulum equation in the cylindrical phase space. The figure on the left shows the level lines of energy of the pendulum equation $\dot{x} = y, \ \dot{y} = -\sin(x)$ on the cylindrical phase space. The figure on the right is the projection of the level lines from the cylinder onto a plane in $\mathbb{R}^2$, showing the

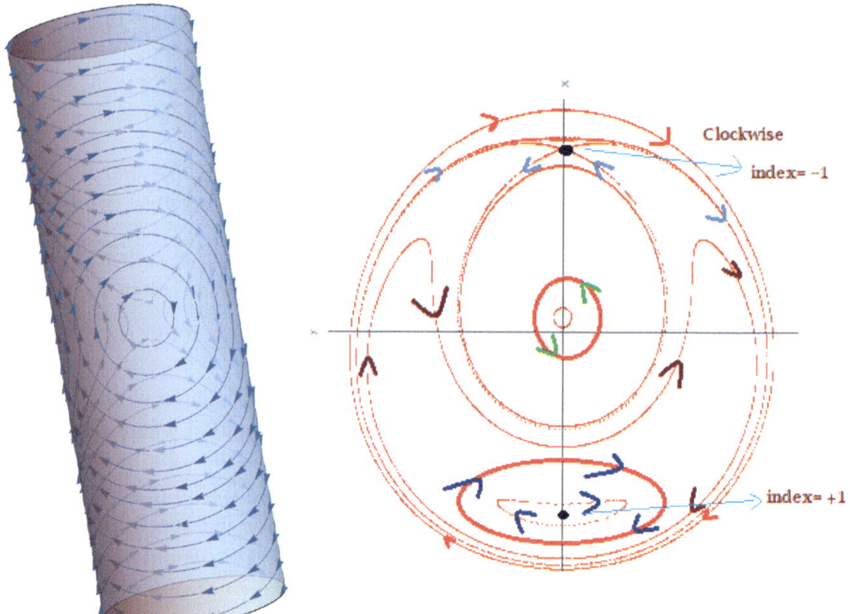

Fig. 2.2 Portrait of the pendulum equation in the cylindrical phase space and its projection

[4] A Poincaré map is always homotopic to the Identity map. A homotopy given by $\lambda \mapsto \tilde{x}(\lambda T; z)$, where $\tilde{x}(t; z)$ is the solution of the differential system with initial point $x(0) = z$. Clearly, for $\lambda = 1$ we have the Poincaré map, whilst for $\lambda = 0$ we have the Identity map.

sum $= 0$ of the fixed point indices of the two equilibrium points, the saddle point with index $= -1$, and the centre with index $= 1$, $(1 + (-1) = 0)$.

2.10 Computation of the Degree with Some Examples

In this section, we give some examples of applications of the fixed point index and degree theory to nonlinear differential equations. As a first example we derive the classical Schauder fixed point theorem and its application to the Peano local existence theorem for the initial-value problems of ODEs. As a second example we introduce Mawhin's coincidence degree and give some applications to the periodic problem associated to the Duffing equation.

2.10.1 The Schauder Fixed Point Theorem

We give an extension of the Brouwer fixed point theorem to normed spaces due to Schauder, [173], (see also, [120]). For convex sets we have the following analogue of the Brouwer fixed point theorem.

Theorem 2.8 *Let $\Omega \subset X$ be a bounded, open and convex set with $0 \in \Omega$ and let $g : \overline{\Omega} \subset X \to X$ be a compact mapping. If $g(\overline{\Omega}) \subset \overline{\Omega}$ then g has at least one fixed point.*

Proof We argue by contradiction. Suppose $g(x) \neq x$ for all $x \in \overline{\Omega}$. Consider the homotopy $h_t(x) = x - tg(x)$. In particular, $g(x) \neq x$ for $x \in \partial\Omega$. This implies that $0 \notin h_1(x)$. Observe, since Ω is open, contains 0 and is convex, that $tg(x) \in \Omega$ for all $0 \le t < 1$ and for all $x \in \overline{\Omega}$.

Consequently, if $x \in \partial\Omega$, then $h_t(x) \neq 0$ for all $0 \le t < 1$. The Leray-Schauder degree $deg_{LS}(h_t, \Omega, 0)$ is well defined and independent of $t \in [0, 1]$. For t = 0, $deg_{LS}(h_0, \Omega, 0) = deg_{LS}(I, \Omega, 0) = 1$. Hence $deg_{LS}(h_1, \Omega, 0) = 1$. On the other hand, since $g(x) \neq x$ for all $x \in \overline{\Omega}$, we have that $h_1^{-1}(0) = \emptyset$, which implies that $deg_{LS}(h_1, \Omega, 0) = 0$, a contradiction. □

Remark 2.8 The Schauder fixed point theorem formulated above is also true if we replace the condition $0 \in \Omega$, by $\Omega \neq \emptyset$. In these let $x_0 \in \Omega$ be an interior point. Define $y = x - x_0$ and $\tilde{g}(y) = g(y + x_0) - x_0$. If we now apply Theorem 2.8 to $\tilde{g}$ on the domain $\tilde{\Omega} = \Omega - x_0$.

If, instead of the Leray-Schauder degree, we use the fixed point index, we can get the result for an arbitrary nonempty convex set C instead of $\overline{\Omega}$.

As an application of the Schauder fixed point theorem we now consider Peano's Theorem which guarantees the existence of solutions to initial-value problems in $\mathbb{R}^n$. Consider the initial-value problem,

$$\dot{y}(t) = f(t, y(t)), \quad y(t_0) = y_0 \in \mathbb{R}^n, \ t \in \mathbb{R}, \tag{2.38}$$

where $y : \mathbb{R} \to \mathbb{R}^n$ and $f : \mathbb{R} \times \mathbb{R}^n \to \mathbb{R}^n$ is a continuous function. Note that we only assume continuity of f.

Theorem 2.9 (Peano's theorem) *Consider Eq.* (2.38) *with f continuous. Let $y_0 \in \mathbb{R}^n$. Then there exists a time $\tau > 0$ and a C^1- function $y : [t_0 - \tau, t_0 + \tau] \to \mathbb{R}^n$ which satisfies Eq.* (2.38).

Proof We will restrict the proof to the case $t_0 = 0$ and $y_0 = 0$ without loss of generality. We start with reformulating Eq. (2.38) in terms of an integral equation. Integration of (2.38) over $[0, t]$ yields the Volterra integral equation:

$$y(t) = \int_0^t f(s, y(s))ds =: g(y)(t), \tag{2.39}$$

which provides a fixed point problem for the mapping g. Let $Y = C^0([-\tau, \tau]; \mathbb{R}^n)$ with norm $\| y \| := \sup_{t \in [-\tau, \tau]} |y(t)|$ and $\overline{\Omega} = \{y \in Y : \| y \| \leq M\}$ is a closed, bounded subset containing 0. The mapping g maps from $\overline{\Omega}$ in to Y. We first need to choose the constants τ and M such that $g(\overline{\Omega}) \subset \overline{\Omega}$. We have for $y \in \overline{\Omega}$

$$|g(y)(t)| \leq \left| \int_0^t f(s, y(s))ds \right| \leq \left| \int_0^t |f(s, y(s))| ds \right| \leq |t||f(s, y(s))| \leq \tau N,$$

where $N = \max_{(s, y) \in [-\tau, \tau] \times [-M, M]} |f(s, y)|$. This implies $\| g(y) \| \leq \tau N$. If we choose $\tau \leq M/N$, then $g(\overline{\Omega}) \subset \overline{\Omega}$.

Finally the mapping g is compact. Indeed, $g(y)(t)$ is continuously differentiable and $(d/dt)g(y)(y) = f(t, y(t))$. We now use the fact that the embedding $C^1([-\tau, \tau]) \hookrightarrow C^0([-\tau, \tau])$ is compact. Therefore, g maps bounded sets to precompact sets in Y. The existence of a solution to $y = g(y)$ is now guaranteed by the Schauder fixed point theorem (Theorem 2.8), which completes the proof. □

Remark 2.9 As mentioned earlier, in order to prove Peano's Theorem for every t_0 and y_0 we can use the extension of the Schauder fixed point theorem discussed in Remark 2.8.

With the fixed point index, the Schauder fixed point theorem may be proved easily, as observed previously, we detail this in Theorem 2.10.

Theorem 2.10 *Let C be a nonempty, convex closed set of a normed space and $\phi : C \to C$ be a compact map. Then ϕ has a fixed point.*

Proof Let $X = C$ be ANR, which would be actually an AR by the Dugundji extension theorem. Then, since U, the open set is taken to be $U = X = C$, $\partial U = \emptyset$. Let $p \in C$ (a given point) and a homotopy $\psi(X, \lambda) = \lambda\phi(X) + (1 - \lambda)p$. We have $\psi : C \times [0, 1] \to C$ is compact (also there are no fixed points on the boundary, as

$\partial U = \emptyset$.) Then, $i_C(\phi, C) = i_C(p) = 1$ (by homotopic invariance and the weak form of normalization) and $i_C(\phi, C) \neq 0$, $\implies$ ϕ has a fixed point in C. □

The difference between this proof and the previous proof is that we have $C = \overline{\Omega}$ with Ω being convex and open in the previous case. In this case, C is any closed convex set.

2.10.2 *The Coincidence Degree and Some Applications*

The Coincidence Degree is an extension of the Leray-Schauder degree to the study of abstract operator equations, in normed spaces of the form,

$$L(x) = N(x), \quad x \in dom(L). \tag{2.40}$$

where L is a linear operator and N is a nonlinear one. When L is invertible, Eq. (2.40) can be transformed to the fixed point equation

$$x = \phi(x), \quad \text{with} \quad \phi = L^{-1} \circ N$$

and, under suitable continuity and compactness assumptions on $L^{-1} \circ N$, one can apply the Leray-Schauder degree theory to prove the existence of a fixed point for ϕ and hence the existence of a solution to Eq. (2.40). A problem arises when L is not invertible. In this case it is necessary to use a different approach (by going around the equation) to solve (2.40), which is usually named as a *coincidence equation.* A solution to this problem was proposed by J. Mawhin in [127] and subsequently fully developed in the setting of locally convex topological vector spaces in [128]. In the long and exhaustive introduction of [128] the interested reader will find several relevant references related to previous approaches to deal with equations of the type (2.40). Of late the term coincidence degree reveals more than 1600 references, indicating also its widespread use in a large variety of nonlinear problems. This theory is especially well suited to the study of the periodic problem. Indeed, if we look at (2.6), we can consider the linear differential operator $L : x \mapsto \dot{x}$ and the Nemytskii operator $N : x \mapsto f(\cdot, x(\cdot))$ in order to transform (2.6) to the operator equation (2.40) in a suitably chosen space X of periodic functions (in this case typically X is the space of continuous functions with periodic boundary conditions). If X is an absolutely continuous function, with $x(0) = x(T)$, then $\dot{x} = 0$ if and only if x is a constant. Hence the kernel of L is an m-dimensional subspace and L is not invertible. A systematic development of the coincidence degree with several applications may be seen in [91, 132, 134]. An extension of the theory to the case of nonlinear differential operators may be seen in [121].

To present the full range of topics of coincidence degree theory is well beyond the scope of this manuscript; moreover, this theory is represented exhaustively in the references just cited. We will confine ourselves to the applications of this theory to the Duffing equation.

Let us consider the differential system (2.5) that we recall is

$$\dot{x} = f(t, x),$$

where $f = f(t, x) : \mathbb{R} \times \mathbb{R}^m \to \mathbb{R}^m$ is a continuous function which is periodic of period $T > 0$ in the t-variable. The search of T-periodic solutions for this differential system will be transformed into an equivalent coincidence equation in the space $X = \mathscr{C}_T$ of the continuous and T-periodic functions $\mathbb{R} \to \mathbb{R}^m$. In $\mathscr{C}_T$ we take the natural norm, namely,

$$||x||_{\mathscr{C}_T} = ||x||_\infty = \sup_{t\in[0,T]} ||x(t)||,$$

where $||\cdot||$ is the Euclidean norm in $\mathbb{R}^m$. To this end we set

$$L(x) := \dot{x}, \qquad N(x) := f(\cdot, x(\cdot)). \tag{2.41}$$

The linear operator L is defined on the subspace

$$\mathscr{C}_T^1 = \{x \in \mathscr{C}_T : \ x \text{ is of class } C^1\}.$$

By the above definitions it is clear that $x(\cdot)$ is a T-periodic solution of the Eq. (2.5) if and only if x is a solution of a coincidence-type equation in the form of (2.40).

Within the above framework, the kernel of L consists of constant functions as mentioned earlier. Therefore we set

$$\text{Ker } L = \mathbb{R}^m$$

because we identify constant functions with the points of $\mathbb{R}^m$. The range of L consists of T-periodic functions with zero mean value. Now it is convenient to introduce the projection

$$P : \mathscr{C}_T \to \mathbb{R}^m, \qquad P(x) := \frac{1}{T}\int_0^T x(t)dt. \tag{2.42}$$

By definition

$$P : \mathscr{C}_T \to \text{Ker } L, \quad I - P : \mathscr{C}_T \to Im(L).$$

The linear operator L even when noninvertible admits a so-called *right inverse*, namely, there exists

$$\mathscr{K} : Im(L) \to \text{Ker } P = \left\{x \in \mathscr{C}_T : \int_0^T x(t)dt = 0\right\}$$

such that

$$L \circ \mathscr{K} = \text{Identity on } Im(L), \qquad \mathscr{K} \circ L = \text{Identity on } dom(L) \cap \text{Ker } P.$$

Then we have that $x \in dom(L)$ is a solution of (2.40) if an only if

$$\begin{cases} (I - P)x = \mathscr{K} \circ (I - P)N(x) \\ PN(x) = 0. \end{cases} \tag{2.43}$$

The first equation in (2.43) is defined on Ker $P \cap dom(L)$, whilst the second equation is defined in its complement (finite dimensional) space Ker L. Since the two equations lie in orthogonal spaces, we can merge them into a unique fixed point equation of the form

$$x = \phi(x) := Px + JPN(x) + \mathscr{K} \circ (I - P)N(x), \quad x \in X, \tag{2.44}$$

where $J : Im(P) \equiv \mathbb{R}^m \to \text{Ker } L \equiv \mathbb{R}^m$ is a fixed linear isomorphism. In the sequel, we take

$$J = \text{Identity on Ker } L = I_{\mathbb{R}^m} .$$

Observe that $x \in X$ is a fixed point of ϕ if and only if $x \in dom(L)$ is a solution of the coincidence equation (2.40). As explained in [128], the above transformation works in full generality whenever $L : dom(L) \subset X \to Z$ is a linear operator with finite-dimensional kernel and a closed range having a complementary subspace of the same dimension as the kernel. Such linear operators are called *linear Fredholm maps of index zero*. Moreover $N : dom(N) \subset X \to Z$ need not necessarily be a linear operator. In the general setting, one takes *two linear projections* $P : X \to \text{Ker } L$ and $I - Q : Z \to Im(L)$ and (2.44) takes the form

$$x = \phi(x) := Px + JQN(x) + \mathscr{K} \circ (I - Q)N(x), \quad x \in dom(N), \tag{2.45}$$

where $\mathscr{K} : Im(L) \to \text{Ker } P$ is the right inverse of L and $J : Im(Q) \to \text{Ker } L$ is a linear isomorphism between finite-dimensional spaces of the same dimension.

In our setting, concerning the T-periodic problem for (2.5) and $X = Z = \mathscr{C}_T$, the operator ϕ defined in (2.44) is completely continuous. This is followed by the continuity of the vector field f. If we assume on f the more general Carathéodory conditions, then we can still have the complete continuity of the operator ϕ in the more general setting of (2.45) if we take $X = \mathscr{C}_T$ and as Z the space of T-periodic functions $u : \mathbb{R} \to \mathbb{R}^m$ with $u \in L^1([0, T])$. In order to simplify our presentation, we restrict ourselves to the case in which f is continuous and refer to [131] for the Carathéodory setting.

Having set the equivalence between (2.40) and (2.44), we are now in a position to apply the topological degree. In fact, according to [128] we can define the *Coincidence Degree* for $L - N$ as

$$D(L - N, \Omega) := deg_{LS}(I - \phi, \Omega, 0), \tag{2.46}$$

where "deg_{LS}" is the previously introduced Leray-Schauder degree and $\Omega \subset X$ is an open and bounded set such that

$$L(x) \neq N(x) \quad \forall x \in \partial\Omega \cap dom(L). \tag{2.47}$$

In this manner, the properties of topological degree, such as *Additivity/Excision, Invariance under homotopy, Existence of solutions*, are transferred to the coincidence degree. In particular, if $D(L - N, \Omega) \neq 0$, then (2.5) has a T-periodic solution in Ω.

Now we present a few examples which link the calculation of the coincidence degree to the Brouwer degree of the vector field f. Typically these kind of results belong to the family of the so-called *Principles of Relatedness* (or *Duality Principles*), following the terminology adopted in [107, Chap. 3].

Theorem 2.11 *Let $\Omega \subset \mathscr{C}_T$ be an open and bounded set and assume the following two conditions:*

$(A1)$ *the equation*

$$\dot{x} = \lambda f(t, x),$$

has no T-periodic solutions on $\partial\Omega$, for each $\lambda \in (0, 1]$;

$(A2)$ *for*

$$\bar{f}(z) := \frac{1}{T}\int_0^T f(t, z)\, dt, \quad z \in \mathbb{R}^m,$$

we have that

$$\bar{f}(z) \neq 0, , \quad \forall z \in \partial\Omega \cap \mathbb{R}^m.$$

Then,

$$D(L - N, \Omega) = (-1)^m d(\bar{f}, \Omega \cap \mathbb{R}^m, 0).$$

We recall that $d(\bar{f}, \Omega \cap \mathbb{R}^m, 0)$ is the Brouwer degree as introduced in Sect. 2.6.

Proof We consider (2.44) and perform the homotopy

$$x - h(x, \lambda) := x - Px - JPN(x) - \lambda \mathscr{K} \circ (I - P)N(x), \quad \lambda \in [0, 1], \quad x \in \overline{\Omega}$$

such that

$$x - h(x, 1) = x - \phi(x), \quad \text{and } x - h(x, 0) = x - Px - JPN(x).$$

Now we have that $x = h(x, \lambda)$ with $\lambda \in (0, 1]$ if and only if $L(x) = \lambda N(x)$ for $x \in dom(L)$ and $\lambda \in (0, 1]$. On the other hand, for $\lambda = 0$, we have $x = h(x, 0)$ if and only if $x = Px + JPN(x)$. In this case, $x \in Im(P) = \text{Ker } L = \mathbb{R}^m$ and therefore

$x = Px + JPN(x)$ reduces to $JPN(x) = 0$ with $x \in \text{Ker } L$. By (2.41) we have that $L(x) = \lambda N(x)$ with $x \in dom(L)$ if and only if $x(\cdot)$ is a T-periodic solution of $\dot{x} = \lambda f(t, x)$. Moreover, by the definition of the projection P in (2.42) we have that $JPN(x) = 0$ with $x \in \text{Ker } L$, if and only if $x(\cdot) = \text{constant} = z$ with $\bar{f}(z) = 0$. As a consequence, we have that conditions $(A1)$ and $(A2)$ imply that

$$x - h(x, \lambda) \neq 0, \quad \forall x \in \overline{\Omega}, \; \lambda \in [0, 1]$$

and therefore the homotopy is admissible. The homotopy invariance of the Leray-Schauder degree and the reduction formula (see [49, Sect. 1.3.4]) imply that

$$\begin{aligned} D(L - N, \Omega) &= deg_{LS}(I - \phi, \Omega, 0) = deg_{LS}(-JPN(x), \Omega \cap \mathbb{R}^m, 0) \\ &= d(-\bar{f}, \Omega \cap \mathbb{R}^m, 0) = (-1)^m d(\bar{f}, \Omega \cap \mathbb{R}^m, 0), \end{aligned}$$

because we have taken as J the identity in $\mathbb{R}^m$. This concludes the proof of the theorem. □

A first consequence of Theorem 2.11 is Mawhin's continuation theorem which can be expressed as follows.

Theorem 2.12 *Let $\Omega \subset \mathscr{C}_T$ be an open and bounded set and assume the following two conditions:*

$(B1)$ *the equation*

$$\dot{x} = \lambda f(t, x),$$

has no T-periodic solutions on $\partial\Omega$, for each $\lambda \in (0, 1)$;

$(A2)$ *for*

$$\bar{f}(z) := \frac{1}{T} \int_0^T f(t, z)\, dt, \quad z \in \mathbb{R}^m,$$

we have that

$$\bar{f}(z) \neq 0, , \quad \forall z \in \partial\Omega \cap \mathbb{R}^m.$$

If

$$d(\bar{f}, \Omega \cap \mathbb{R}^m, 0) \neq 0,$$

then system (2.5) *has a T-periodic solution $\tilde{x}(\cdot) \in \overline{\Omega}$.*

Proof If there is a T-periodic solution $\tilde{x}(\cdot) \in \partial\Omega$ of (2.5), we are done. Otherwise, if there are no T-periodic solutions of (2.5) on $\partial\Omega$, we have that condition $(B1)$ holds for all $\lambda \in (0, 1]$ and therefore $(A1)$ of Theorem 2.11 is satisfied. From $(A1)$ and $(A2)$, we obtain

$$D(L - N, \Omega) = (-1)^m d(\bar{f}, \Omega \cap \mathbb{R}^m, 0) \neq 0$$

and therefore $D(L - N, \Omega) \neq 0$, which in turn implies that $L(x) = N(x)$ has at least one solution in Ω. Hence, in any case, (2.5) has a T-periodic solution $\tilde{x}(\cdot) \in \overline{\Omega}$. □

A second consequence can be obtained by combining Theorem 2.11 with the Leray-Schauder continuation Theorem 2.6, as follows.

Theorem 2.13 *Let $\Omega \subset \mathscr{C}_T$ be an open and bounded set and assume* (A2) *with*

$$d(\bar{f}, \Omega \cap \mathbb{R}^m, 0) \neq 0.$$

Then there is a continuum Σ of solution-pairs $(\lambda, \tilde{x})$, with $\tilde{x}$ a T-periodic solution of

$$\dot{x} = \lambda f(t, x), \quad \lambda \in [0, 1]$$

emanating from

$$\Sigma_0 \subset \{z \in \Omega \cap \mathbb{R}^m : \bar{f}(z) = 0\}$$

and such at least one of the following conditions hold:

- $\Sigma_\lambda \cup \partial\Omega \neq \emptyset$ *for some* $\lambda > 0$
- Σ *takes all the values in* $[0, 1]$.

As usual, we have denoted by Σ_λ the section of Σ at λ, namely, $\Sigma_\lambda = \{x \in X : (\lambda, x) \in \Sigma\}$.

Applications of this result to the periodic problem associated with (2.5) can be found in [31] and in [137].

An immediate consequence of the above results is the following.

Corollary 2.5 *Let $\Omega \subset \mathscr{C}_T$ be an open and bounded set and assume* (A2) *with*

$$d(\bar{f}, \Omega \cap \mathbb{R}^m, 0) \neq 0.$$

Then there is $\lambda_0 > 0$ such that for each $\lambda \in (0, \lambda_0]$ the system

$$\dot{x} = \lambda f(t, x)$$

has a T-periodic solutions in $\overline{\Omega}$.

Proof We apply Theorem 2.13 and consider the continuum Σ of solutions pairs $(\lambda, \tilde{x})$ emanating from Σ_0. If $\Sigma_\lambda \subset \Omega$ for each $\lambda \in [0, 1]$, our thesis is achieved with $\lambda_0 = 1$. Otherwise, there is a $\lambda > 0$ such that $\Sigma_\lambda \cup \partial\Omega \neq \emptyset$. In this case, we define

$$\lambda_0 = \min\{\lambda > 0 : \Sigma_\lambda \cup \partial\Omega \neq \emptyset\}$$

(such a minimum does exist) and we can conclude. □

It is interesting to observe that all the above results concerning the coincidence equation, can be applied to different scenarios provided that we can enter in the above setting. For instance, instead of

$$\dot{x} = f(t, x)$$

and the corresponding parameter-dependent equation

$$\dot{x} = \lambda f(t, x), \quad \lambda > 0,$$

we can consider an nth-order differential equation

$$\mathscr{D}(x) = f\Big(t, x, \dot{x}, \cdots, \frac{d^{n-1}}{dt^{n-1}}x\Big),$$

where $\mathscr{D}$ is a nth-order linear differential operator. To show a specific example dealing with a Duffing-type equation, let us consider the T-periodic problem for the scalar second-order equation

$$\ddot{x} + c\dot{x} + h(t, x) = 0, \tag{2.48}$$

where $c \geq 0$ is a fixed constant (this covers the friction case $c > 0$ as well as the frictionless case $c = 0$) and $h = h(t, x) : \mathbb{R} \times \mathbb{R} \to \mathbb{R}$ is a continuous function which is T-periodic in the t-variable. In the space $X = \mathscr{C}_T$ we can define the T-periodic problem associated with (2.48) as a coincidence Eq. (2.40) with

$$L(x) := -\ddot{x} - c\dot{x}, \quad x \in dom(L) = \mathscr{C}_T^2,$$

with

$$\mathscr{C}_T^2 = \{x \in \mathscr{C}_T : \ x \text{ is of class } C^2\}$$

and

$$N(x) := h(\cdot, x(\cdot)).$$

Observe that Ker $L = \mathbb{R}$ (the constant functions) and the range of L consists of (continuous and) T-periodic functions with zero mean value. Now we can repeat all the preliminary construction leading finally to the fixed point equation $x = \phi(x)$ in (2.44), with the same projection P and the right-inverse $\mathscr{K}$ defined accordingly. In this case $\mathscr{K}(y)$ is the unique T-periodic solution with zero mean value of the equation

$$\ddot{x} + c\dot{x} = y(t), \quad \text{with} \quad y \in \mathscr{C}_T, \text{ and } \quad \int_0^T y(t)\, dt = 0.$$

As a last step, we have to choose an open bounded set $\Omega \in \mathscr{C}_T$ in order to apply Theorems 2.11, 2.12, or Theorem 2.13. Since we are dealing with a scalar equation, a convenient choice may be the following. Let

$$\Omega = \Omega(a, b) := \{u(\cdot) \in \mathscr{C}_T : a < u(t) < b, \ \forall t\},$$

where a, b with $a < b$ are fixed constants. In this case, $u \in \partial\Omega$ is and only if $a \leq u(t) \leq b$, for all t and there exists t_0 such that $u(t_0) = a$ or $u(t_0) = b$. Observe also that

$$\Omega \cap \operatorname{Ker} L = \Omega \cap \mathbb{R} = (a, b).$$

Finally, we define

$$\bar{h}(z) := \frac{1}{T} \int_0^T h(t, z)\, dt, \quad \text{for } z \in \mathbb{R}.$$

In this framework, Mawhin's continuation theorem 2.12 reads as

Theorem 2.14 *Let $a, b \in \mathbb{R}$ with $a < b$ such that*

$$\bar{h}(a) < 0 < \bar{h}(b), \quad \text{or} \quad \bar{h}(a) > 0 > \bar{h}(b)$$

and assume that, for each $\lambda \in (0, 1)$, the equation

$$\ddot{x} + c\dot{x} + \lambda h(t, x) = 0, \tag{2.49}$$

has no T-periodic solutions on $\partial\Omega$, for $\Omega = \Omega(a, b)$. Then (2.48) has at least one T-periodic solution in $\overline{\Omega}$.

Proof We just observe that the hypothesis of no solutions on $\partial\Omega$ for (2.49) corresponds to $(B1)$ of Theorem 2.12 in our new setting for Eq. (2.48). Condition $(A2)$ now reads as $\bar{h}(z) \neq 0$ for $z = a$ or $z = b$, because $\partial\Omega \cap \mathbb{R} = \{a, b\}$. Finally, as shown in Example (2.7) we have $d(\bar{h}, \Omega \cap \mathbb{R}, 0) \neq 0$, due to the fact that $\bar{h}(a)\bar{h}(b) < 0$. Then we conclude with Theorem 2.12 adapted to the linear differential operator $L(x) := -\ddot{x} - c\dot{x}$. □

Along the same lines, we can provide a version of Corollary 2.5 to the case of (2.48).

Corollary 2.6 *Let $a, b \in \mathbb{R}$ with $a < b$ such that*

$$\bar{h}(a) < 0 < \bar{h}(b), \quad \text{or} \quad \bar{h}(a) > 0 > \bar{h}(b).$$

Then there is $\lambda_0 > 0$ such that for each $\lambda \in (0, \lambda_0]$ the equation

$$\ddot{x} + c\dot{x} + \lambda h(t, x) = 0,$$

has a T-periodic solutions $\tilde{x}_\lambda$ with $a \leq \tilde{x}_\lambda(t) \leq b$, for all t.

We conclude this section by recalling another duality principle which has been successfully applied in different contexts. It was developed in [32], with applications to differential systems, and further elaborated in [7] with applications to delay differential systems. We present here the main results without proof.

Let $f_0 : \mathbb{R}^m \to \mathbb{R}^m$ be a continuous vector field and let us consider the autonomous differential system

$$\dot{x} = f_0(x). \tag{2.50}$$

For a fixed period $T > 0$, we consider the T-periodic problem associated with (2.50). Moreover, within the same coincidence degree framework as described above, leading to (2.40), we set

$$L(x) := \dot{x}, \qquad N_0(x) := f_0(x(\cdot)). \tag{2.51}$$

Then, from [32, Theorem 1], the following result is valid.

Theorem 2.15 *Let $\Omega \subset \mathscr{C}_T$ be an open and bounded set such that there is no T-periodic solution $x(\cdot)$ of* (2.50) *with $x(\cdot) \in \partial\Omega$. Then,*

$$D(L - N_0, \Omega) = (-1)^m d(f_0, \Omega \cap \mathbb{R}^m, 0).$$

Now, applying the homotopy invariance of the Coincidence Degree like in the development of Theorem 2.12 from Theorem 2.11, we obtain the next result, where we introduce an auxiliary continuous vector field

$$f^* = f^*(t, x, \lambda) : \mathbb{R} \times \mathbb{R}^m \times [0, 1] \to \mathbb{R}^m,$$

depending on the parameter λ and such that f^* is T-periodic in the t-variable and, moreover,

$$f^*(t, x, 0) = f_0(x), \quad \text{and} \quad f^*(t, x, 1) = f(t, x),$$

with $f(t, x)$ as in (2.5) and $f_0(x)$ as in (2.50).

Theorem 2.16 *Let $\Omega \subset \mathscr{C}_T$ be an open and bounded set such that*

$(C1)$ *the equation*

$$\dot{x} = f^*(t, x, \lambda),$$

has no T-periodic solutions on $\partial\Omega$, for each $\lambda \in [0, 1)$.

If

$$d(f_0, \Omega \cap \mathbb{R}^m, 0) \neq 0,$$

then system (2.5) *has a T-periodic solution $\tilde{x}(\cdot) \in \overline{\Omega}$.*

Proof If there is a T-periodic solution $\tilde{x}(\cdot) \in \partial\Omega$ of (2.5), we are done. Otherwise, if there are no T-periodic solutions of (2.5) on $\partial\Omega$, we have that condition $(C1)$ holds for all $\lambda \in [0, 1]$. From the homotopic invariance of the coincidence degree, we obtain

$$D(L - N^*(\cdot, 1), \Omega) = D(L - N^*(\cdot, 0), \Omega) = D(L - N_0, \Omega)$$

and, then, from Theorem 2.15,

$$D(L - N_0, \Omega) = (-1)^m d(\bar{f}, \Omega \cap \mathbb{R}^m, 0) \neq 0$$

where we have set

$$N^*(x, \lambda) := f^*(\cdot, x(\cdot), \lambda).$$

Therefore $D(L - N^*(\cdot, 1), \Omega) \neq 0$, namely, the coincidence degree associated with (2.5) in the space of T-periodic functions is non-trivial, which in turn implies that $L(x) = N^*(x, 1)$ has at least one solution in Ω. Hence, in any case, (2.5) has a T-periodic solution $\tilde{x}(\cdot) \in \overline{\Omega}$. □

The meaning of Theorem 2.16 is that whenever we have a perturbation of an autonomous differential system without T-periodic solutions on the boundary of an open bounded set Ω, if the degree of the vector field is nonzero and, along the perturbation, we do not create new solutions on $\partial\Omega$ then a T-periodic solution (in Ω) exists also for the perturbed system. A typical application to the periodically perturbed Duffing equation

$$\ddot{x} + g(x) = \varepsilon p(t), \tag{2.52}$$

with $p(t)$ a T-periodic function is the following (see [32, Proposition 1]).

Corollary 2.7 *Let $g : \mathbb{R} \to \mathbb{R}$ be a continuous function such that*

$$g(x)x > 0 \ \textit{for}\ x \neq 0 \ \textit{and} \ \lim_{x \to \pm\infty} G(x) = +\infty,$$

for $G(x) = \int_0^x g(s)\,ds$. If the centre of the associated planar system

$$\dot{x} = y, \qquad \dot{y} = -g(x)$$

is not isochronous, then there exists $\varepsilon_0 > 0$ such that, for each $0 < \varepsilon \leq \varepsilon_0$, (2.52) has a T-periodic solution.

For further continuation theorems along this line and applications, see also [32, 54] and, for a more recent article, also [74].

As a final remark, we also mention that, concerning the T-periodic problem for the autonomous system (2.50), there is also a formula for the degree associated with the Poincaré map (see [32, Corollary 2]). To this end, we denote by Φ_0 the Poincaré map for system (2.50).

Theorem 2.17 *Let $\mathcal{O} \subset \mathbb{R}^m$ be an open and bounded set such that Φ_0 is defined on $\overline{\mathcal{O}}$ and*

$$z \neq \Phi_0(z), \quad \forall z \in \partial\mathcal{O}.$$

Then,

$$d(I - \Phi_0, \mathcal{O}, 0) = (-1)^m d(f_0, \mathcal{O}, 0).$$

See [32] for a proof. The formula in Theorem 2.17 allows to obtain existence results of T-periodic solutions for the non-autonomous system (2.5), by relating through an admissible homotopy the two Poincaré maps of (2.5) and (2.50), respectively.

2.11 The Poincaré-Birkhoff Twist Theorem

The Poincaré-Birkhoff fixed point theorem, named also the "twist theorem" or the "Poincaré's last geometric theorem", in the original formulation [160], claims that there exist at least two fixed points for a homeomorphism, ψ, which is area preserving, of a planar circular annulus $\mathcal{A} = B[0, R_2] \setminus B(0, R_1)$ onto itself, such that the points of the inner boundary Γ_i are advanced along Γ_i in the clockwise sense and the points of the outer boundary Γ_o are advanced along Γ_o in the counter-clockwise sense (or vice versa). This remarkable result, conjectured by Henri Poincaré, was published by him with some reluctance in 1912, the year of his death. His celebrated paper was titled "Sur un thèorème de géométrie" and published in Rendiconti del Circolo Matematico di Palermo. In the words of Poincaré (translated from French) according to [77]:
I have never presented such an incomplete work to the public; therefore, I think it necessary to briefly explain the reasons which convinced me to publish it, and, above all, those which drove me to start it. $[\cdots]$.
So, I was led to research the veracity of this theorem, but I met some unexpected difficulties. $[\cdots]$ *It seems that, in such a situation, I should refrain from any publication until I have solved the problem; but, after all the pointless efforts made over many months, I thought that the wiser choice was to leave the problem to mature, whilst resting for some years; this would have been fine if I had been sure to be able to take it up again one day; but at my age, I cannot be so sure. On the other hand, the importance of the subject is too great, and the quantity of results so far obtained too considerable, to resign myself to let them definitively unfruitful.* $[\cdots]$
I think that these considerations are sufficient to justify me.

Poincaré also verified the validity of his conjecture in many special cases, but a complete proof was only provided by George D. Birkhoff [13], in 1913, with the proof of the existence of at least one fixed point. For the existence of a second fixed point, Birkhoff refers to a remark by Poincaré, according to which [160, p. 377]

> *There will be at least two of them, since the Analysis situs (and in particular Kronecker theorem) shows us immediately that their number must be even.*

The above-quoted remark can be better understood within the framework of fixed point index theory. Indeed, we have already observed that a map defined on an annular domain and homotopic to the identity has zero as its fixed point index. Therefore, if we find a fixed point with an associated nonzero index, there must be at last another fixed point. This is the so-called "non-degenerate situation". However, this argument does not prevent a priori that there is only a "degenerate" fixed point with zero index. This subtle issue was settled only in 1993 by Slaminka in [178] who proved that an isolated fixed point of zero index for an area-preserving homeomorphism could be removed (see [178] for the precise statement of the theorem). Hence, as observed by Birkhoff himself [16, p. 299], the proof for a second fixed point needed to be precisely derived. Thereafter, in 1926, Birkhoff gave a proof of the existence of a second fixed point in any case, and also replaced the area-preserving condition with a more general assumption of topological nature. Some mathematicians were skeptical about the validity of Birkhoff's original proof. M. Brown and W.D. Neumann [28], in 1977 precisely verified Birkhoff's work, and showed that it was indeed correct and mathematically rigorous. Actually Brown and Neummann proved that a modification of the argument used by Birkhoff in his paper of 1913 shows that a second fixed point may be obtained.

Applications of the twist theorem to problems in dynamical systems arising from nonlinear mechanics and geometry were already suggested by Poincaré in [160] and studied by Birkhoff in [14, 15].

In planar non-autonomous ordinary differential equations, to find the existence of periodic solutions or subharmonic solutions via the search of the fixed points of the Poincaré map or of its iterates, respectively, there is a major difficulty in the application of the Poincaré-Birkhoff theorem in the version stated above. This is the construction of annular regions which are invariant under these transformations. A modification of this fixed point theorem is therefore necessary, in which the invariance conditions for the annulus and its inner and outer boundaries are not assumed. Birkhoff himself, motivated by different dynamical applications, was interested in proving some extensions of the theorem along these directions. For instance, in, [14, 16], he showed that his proof was valid also when the annulus was not necessarily invariant under ψ but when its inner boundary was still rotated onto itself. In 1976, H. Jacobowitz [102, 103], following a suggestion of J. Moser [142], proposed a modified version of the Poincaré-Birkhoff theorem for a topological pointed disc, showing an application to the search of periodic solutions to some superlinear second-order differential equations. Further applications in this direction were also given by P. Hartman [99] and G. J. Butler [29]. In view of these results, W. Y. Ding [55, 56], J. Franks [85, 86], C. Rebelo [164], D. Qian and P. J. Torres [161], P. Le Calvez, and J. Wang [110] offered new versions of the Poincaré-Birkhoff theorem where the boundary invariance assumption is removed. The article of Dalbono and Rebelo [41] gives a survey on these developments till 2001. For an updated survey see the introduction in [77] and the book of A. Fonda [69].

After the introductory overview, we now give a more systematic and detailed elaboration of the Poincaré-Birkhoff fixed point theorem. As we already observed, in the original formulation the theorem asserts the existence of at least two fixed points for an area-preserving homeomorphism of a planar circular annulus

$$\mathcal{A}[a, b] = \{(x, y) : a^2 \leq x^2 + y^2 \leq b^2\}$$

onto itself, such that the points of the inner boundary C_a are advanced along C_a in the clockwise sense and the points of the outer boundary C_b are advanced along C_b in the counter-clockwise sense (or, vice versa). We include a geometric example to

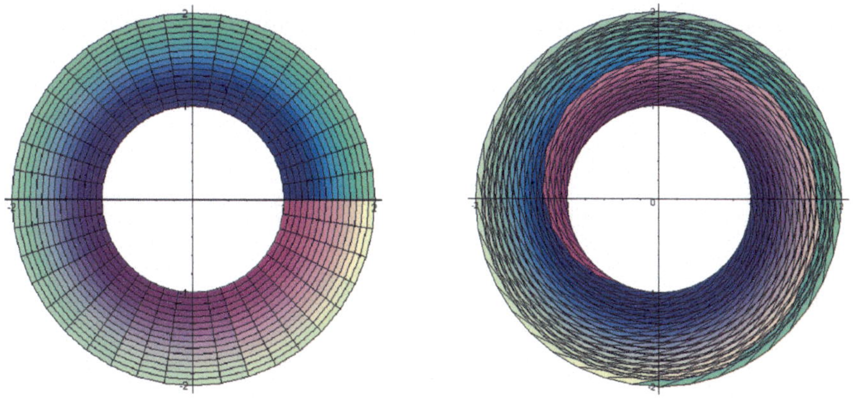

Fig. 2.3 The effect of the twist on the annulus $\mathcal{A}[1,2]$ under the action of the dynamical system (S) for $p > 1$, on a short time interval

illustrate the Poincaré-Birkhoff fixed point twist theorem. Our example considers the planar dynamical system defined by

$$\begin{cases} \dot{x} = 2py(x^2+y^2)^{p-1} \\ \dot{y} = -2px(x^2+y^2)^{p-1} \end{cases} \tag{S}$$

Figure 2.3 shows the effect of the twist on the annulus. The Poincaré map moves all the points in the clockwise sense along the circumferences, centred at the origin. However the visual effect is that of the inner and outer boundaries being rotated in different directions.

We present now a framework where the concept of boundary rotation can be defined in a more rigorous manner. Let $\phi : \mathcal{A} \to \mathcal{A}$ be an area-preserving homeomorphism of the closed annulus $\mathcal{A} := \mathcal{A}[a,b]$ onto itself.

Passing to the polar coordinates we assume that ϕ can be expressed as

$$\tilde{\phi} : (\theta, \rho) \mapsto (\theta_1, \rho_1),$$

with

$$\begin{cases} \theta_1 = \theta + \Theta(\theta, \rho) \\ \rho_1 = R(\theta, \rho). \end{cases}$$

More precisely, on the universal covering space

$$H := \{(\theta, \rho) : \theta \in \mathbb{R},\ \rho > 0\},$$

of $\mathbb{R}^2 \setminus \{0\}$ with the standard covering projection

$$\Pi : (\theta, \rho) \mapsto (\rho\cos\theta, \rho\sin\theta), \tag{2.53}$$

we consider a lifting of $\phi|_{\mathcal{A}}$ of the form $\tilde{\phi}$ (as above), defined on the strip $\tilde{\mathcal{A}} = \mathbb{R} \times [a, b]$. More generally, given a continuous map $\phi : D \subset \mathbb{R}^2 \setminus (0, 0) \to \mathbb{R}^2 \setminus (0, 0)$, a map $\tilde{\phi} : \Pi^{-1}(D) \to H$, is called a lifting of ϕ if

$$\Pi \circ \tilde{\phi} = \phi \circ \Pi.$$

Furthermore for each set $D \subset \mathbb{R}^2 \setminus (0, 0)$ we set $\tilde{D} := \Pi^{-1}(D)$.

Remark 2.10 Some authors, prefer to consider as a covering projection the map

$$\Pi : (\theta, \rho) \mapsto (\sqrt{2\rho} \cos \theta, \sqrt{2\rho} \sin \theta) \tag{2.54}$$

(see, for instance, [69, p. 227]). In fact, the map Π defined in (2.53) transforms the element of area $dxdy$ to $\rho d\theta d\rho$, whilst (2.54) transforms the element of area $dxdy$ to $d\theta d\rho$, which is much better in order to have the lifting $\tilde{\phi}$ as an area-preserving homeomorphism of the covering space H. Since we do not provide here a specific proof of the Poincaré-Birkhoff theorem, we prefer to rely on (2.53).

Now we are ready to state precisely the hypotheses of the Poincaré-Birkhoff theorem.

Theorem 2.18 *Assume*
Boundary Invariance: $R(\theta, a) = a, \quad R(\theta, b) = b, \quad \forall \theta \in \mathbb{R}$
and the
Twist Condition:
$\Theta(\theta, a) > 0 > \Theta(\theta, b), \quad \forall \theta \in \mathbb{R}$ *or* $\Theta(\theta, a) < 0 < \Theta(\theta, b), \quad \forall \theta \in \mathbb{R}.$
Then, ϕ has at least two fixed points z_1, z_2 in the interior of $\mathcal{A}$, such that $\Theta(\Pi^{-1}(z_1)) = \Theta(\Pi^{-1}(z_2)) = 0$.

An illustration of the two conditions in the theorem is given in Fig. 2.4. The map $\tilde{\phi}$ moves the strip $\tilde{\mathcal{A}}$ onto itself leaving the two components of its boundary invariant. The twist condition is expressed by the fact that the points of the lines $\rho = a$ and $\rho = b$ are moved in opposite directions.

Equivalently, we can suppose that the twist condition can be expressed as follows: *There exists an integer $j \in \mathbb{Z}$ such that*

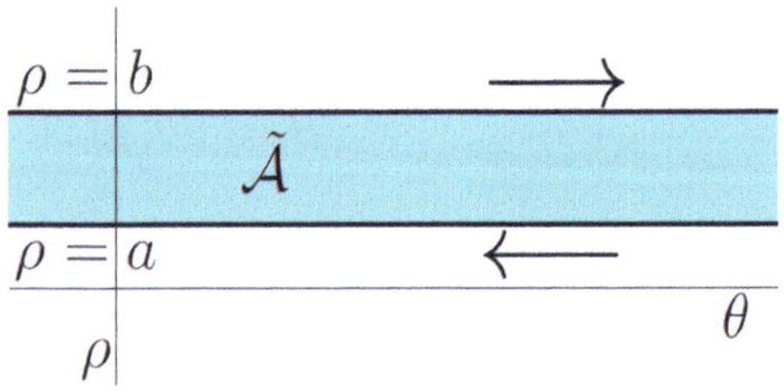

Fig. 2.4 Lifting of $\phi|_{\mathcal{A}}$ of the form $\tilde{\phi}$, defined on the strip $\tilde{\mathcal{A}} = \mathbb{R} \times [a, b]$

$$\Theta(\theta, a) > 2j\pi > \Theta(\theta, b), \quad \forall \theta \in \mathbb{R}$$

or

$$\Theta(\theta, a) < 2j\pi < \Theta(\theta, b), \quad \forall \theta \in \mathbb{R}.$$

Then, we can conclude that ϕ has at least two fixed points z_1, z_2 in the interior of $\mathcal{A}$, such that

$$\Theta(\Pi^{-1}(z_1)) = \Theta(\Pi^{-1}(z_2)) = 2j\pi \, .$$

For a detailed proof of this theorem, we refer to [28].

Poincaré himself claimed that, once the theorem is proved for the annulus, one should obtain the same result for the *pointed disk*

$$B[0, R] \setminus \{0\}.$$

In his article [102] Jacobowitz (see also [103]), gave a version of the twist fixed point theorem in which the area-preserving twist homeomorphism is defined on an annulus whose internal boundary degenerates into a point, whilst the external one is a simple curve around it. More generally, we are interested in studying the case in which an area-preserving homeomorphism is defined on a closed topological annulus, namely, a planar region between two simple closed curves say Γ_1 (the inner boundary of the annulus) and Γ_2 (the outer boundary of the annulus). The assumption that boundaries of the annulus are star shaped is a delicate hypothesis. In [56], W. Y. Ding assumed that only Γ_1 was star shaped. Based on the Jacobowitz version of the twist theorem he considered this as a technical assumption, although crucial to his proof, was probably unnecessary. W. Y. Ding in [55] and [56] treated the case in which also the inner boundary can vary under the area-preserving homeomorphism. In [56], using the result by Jacobowitz, he proved the existence of two fixed points for the inner boundary C_1 being star shaped. As remarked by W. Y. Ding himself, the condition of the star-shapedness of the inner boundary was crucial for the proof. However, its necessity for the theorem was doubtful. Later on, it has been shown by R. Martins and A.J. Ureña in [124] that the assumption of star-shapedness on the inner boundary cannot be removed.

More recently, P. Le Calvez and J. Wang in [110] showed with an example that the star-shapedness assumption was necessary for the outer boundary too, thus correcting both Jacobowitz [102, 103] and Ding [56] versions of the theorem. In the same paper, corrections to some previous proofs were also given. In any case, the version of W. Y. Ding theorem in [55] remains undisputed today, as it deals with the case in which the (not necessarily invariant) boundaries are circumferences. For the proof of Theorem 2.19 below, we refer to [164], where a direct reduction to the classical Poincaré-Birkhoff theorem for the standard annulus, proved in [28], is given. This approach appears a safer approach (in our opinion) without using the Jacobowitz result. The version of the theorem for non-invariant boundaries is extremely useful for the applications and has been utilized by many authors to prove the existence and multiplicity of periodic solutions of some non-autonomous Hamiltonian systems in a variety of situations, see, for, e.g. [29, 30, 41, 47, 52, 53, 65, 68, 70, 71, 75, 78, 84, 100, 102, 122, 164, 165, 195]. Most of these papers rely on Ding's version of the Poincaré-Birkhoff theorem, safely applying it to annuli for which both boundaries are star shaped. It can be seen that in recent years, this powerful tool has been attracting a rapidly growing interest in applications.

A modern version of the Poincaré-Birkhoff theorem (see [164, Corollary 2 and Remark 2]) is now stated here. For a simple closed curve γ in $\mathbb{R}^2$, we denote by $\mathcal{I}(\Gamma)$ the open domain bounded by Γ, and by $\overline{\mathcal{I}}(\Gamma)$ its closure.

Theorem 2.19 *[77, 164] Let $\mathcal{A} = \overline{\mathcal{I}}(\Gamma_2) \setminus \mathcal{I}(\Gamma_1)$ with $0 \in \mathcal{I}(\Gamma_1)$, be an annular region bounded by two curves Γ_1 and Γ_2, which are strictly star shaped with respect to the origin. Let $\Psi : \overline{\mathcal{I}}(\Gamma_2) \to \Psi(\overline{\mathcal{I}}(\Gamma_2))$ be an area-preserving homeomorphism such that $0 \in \Psi(\mathcal{I}(\Gamma_1))$. On the universal covering space $(\theta, \rho) : \theta \in \mathbb{R}, \rho > 0$, with the standard covering projection $\Pi : (\theta, \rho) \mapsto (\rho \cos\theta, \rho \sin\theta)$, consider a lifting of $\Psi|_{\mathcal{A}}$of the form*

$$h(\theta, \rho) = (\theta + \gamma(\theta, \rho), \eta(\theta, \rho)),$$

where $\gamma(\theta, \rho)$ and $\eta(\theta, \rho)$ are continuous, and $2\pi-$periodic in their first variable. Correspondingly, for $\widetilde{\Gamma}_1 = \Pi^{-1}(\Gamma_1)$ and $\widetilde{\Gamma}_2 = \Pi^{-1}(\Gamma_2)$, assume the twist condition

$$\gamma(\theta, \rho) > 2\pi j \text{ on } \widetilde{\Gamma}_1, \quad \gamma(\theta, \rho) < 2\pi j \text{ on } \widetilde{\Gamma}_2,$$

or

$$\gamma(\theta, \rho) < 2\pi j \text{ on } \widetilde{\Gamma}_1, \quad \gamma(\theta, \rho) > 2\pi j \text{ on } \widetilde{\Gamma}_2,$$

for some integer j.

Then, Ψ has at least two fixed points z_1, z_2 in the interior of $\mathcal{A}$, such that

$$\gamma(\Pi^{-1}(z_1)) = \gamma(\Pi^{-1}(z_2)) = 2\pi j.$$

A version for planar Hamiltonian systems

We consider a planar Hamiltonian system

$$\dot{x} = \frac{\partial}{\partial y}\mathcal{H}(t, x, y), \qquad \dot{y} = -\frac{\partial}{\partial x}\mathcal{H}(t, x, y) \tag{2.55}$$

and suppose that the vector field $\mathbf{F}(t, z) := (\frac{\partial}{\partial y}\mathcal{H}(t, x, y), -\frac{\partial}{\partial x}\mathcal{H}(t, x, y))$, for $z = (x, y)$, is continuous and T-periodic in the t-variable.

We also assume the uniqueness and the global continuability for the solutions of the initial-value problems.

For any initial point $w \in \mathbb{R}^2$ let $\zeta(\cdot, w) = (\zeta_1(\cdot, w), \zeta_2(\cdot, w))$ be the solution of (2.55) with $\zeta(0, w) = w$.

If

$$\mathbf{F}(t, 0) \equiv 0, \tag{2.56}$$

we have that $\zeta(t, w) \neq 0$ for every t, provided that $w \neq 0$. Hence we can pass to polar coordinates and determine the normalized angular displacement of the solution $\zeta(\cdot, w)$ on a time interval $[0, \tau]$ as

$$\mathrm{rot}_w(\tau) := \frac{1}{2\pi}\int_0^\tau \frac{\frac{\partial}{\partial y}\mathcal{H}(t, \zeta(t, w))\zeta_2(t, w) + \frac{\partial}{\partial x}\mathcal{H}(t, \zeta(t, w))\zeta_1(t, w)}{|\zeta(t, w)|^2}\, dt,$$

where $|\cdot|$ denotes the Euclidean norm of a vector in the plane. The *rotation number* $\mathrm{rot}_w(\tau)$ is the algebraic count of the clockwise turns of the solution $\zeta(t, w)$ around the origin in the time interval $[0, \tau]$. (See also Sect. 2.4 for the definition of rotation number). Using the fact that (as a consequence of Liouville's theorem [5, 176])[5] for every $\tau > 0$, the mapping $w \mapsto \zeta(\tau, w)$ is an area-preserving homeomorphism of the plane onto itself, we can apply a consequence of the Poincaré-Birkhoff fixed point theorem[6] which reads as follows.

Theorem 2.20 *[25, 122] Assume* (2.56) *and let $k \geq 1$ be a fixed integer. Suppose that there are $0 < r < R$ and a positive integer j such that*

$$\mathrm{rot}_w(kT) > j,\ \forall\,|w| = r \quad \textit{and} \quad \mathrm{rot}_w(kT) < j,\ \forall\,|w| = R, \tag{2.57}$$

or vice versa. Then there exist at least two initial points $w_1 \neq w_2$, with

$$r < |w_1|,\ |w_2| < R$$

such that

$$\mathrm{rot}_{w_1}(kT) = \mathrm{rot}_{w_2}(kT) = j \tag{2.58}$$

and the solutions $\zeta(\cdot, w_1)$ and $\zeta(\cdot, w_2)$ of (2.55) *are kT-periodic.*

---

[5] **Liouville's Theorem** [176, p. 150] Let $f(x, t)$ be a C^2−vector field on $W \times \mathbb{R}$ where W is an open set in $\mathbb{R}^n$ and assume that $div_x f = \sum_{i=1}^n \frac{\partial f_i}{\partial x_i} = 0$ for all (x, t) in $W \times \mathbb{R}$. Let D be a Lebesgue measurable set in W and let $D_t = \{\phi(t, x, f) : x \in D\}$ be the translate of D by solutions of $\dot{x} = f(x, t)$. Assume that every solution starting in D can be continued to D_t. Then $\mu(D) = \mu(D_t)$ where μ denotes the Lebesgue measure. In other words, the flow is volume-preserving, or incompressible. In our applications to planar Hamiltonian systems the regularity assumptions on the vector field can be relaxed, by means of a standard smoothing procedure.

[6] In the version of Ding [Acta Mathematica Sinica (1982)] for the case of a standard annulus or, respectively, in the version of Ding-Rebelo as stated in Theorem 2.19.

Instead of the circumferences of radius r and R we can take two *strictly star-shaped curves* around the origin bounding a *topological annulus*. The fixed points belong to the *interior* of such annulus.

The points w_1 and w_2 are obtained as fixed points of the k-th iterate of the Poincaré map $\Phi : w \mapsto \zeta(T, w)$. Hence, they determine two orbits

$$\mathcal{O}_1 := \{w_1, \Phi(w_1), \ldots, \Phi^{k-1}(w_1)\} \quad \text{and} \quad \mathcal{O}_2 := \{w_2, \Phi(w_2), \ldots, \Phi^{k-1}(w_2)\}$$

of fixed points of Φ^k. By a remark in [148] we know that it is always possible to choose w_1 and w_2 so that their orbits are disjoint. This means that the corresponding kT-periodic solutions are not in the same periodicity class.

In the special case of

$$\ddot{u} + f(t, u) = 0 \tag{2.59}$$

which writes as a planar Hamiltonian system of the form

$$\dot{x} = y, \qquad \dot{y} = -f(t, x)$$

and if we assume

$$f(t, 0) \equiv 0,$$

the condition (2.58) corresponds to the fact that the solutions $v^{(i)}(\cdot) := \zeta_1(\cdot, w_i)$ ($i = 1, 2$) of (2.59) possess exactly $2j$ zeros in the interval $[0, kT[$. From the relationships between k and j it is easy to produce conditions ensuring the minimality of the period, and therefore to obtain subharmonic solutions with a given minimal period.

Remark 2.11 One of the remarkable aspects in the version of the of Poincaré-Birkhoff theorem due to Fonda and Ureña for Hamiltonian systems (see [82, 83] and also [69]) is that the uniqueness hypothesis for the Cauchy problems can be avoided. However, in the setting of (2.56) one still needs to assume that the solutions with initial points different from the origin do not pass through the origin.

In recent years the applications of the Poincaré-Birkhoff theorem to Hamiltonian systems has grown to include many new results in higher dimensional systems. See for instance the works in [23, 72, 80–83]. Recent advancements include the case of infinite dimensional systems as in [22, 76].

The following examples are taken from [24, 25], respectively.

A first application

Let us consider the Duffing-type equation

$$\ddot{u} + \mu u - p(t)u^3 = 0, \tag{2.60}$$

where $\mu > 0$ and $p : \mathbb{R} \to \mathbb{R}$ is a strictly positive, continuous and T-periodic function. The example is inspired by a paper of Belmonte-Beitia and Torres [10] where the authors study the existence of heteroclinic solutions for a nonlinear Schrödinger equation. An analysis of the phase portrait of the system

Fig. 2.5 The phase portrait of the system $\dot{x} = y,\ \dot{y} = -\mu x + pu^3$

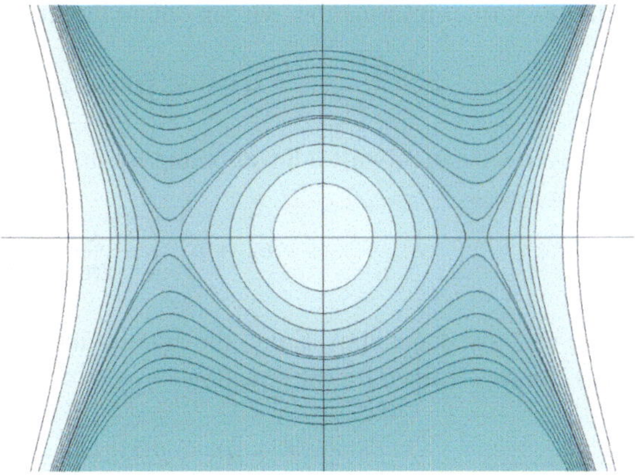

$$\begin{cases} \dot{x} = y \\ \dot{y} = -\mu x + pu^3 \end{cases}$$

with $p(t) = \text{constant} = p > 0$, shows the presence of a large number of periodic solutions. See Fig. 2.5 for the phase portrait.

Following [10] we introduce the constants

$$\xi_1 := \sqrt{\frac{\mu}{\min p(t)}}, \quad \xi_2 := \sqrt{\frac{\mu}{\max p(t)}}.$$

A simple computation shows that ξ_2 is a constant lower solution and ξ_1 is a constant upper solution for Eq. (2.60). By symmetry, also the constant functions $-\xi_1$ and $-\xi_2$ are, respectively, a lower and an upper solution for the same equation.

Using a theorem of Boscaggin, [21] the following result holds. *With the above positions, there exists $k^* \in \mathbb{N}_0$ such that, for every integer $k \geq k^*$, there exists an integer m_k such that, for every integer j relatively prime with k and such that $1 \leq j \leq m_k$, equation* (2.60) *has at least two subharmonic solutions $u^{(1)}_{j,k}(t)$, $u^{(2)}_{j,k}(t)$ of order k (not belonging to the same periodicity class) such that, for $i = 1, 2$, $u^{(i)}_{j,k}(t)$ has exactly $2j$ zeros in the interval $[0, kT[$ and*

$$-\xi_1 \leq u^{(i)}_{j,k}(t) \leq \xi_1, \quad \text{for every } t \in \mathbb{R}. \tag{2.61}$$

If we like to improve the estimate (2.61), we can take advantage of the fact that, as proved in [10] there exist two T-periodic solutions to (2.60), say $\rho_-(t)$ and $\rho_+(t)$, such that, for every $t \in [0, T]$,

$$-\xi_1 \leq \rho_-(t) \leq -\xi_2, \quad \xi_2 \leq \rho_+(t) \leq \xi_1$$

With this further information, we can obtain the sharper estimate

$$\rho_-(t) < u^{(i)}_{j,k}(t) < \rho_+(t).$$

Moreover, we can provide a lower bound for m_k as follows:

$$m_k \geq \mathscr{E}^- \left(\frac{kT\sqrt{\mu}}{2\pi} \right), \tag{2.62}$$

where we have denoted by $\mathscr{E}^-(x)$ the greatest integer strictly less then x. A similar result can be obtained for the equation

$$\ddot{u} + \mu u - p(t)g(u) = 0,$$

with $g(0) = 0$ and $g(x)/x$ sufficiently large at $\pm\infty$ (g not necessarily odd), as well as for

$$\ddot{u} + f(t, u) = 0$$

with

$$f(t, 0) \equiv 0, \quad \int_0^T f_x(t, 0)dt > 0$$

and $f(t, x)$ suitably "confined" at $\pm\infty$ (either by assuming a certain decay at infinity or the presence of lower and upper solutions).

For the technical details in the proof, see [25].

A second application

The previous example considered a case in which $f(t, x)/x \not\to 0$ as $x \to 0$. Now we continue the investigation with respect to the existence and multiplicity of *nodal periodic solutions* of

$$\ddot{u} + \lambda f(t, u) = 0 \tag{2.63}$$

namely, periodic solutions with a prescribed number of simple zeros in $[0, T[$, assuming that $f(t, x)/x \to 0$ for $x \to 0$, where, $\lambda > 0$ is a given parameter. For the Dirichlet (two-point) boundary value problem, a previous result in this direction was obtained by P. H. Rabinowitz in [163] proving the existence of unbounded connected components of nodal solution $\mathcal{C}_k^{\pm} \subset \mathbb{R} \times C^1([0, T])$. More in detail, for every positive integer k, there exists $\Lambda_k > 0$ such that for every $\lambda > \Lambda_k$ there exists at least *two* distinct points $(\lambda, u), (\lambda, v) \in \mathcal{C}_k^{\pm}$ with $u(t), v(t)$ solutions of (2.63) vanishing at $t = 0$ and $t = T$ and having precisely k zeros in $]0, T[$. Moreover $\dot{u}(0), \dot{v}(0) > 0$ for $(\lambda, u), (\lambda, v) \in \mathcal{C}_k^+$ and $\dot{u}(0), \dot{v}(0) < 0$ for $(\lambda, u), (\lambda, v) \in \mathcal{C}_k^-$. Rabinowitz's theorem also applies to the equation

$$\mathcal{L}u \equiv -\frac{d}{dt}(p(t)\dot{u}) + a(t)u = \lambda f(t, u)$$

with $p(t)$ positive and continuously differentiable in $[0, T]$ and $a(t)$ non-negative and continuous. We prefer to consider only the case $\mathcal{L}u \equiv -\ddot{u}$ for uniformity in the exposition. The assumptions on $f(t, s)$ in [163] are essentially the same as those considered in [162] for the case of positive solutions, except for the fact that now two-sided conditions on $f(t, z)$, are considered with $z \in \mathbb{R}$ and they are also uniform in $t \in [0, T]$. Such hypotheses require a *superlinear condition at zero*, as

$$\lim_{z \to 0} f(t, z)/z = 0, \quad \text{uniformly in } t \in [0, T], \tag{2.64}$$

a sign condition, as

$$f(t, z)z > 0, \quad \text{for } 0 < |z| < r_0, \ t \in [0, T] \tag{2.65}$$

and a condition at infinity given by

$$f(t, z)z < 0, \quad \text{for } |z| > r_1, \ t \in [0, T]. \tag{2.66}$$

Regarding this latter condition, we remark that it could be replaced by other hypotheses ensuring the existence of a priori bounds for the solutions. For instance (as observed above) one could assume $f(t, z)/z \to 0$ as $|z| \to +\infty$. Another aim considered in [24] was to relax the above conditions (2.64), (2.66) by assuming *one-sided* growth restrictions on $f(t, z)$ at zero and infinity (namely, only for $z > 0$ or $z < 0$ and for z near zero or near infinity, in all the possible combinations), as well as to replace the sign condition (2.65), which is uniform with respect to $t \in [0, T]$, by a local type one. This, in particular, will permit us to obtain some new applications to (2.63) when the weight function is of nonconstant sign. For the proofs a classical ODE approach, which consists in studying the qualitative properties of the trajectories in the phase-plane for the system

$$\begin{cases} \dot{x} = -y \\ \dot{y} = \lambda f(t, x) \end{cases} \tag{2.67}$$

is used. More specifically, again one is led to look for the fixed points and periodic points of the Poincaré map $\Phi^\tau : z \mapsto \zeta(t_0 + \tau; t_0, z)$, where $\zeta(\cdot; t_0, z)$ is the solution of (2.67) with $\zeta(t_0) = z$.

Figure 2.6 explains the typical dynamics associated with Eq. (2.67). Whilst points near and far from the origin moves slowly, on the other hand, taking $\lambda > 0$ sufficiently large, intermediate points can be seen to rotate faster (left panel). The right panel shows a typical deformation through the Poincaré map of a line passing through the origin.

Figure 2.7 considers the image of an interval on the positive x-axis and taken

$$f(t, x) = q(t)\mathtt{f}(x)$$

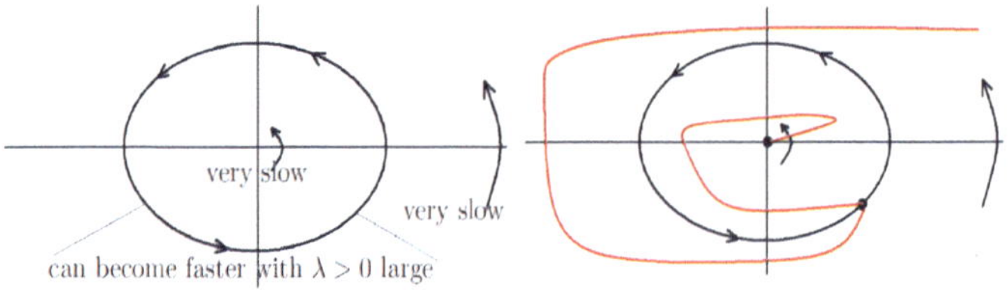

Fig. 2.6 The slow rotational movement of points near and far from the origin

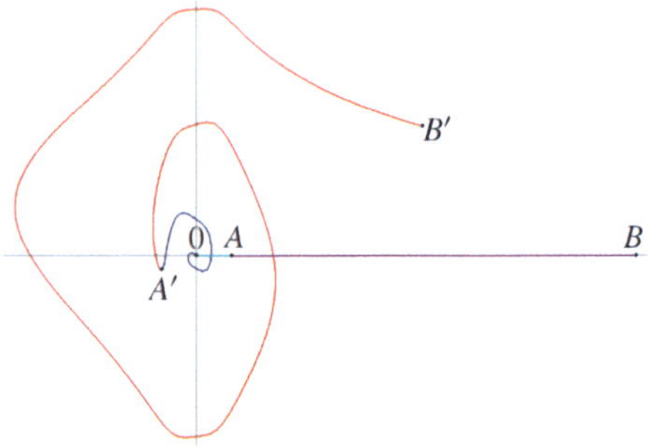

Fig. 2.7 Example of the deformation of a half-line from the origin by the Poincaré map Φ associated with (2.67) [24]. For $T = 1$ and $\lambda = 50$ the positive x-axis is transformed to a double spiral winding around the origin

with

$$q(t) \equiv 1, \quad \mathtt{f}(x) = 4x^3/(1 + x^4).$$

The six non-trivial intersections of $\Phi(\overline{0B})$ with the x-axis correspond to six nodal solutions of the Neumann problem starting with a positive initial value $x(0)$ with $(x(0), 0)$ on the segment $\overline{0B}$. Indeed, we have two solutions with respectively: one zero, two zeros, and three zeros in $]0, T[$ (exactly). For each pair of such solutions, one is with a "small" initial value (namely, with $(x(0), 0) \in \overline{0A}$) and the other is with a "large" initial value (namely, with $(x(0), 0) \in \overline{AB}$). Other six solutions can be obtained from initial points on the negative x-axis. The spiral will wind more turns around the origin as λ grows, thus producing more pairs of solutions to the Neumann problem. An analogous qualitative portrait can be provided for the Dirichlet problem, by shooting from the y-axis. In this example, the situation is particularly simple since we have considered an autonomous equation with f odd. However, the results in [24] show that the same kind of conclusion is true for a general weight function $q(t)$ (possibly of nonconstant sign) and requiring on f only one-sided growth conditions in zero and infinity. Figure 2.8 provides an example of energy level lines for

$$\frac{1}{2}y^2 + \mathtt{F}(x), \tag{2.68}$$

where $\mathtt{F}$ is a primitive of $\mathtt{f}$. The function $\mathtt{f}$ in the present example is defined as

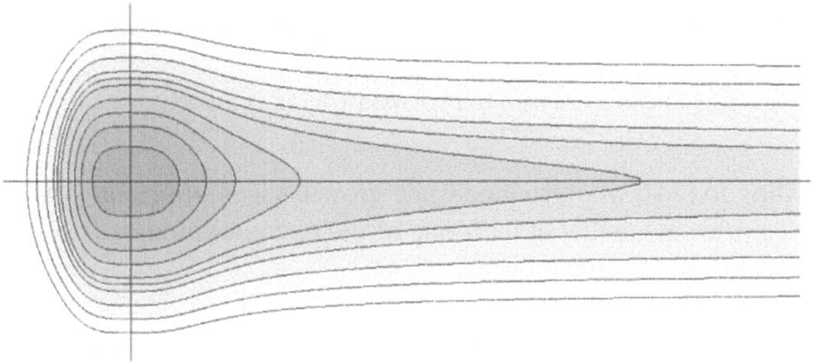

Fig. 2.8 Example of energy level lines for (2.68), with $\mathtt{f}(x)$ as in (2.69) [24]

$$\begin{cases} \mathtt{f}(x) = & 5x^3/(1+x^4) \quad \text{for } x \leq 0 \\ \mathtt{f}(x) = & x^2/(1+x^4) \quad \text{for } x \geq 0. \end{cases} \tag{2.69}$$

In this application, it is shown that

- One-sided conditions on
$$x \mapsto f(t,x),$$
- local conditions on
$$t \mapsto f(t,x)$$

are enough. More precisely the assumptions are as follows (see [24]): A first hypothesis is a sign condition on the continuous function $f = f(t,x) : [0,T] \times \mathbb{R} \to \mathbb{R}$ which is of local nature in x and nonuniform in t. Indeed, it is assumed:

$(f_{g,h})$ *there exist* $\delta > 0$, *two continuous functions* $g, h : [-\delta, \delta] \to \mathbb{R}$, *with*

$$0 < g(x)x \leq h(x)x, \quad \text{for every } x \in [-\delta,\delta] \setminus \{0\},$$

and a non-degenerate closed interval $I_0 = [a,b] \subset [0,T]$ *such that*

$$g(x)x \leq f(t,x)x \leq h(x)x, \quad \text{for every } t \in I_0,\ |x| \leq \delta. \tag{2.70}$$

Note that in the special case $f(t,x) = \mathtt{f}(x)$, the above condition is fulfilled if and only if $\mathtt{f}(x)x > 0$ for $x \in [-\delta,\delta] \setminus \{0\}$ and we can take $g = h = \mathtt{f}$ on $[-\delta,\delta]$. If, moreover, the function f takes the form

$$f(t,x) = q(t)\mathtt{f}(x), \tag{2.71}$$

with $q : [0,T] \to \mathbb{R}$ continuous, then the condition $(f_{g,h})$ is satisfied provided that $q(t_0) > 0$ for some $t_0 \in [0,T]$ and $\mathtt{f}(x)x > 0$ for all $x \in [-\delta,\delta] \setminus \{0\}$. Indeed, in such a situation, one can choose as $[a,b]$ a suitable interval (containing t_0) such that

$0 < \min_{[a,b]} q(t)$ and take

$$g(x) := \mathtt{f}(x) \min_{[a,b]} q(t), \quad h(x) := \mathtt{f}(x) \max_{[a,b]} q(t).$$

Next, according to [24], we introduce some growth assumptions on f at zero and at infinity. By (f_0) we mean that at least one of the two conditions

$$(f_0^+) \qquad \limsup_{x \to 0^+} \frac{f(t,x)}{x} \leq 0, \quad \textit{uniformly in } t \in [0,T],$$

$$(f_0^-) \qquad \limsup_{x \to 0^-} \frac{f(t,x)}{x} \leq 0, \quad \textit{uniformly in } t \in [0,T],$$

holds. Similarly, by (f_∞) we express the fact that at least one of the two conditions

$$(f_\infty^+) \qquad \limsup_{x \to +\infty} \frac{f(t,x)}{x} \leq 0, \quad \textit{uniformly in } t \in [0,T],$$

$$(f_\infty^-) \qquad \limsup_{x \to -\infty} \frac{f(t,x)}{x} \leq 0, \quad \textit{uniformly in } t \in [0,T],$$

holds. To conclude with the list of the hypotheses of the main result in [24], we add a technical condition which is required by the particular approach. Indeed, we need to have a well-defined Poincaré map. Accordingly, it is supposed that

- *the uniqueness and the global continuability of the solutions to the Cauchy problems associated with*

$$\ddot{u} + \lambda f(t,u) = 0, \tag{2.72}$$

  are ensured.

We enter in this setting if, for instance, $f(t,x)$ *is locally Lipschitz continuous in* x *and it grows at most linearly at infinity* (see also [24] for more information and comments on this issue). As mentioned in Remark 2.11, one could also slightly relax this assumption to a uniqueness condition only near the origin.

In such a framework the following results hold. As usual, by a T-periodic solution of (2.72) we mean a solution $u(t)$ defined on $[0,T]$ and such that

$$u(T) - u(0) = \dot{u}(T) - \dot{u}(0) = 0.$$

When $f(t,x)$ is extended by T-periodicity for all $t \in \mathbb{R}$, the solutions satisfying the above boundary conditions also extend to T-periodic solutions in the classical sense.

Theorem 2.21 *[24] Assume $(f_{g,h})$, (f_0), (f_∞). Then, for every positive integer m, there exists $\Lambda_m^* > 0$ such that, for each $\lambda > \Lambda_m^*$, Eq. (2.72) has at least $4m$ T-periodic solutions. Precisely, for every $j = 1, \ldots, m$, there are at least four T-periodic solutions having exactly $2j$ zeros in $[0, T[$; for two of such solutions $|u(t)| + |\dot{u}(t)|$ is small and for the other two $|u(t)| + |\dot{u}(t)|$ is large.*

Subharmonic solutions can be obtained, as well. We give a sketch of the proof, along the lines of [24], without going into too many details.

Proof Via a simple change of variable, $\lambda > 0$ and large can be replaced by $\lambda = 1$ and by an expansion of the time interval. Hence, instead of using in the proof λ large, we allow the time to grow. Let $[t_0, t_1] \subset \mathbb{R}$ be an interval and let $\zeta : [t_0, t_1] \to \mathbb{R}^2 \setminus \{0\}$ be a C^1-path with $\zeta(t) = (x(t), y(t))$. The *rotation number*, of ζ, as mentioned earlier, counts the number of algebraic turns (in the counter-clockwise sense) around the origin $0 = (0, 0)$ in the given time interval. It is defined as

$$\mathrm{Rot}\,(\zeta; [t_0, t_1]) := \frac{1}{2\pi} \int_{t_0}^{t_1} \frac{x(s)\dot{y}(s) - y(s)\dot{x}(s)}{x(s)^2 + y(s)^2}\, ds,$$

or, equivalently, as

$$\mathrm{Rot}\,(\zeta; [t_0, t_1]) := \frac{\theta(t_1) - \theta(t_0)}{2\pi}, \tag{2.73}$$

once we have expressed $\zeta(t) = (x(t), y(t))$ in polar coordinates:

$$x(t) = \rho(t) \cos\theta(t), \qquad y(t) = \rho(t) \sin\theta(t),$$

so that

$$\dot{\theta}(t) = \frac{1}{\rho(t)} (\dot{y}(t) \cos\theta(t) - \dot{x}(t) \sin\theta(t)).$$

In some applications $\zeta(t)$ is the solution of a planar differential system

$$\begin{cases} \dot{x} = X(t, x, y) \\ \dot{y} = Y(t, x, y) \end{cases} \tag{2.74}$$

with $\zeta(t_0) = z_0$. In such a situation and when $\zeta(\cdot)$ is uniquely determined by z_0 as a solution of (2.74) defined on $[t_0, t_1]$, we also write $\mathrm{Rot}_{(2.74)}\,(z_0; [t_0, t_1])$ in place of $\mathrm{Rot}\,(\zeta; [t_0, t_1])$, so that we have

$$\mathrm{Rot}_{(2.74)}\,(z_0; [t_0, t_1]) = \frac{1}{2\pi} \int_{t_0}^{t_1} \frac{x(s)Y(s, x(s), y(s)) - y(s)X(s, x(s), y(s))}{x(s)^2 + y(s)^2}\, ds.$$

Without ambiguity, we omit to mention the reference to the differential equation and use the simplified notation $\mathrm{Rot}\,(z_0; [t_0, t_1])$. Note that in order to make the above formula meaningful one has to assume that the solution never vanishes. For instance, one can enter in such a situation by assuming

$$X(t, 0, 0) = Y(t, 0, 0) \equiv 0. \tag{2.75}$$

In this case, the uniqueness of the solutions for the initial-value problems ensures that $\zeta(t) \neq 0$ for all t, whenever $z_0 \neq 0$.

- Let $\delta > 0$ and let $g, h : [-\delta, \delta] \to \mathbb{R}$ be two continuous functions such that

$$0 < g(x)x \leq h(x)x, \quad \forall x \neq 0 \tag{2.76}$$

and let $I \subset \mathbb{R}$ be a compact interval. We denote by $\mathcal{W}(g, h; I)$ the set of all continuous functions $w : \mathbb{R} \times \mathbb{R} \to \mathbb{R}$, with $w(t, 0) \equiv 0$, which satisfy the inequality

$$g(x)x \leq w(t, x)x \leq h(x)x, \quad \text{for every } t \in I,\ x \in [-\delta, \delta] \tag{2.77}$$

and such that the Cauchy problems associated with $\ddot{v} + w(t, v) = 0$ have a unique solution.

By the above assumptions on $w(t, x)$, it follows that for every $z_0 \in \mathbb{R}^2 \setminus \{0\}$, the rotation number $\mathrm{Rot}\,(z_0; I)$ is well defined with reference to the differential system

$$\begin{cases} \dot{x} = -y \\ \dot{y} = w(t, x), \end{cases} \tag{2.78}$$

provided that the solution $\zeta(t) = (x(t), y(t))$ of (2.78) with value z_0 at the time $t_0 = \min I$ is defined on the whole interval I. With these positions, the following lemma holds.

Let g, h be as above. Then, for every positive integer j, there exist $\tau_j^* > 0$ and $r_j^* \in]0, \delta[$ such that, for every compact interval I with $|I| > \tau_j^*$ and for every continuous function $w \in \mathcal{W}(g, h, I)$, it holds that

$$\mathrm{Rot}_{(2.78)}\,(z_0; I) > j, \quad \forall z_0 \text{ with } |z_0| = r_j^*, \tag{2.79}$$

being $|z_0|$ the Euclidean norm of the point $z_0 \in \mathbb{R}^2$.

On the other hand, the condition of sublinear growth at infinity for $f(t, x)$ implies that the rotation for large solutions is less than 1. Thus we enter in the setting of Theorem 2.20 and we prove the existence of a pair of T-periodic solutions with initial points having norm larger than r_j^* and with rotation number equal to j. Similarly the condition on $f(t, x)$ at 0 implies that the rotation for small solutions is less than 1. Thus we enter once more in the setting of Theorem 2.20 and we prove the existence of a pair of T-periodic solutions with initial points having norm smaller than r_j^* and with a rotation number equal to j. Thus altogether, we find four T-periodic solutions with j as a rotation number. Finally, as explained in the beginning of the proof, we can convert the condition on the length of the interval $|I| > \tau_j^*$ to an equivalent condition of the form $\lambda > \Lambda_m^*$ where m is a fixed positive integer with $1 \leq j \leq m$. See [24] for more technical details. □

The following figures illustrate the remarks made. More precisely, for the system $\dot{x} = -y, \ \dot{y} = w(t, x)$, the solutions move in an anti-clockwise sense, i.e. from the right to the left in the upper half of the plane and in the lower half-plane the solutions move from the left to the right. In this figure, if the solutions are at the point P, the alternating of blue and red lines constitute a barrier from P to the region $\mathcal{R}$. Figure 2.9shows that for a solution to enter a neighbourhood of the origin, it is required to execute several turns around the origin, or the solution has several rotations, that is, for solutions which are away from the origin, to enter a region close to the origin need to have a "rotation" number.

Figure 2.10 shows that solutions may go from the point a to the point c, as well as from a to b and b to a, that is along the direction of the vector field. But the solutions cannot go against the direction of the vector field, that is from b to c or c to a by crossing, respectively, the positive or negative y-axis.

In Fig. 2.11, it is indicated that the solutions rotate $> j$ times around the origin, on the other hand, the solutions which start further away from the origin, do not complete a full turn that is, they rotate < 1 times. The same occurs for the solutions which start very close to the origin.

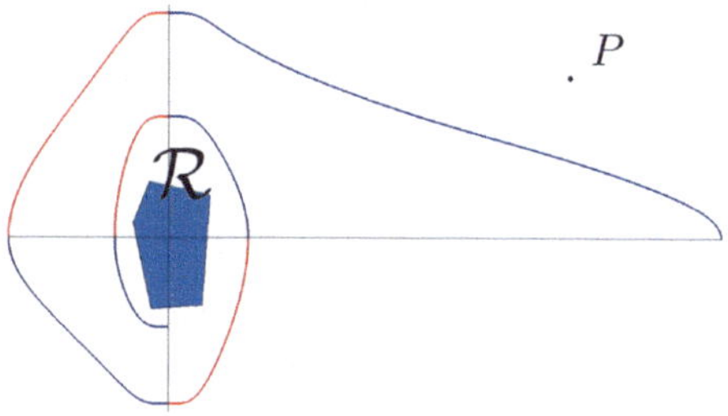

Fig. 2.9 Starting from P the solution must perform some turns before entering $\mathcal{R}$

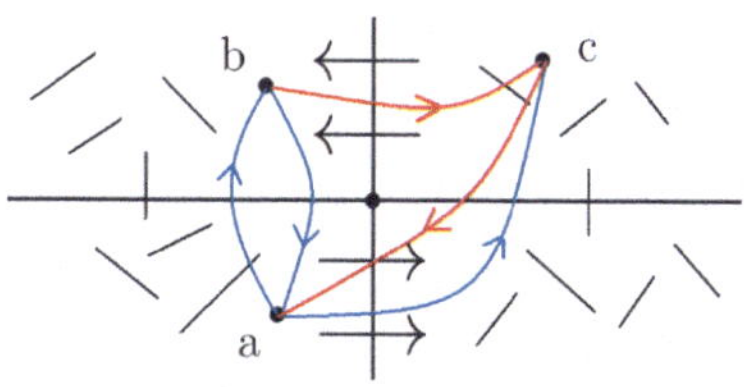

Fig. 2.10 The direction of the vector field is indicated by the arrows in this figure as well as admissible or forbidden moves

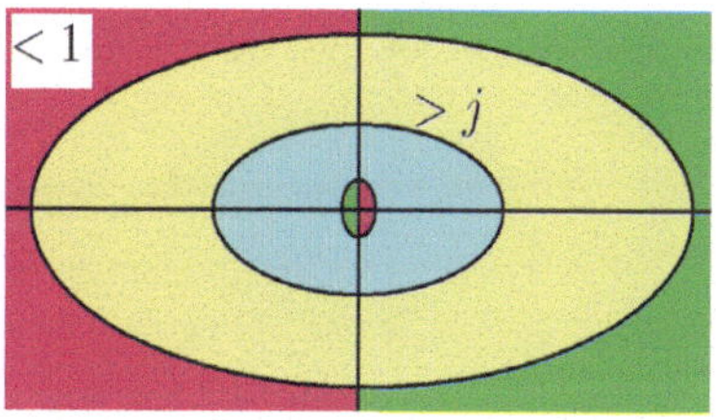

Fig. 2.11 Rotation numbers are indicated in the figure

Some simple corollaries taken from [24]
In the special case when

$$f(t, x) = q(t)\mathtt{f}(x), \tag{2.80}$$

with $\mathtt{f} : \mathbb{R} \to \mathbb{R}$ locally Lipschitz and $q : [0, T] \to \mathbb{R}$ continuous, the following consequences of our main results can be produced for the equation

$$\ddot{u} + \lambda q(t)\mathtt{f}(u) = 0. \tag{2.81}$$

Corollary 2.8 *Suppose that*

$$\mathtt{f}(x)x > 0, \quad \forall x \neq 0,$$

$$\lim_{x\to 0^+} \frac{\mathtt{f}(x)}{x} = 0 \quad or \quad \lim_{x\to 0^-} \frac{\mathtt{f}(x)}{x} = 0$$

and

$$\lim_{x\to +\infty} \frac{\mathtt{f}(x)}{x} = 0 \quad or \quad \lim_{x\to -\infty} \frac{\mathtt{f}(x)}{x} = 0.$$

If, moreover, $q(t) \geq 0$ for all $t \in [0, T]$, with $q \not\equiv 0$, then the same conclusions of Theorem 2.21 hold for Eq. (2.81).

Corollary 2.9 *Suppose that*

$$\exists \delta > 0 : \ x\mathtt{f}(x) > 0, \quad \forall x : 0 < |x| \leq \delta$$

and

$$\lim_{x\to 0^+} \frac{\mathtt{f}(x)}{x} = 0 \quad or \quad \lim_{x\to 0^-} \frac{\mathtt{f}(x)}{x} = 0.$$

Let also exist $x_* < -\delta$ and $x^* > \delta$ such that

$$\mathtt{f}(x_*) = \mathtt{f}(x^*) = 0.$$

Then, if $\max_{t\in[0,T]} q(t) > 0$, *the same conclusions of Theorem 2.21 hold for Eq.* (2.81) *and all the corresponding solutions take value in the open interval* $]x_*, x^*[$.

2.12 The Problem of Boundedness of the Solutions in Duffing-Type Equations

The boundedness of solutions of Hamiltonian systems has been a topic of great interest and has been studied extensively [143–145]. The study of this problem was motivated primarily by celestial mechanics and the stability of the planetary system. This is modelled by the $N-$body problem where N masses move in Euclidean $\mathbb{R}^2$ or $\mathbb{R}^3$ space. The goal of this problem is to find bounded orbits so that collisions of orbits or their escape to infinity may be avoided. In other words, it was required that the distance between two masses be *upper bounded* and *bounded away from zero* for all times. The obvious example of such solutions is periodic solutions. Periodic solutions, being an exceptional set in the phase-plane, are of limited interest from the point of view of dynamical behaviour, until or unless their stability was established. Stability involves finding a region in the phase space in which these orbits are bounded. Thus the concept of boundedness is very closely related to the stability of solutions.

This problem was solved by the KAM theory, which assures the existence of such solutions for Hamiltonian systems. The classical KAM (Kolmogorov-Arnold-Moser) Theory [60, 143] is concerned with the stability of motions in Hamiltonian systems, which are small perturbations of integrable Hamiltonian systems. These integrable systems are characterized by the existence of action-angle coordinates such that the Hamiltonian depends on the action variable alone.[7] Kolmogorov-Arnold-Moser (or KAM) theory was developed for conservative dynamical systems that are nearly integrable. Integrable systems in their phase space usually contain many invariant tori and KAM theory establishes persistence results for such tori, which carry quasi-periodic motions. Kolmogorov discussed the occurrence of multi- or quasi-periodic motions, which in the phase space are confined to invariant tori. This was restricted to conservative (or Hamiltonian) dynamical systems, as these are generally used for modelling in classical mechanics.

Several authors have written about this seminal work of the KAM theory [27]. To quote a few: M. Levi amongst others has several papers in this area. In [112], it is shown that the system of the form $\ddot{x} + V'(x) = p(t)$ with periodic V and p and with mean value $\bar{p} = 0$ is near-integrable for large energies. In particular, most fast solutions are quasi-periodic, provided V is sufficiently smooth, (at least $V \in C^5$) and $p \in L^1$. In [113], Levi studied the $\ddot{x} + V_x(x, t) = 0$. The author showed that under appropriate growth assumptions on V for large x, the system is near-integrable for large amplitudes in the sense that most large amplitude solutions are quasi-periodic and all solutions are bounded for all time with no assumptions related to smallness on the time dependence of $V(x, t)$. In the early 1960s [118, 119], Littlewood studied whether or not all the solutions of the Duffing-type equation $\ddot{x} + g(x) = p(t)$ where $p(t + 1) = p(t)$ stay bounded for all time. The above equation represents the motion

[7] In classical mechanics, Hamilton-Jacobi theory provides a mechanism by which problems may be solved using coordinate transformations. One transformation which has proven useful historically is the transformation from a set of position and moment, to the action (or energy) -angle variables.

of a classical particle in a one-dimensional potential field subjected to external time-periodic force $p(t)$. Littlewood, in [119] suggested that the boundedness of all the solutions of the periodically perturbed Duffing equation, $\ddot{x} + g(x) = e(t)$ need to be examined in the following two cases:

- Superlinear case: $g(x)/x \to \infty$ as $x \to \pm\infty$;
- Sublinear case: $sign(x) \cdot g(x) \to \infty$ and $g(x)/x \to 0$ as $x \to \pm\infty$.

The first result in the superlinear case was obtained by Morris [141] in 1976, who proved the boundedness of all solutions of the equation $\ddot{x} + 2x^3 = p(t)$ where $p(t) \in C^0(T)$. Prompted by the works of Littlewood [119] and Morris [141], Dieckerhoff and Zehnder in [48] extended the result in a more general but still very restricted class of equations for which, in particular the time dependence is involved in the nonlinearity in x. In [12], Liu Bin considers the nonlinear Hill's equation with periodic forcing term $\ddot{x} + \beta x^{2n+1} + (a_1 + \epsilon a(t))x = p(t)$ for $n \geq 1$, where $a(t)$ and $p(t)$ are continuous and l-periodic, a_1 and β are positive constants, and ϵ is a small parameter. The author proves that all solutions of the above-mentioned equation are bounded for $t \in \mathbb{R}$ and that there are infinitely many quasi-periodic solutions and an infinity of periodic solutions of minimal period m, for each positive integer m.

Y. Wang, in [190] considers the sublinear equation with time-dependent potential $\ddot{x} + e(t)|x|^{\alpha-1}x = p(t), \quad 0 < \alpha < 1$, where $p(t), e(t)$ are smooth $1-$periodic functions. He proves a sufficient and necessary condition for the boundedness of all solutions.

Several mathematicians like Moser [143], and Markus [123] focussed a great deal of their attention on the problem of boundedness of the superlinear Duffing equation. Another worthy mention in this area is the work of Tongren Ding [50]. Ding proves that all solutions of Duffing's equation are bounded for $t \in \mathbb{R}$ when $p(t)$ is even (or when $p(t)$ is odd and $g(x)$ is odd). X. Li in [117], deals with the problem of boundedness of solutions for Duffing's equations with semilinear potentials. Rafael Ortega [156], studied the boundedness problem for a Duffing equation of type $\ddot{x} + n^2x + h(x) = p(t)$ with $n = 1, 2, 3 \cdots$ and h a bounded and continuous function. So the function $g(x) = n^2x + h(x)$ has linear growth at infinity.

Just as with the boundedness of solutions, there have been several authors who worked on the unboundedness of solutions. In [114], Levi discussed unbounded motions in super quadratic potentials using as an example, an equation of the form $\ddot{x} + V'(x) = p(t)$ with $V'(x)/x \to \infty$ and yet possessing unbounded solutions, $x(t)$. In [108] the authors prove the existence of invariant curves and thus stability for all time for a class of Hamiltonian systems with time-dependent potentials. For instance, in [115], the authors show that the time-periodic Hamiltonian systems $\dfrac{d^2x}{dt^2} + x^{2n+1} + a(t)x^{2l+1} = 0,\ 2n > 2l > n$, with a discontinuity in $a(t)$, possess unbounded solutions. Zharnitsky in [200] constructs an unbounded solution for a continuous $p(t)$ and a superlinear U' in Littlewood's equation. In [189] Wang examines unbounded solutions in a Duffing Equation with polynomial potentials.

With this brief survey, we have tried to summarize a few aspects concerning the boundedness of the solutions, with a special emphasis on the case of Duffing equations, following the seminal and pioneering work of Littlewood. The full presentation of KAM theory and its applications is beyond the scope of this book. The interested reader will find classical and recent articles on the developments of Littlewood problems following the relevant contributions cited above.

References

1. S. Ahmad, On almost periodic solutions of the competing species problems. Proc. Am. Math. Soc. **102**(4), 855–861 (1988)
2. P. Alexandroff, H. Hopf, *Topologie* (Springer, Berlin, 1935)
3. C. Alvarez, A.C. Lazer, An application of topological degree to the periodic competing species problem. ANZIAM J. **28**(2), 202–219 (1986)
4. H. Amann, S.A. Weiss, On the uniqueness of the topological degree. Mathematische Zeitschrift **130**(1), 39–54 (1973)
5. V.I. Arnold, *Mathematical Methods of Classical Mechanics*, vol. 60 (Springer, 2013)
6. M. Bardi, An equation of growth of a single species with realistic dependence on crowding and seasonal factors. J. Math. Biol. **17**(1), 33–43 (1983)
7. Th. Bartsch, J. Mawhin, The Leray-Schauder degree of S^1-equivariant operators associated to autonomous neutral equations in spaces of periodic functions. J. Differ. Equ. **92**(1), 90–99 (1991)
8. A. Battauz, F. Zanolin, Coexistence states for periodic competitive Kolmogorov systems. J. Math. Anal. Appl. **219**(2), 179–199 (1998)
9. J.W. Bebernes, K. Schmitt, Periodic boundary value problems for systems of second order differential equations. J. Differ. Equ. **13**(1), 32–47 (1973)
10. J. Belmonte-Beitia, P.J. Torres, Existence of dark soliton solutions of the cubic nonlinear Schrödinger equation with periodic inhomogeneous nonlinearity. J. Nonlinear Math. Phys. **15**(sup3), 65–72 (2008)
11. S.R. Bernfeld, V. Lakshmikantham, *An Introduction to Nonlinear Boundary Value Problems* (Elsevier, 1974)
12. L. Bin, Boundedness for solutions of nonlinear Hill's equations with periodic forcing terms via Moser's twist theorem. J. Differ. Equ. **79**(2), 304–315 (1989)
13. G.D. Birkhoff, Proof of Poincaré's geometric theorem. Trans. Am. Math. Soc. 14–22 (1913)
14. G.D. Birkhoff, Dynamical systems with two degrees of freedom. Trans. Am. Math. Soc. **18**(2), 199–300 (1917)
15. G.D. Birkhoff, *Dynamical Systems* (American Mathematical Society, RI, Providence, 1927)
16. G.D. Birkhoff et al., An extension of Poincaré's last geometric theorem. Acta Mathematica **47**(4), 297–311 (1926)
17. G.D. Birkhoff, O.D. Kellogg, Invariant points in function space. Trans. Am. Math. Soc. **23**(1), 96–115 (1922)
18. F. Blanchini, Set invariance in control. Automatica **35**(11), 1747–1767 (1999)
19. L.E. Bobisud, Steady-state turbulent flow with reaction. Rocky Mt. J. Math. 993–1007 (1991)
20. J.-M. Bony, Principe du maximum, inégalite de Harnack et unicité du problème de Cauchy pour les opérateurs elliptiques dégénérés. Ann. Inst. Fourier (Grenoble) **19**(fasc. 1), 277–304 xii (1969)
21. A. Boscaggin, Subharmonic solutions of planar Hamiltonian systems: a rotation number approach. Adv. Nonlinear Stud. **11**(1), 77–103 (2011)
22. A. Boscaggin, A. Fonda, M. Garrione, An infinite-dimensional version of the Poincaré-Birkhoff theorem on the Hilbert cube. Ann. Sc. Norm. Super. Pisa Cl. Sci. (5) **20**(2), 751–770 (2020)
23. A. Boscaggin, R. Ortega, Monotone twist maps and periodic solutions of systems of Duffing type. Math. Proc. Camb. Philos. Soc. **157**(2), 279–296 (2014)

24. A. Boscaggin, F. Zanolin, Pairs of nodal solutions for a class of nonlinear problems with one-sided growth conditions. Adv. Nonlinear Stud. **13**(1), 13–53 (2013)
25. A. Boscaggin, F. Zanolin, Subharmonic solutions for nonlinear second order equations in presence of lower and upper solutions. Discrete Contin. Dyn. Syst. **33**(1), 89–110 (2013)
26. H. Brezis, On a characterization of flow-invariant sets. Commun. Pure Appl. Math. **23**(2), 261–263 (1970)
27. H. Broer, Kam theory: the legacy of Kolmogorov's 1954 paper. Bull. Am. Math. Soc. **41**(4), 507–521 (2004)
28. M. Brown, W.D. Neumann et al., Proof of the Poincaré-Birkhoff fixed point theorem. Michigan Math. J. **24**(1), 21–31 (1977)
29. G.J. Butler, The Poincaré-Birkhoff "Twist" theorem and periodic solutions of second-order nonlinear differential equations, in *Differential Equations* (Elsevier, 1980), pp. 135–147
30. P. Buttazzoni, A. Fonda, Periodic perturbations of scalar second order differential equations. Discrete Contin. Dyn. Syst. **3**, 451–455 (1997)
31. A. Capietto, J. Mawhin, F. Zanolin, A continuation approach to superlinear periodic boundary value problems. J. Differ. Equ. **88**(1), 990 (1990)
32. A. Capietto, J. Mawhin, F. Zanolin, Continuation theorems for periodic perturbations of autonomous systems. Trans. Am. Math. Soc. **329**(1), 41–72 (1992)
33. A. Capietto, F. Zanolin, A continuation theorem for the periodic BVP in flow-invariant ENRs with applications. J. Differ. Equ. **83**(2), 244–276 (1990)
34. A. Capietto, F. Zanolin, A continuation theorem for the periodic BVP in flow-invariant ENRs with applications. J. Differ. Equ. **83**(2), 244–276 (1990)
35. J.A. Cid, J. Mawhin, The Brouwer fixed point theorem and periodic solutions of differential equations. J. Fixed Point Theory Appl. **25**(1), Paper No. 38, 12 (2023)
36. A. Cima, A. Gasull, F. Mañosas, Periods of solutions of periodic differential equations. Differ. Integral Equ. **29**(9–10), 905–922 (2016)
37. B.D. Coleman, Nonautonomous logistic equations as models of the adjustment of populations to environmental change. Math. Biosci. **45**(3–4), 159–173 (1979)
38. M.G. Crandall, A generalization of Peano's existence theorem and flow invariance. Proc. Am. Math. Soc. **36**(1), 151–155 (1972)
39. J. Cronin, *Fixed Points and Topological Degree in Nonlinear Analysis*, vol. 11 (American Mathematical Society, 1995)
40. J.M. Cushing, Periodic time-dependent predator-prey systems. SIAM J. Appl. Math. **32**(1), 82–95 (1977)
41. F. Dalbono, C. Rebelo, Poincaré-Birkhoff fixed point theorem and periodic solutions of asymptotically linear planar Hamiltonian systems. Rend. Sem. Mat. Univ. Politec. Torino **60**(4), 233–263 (2003), 2002. Turin Fortnight Lectures on Nonlinear Analysis (2001)
42. E.N. Dancer, Boundary-value problems for weakly nonlinear ordinary differential equations. Bull. Aust. Math. Soc. **15**(3), 321–328 (1976)
43. C. De Coster, *La méthode des sur et sous solutions dans l'étude de problèmes aux limites*. PhD thesis, UCL-Université Catholique de Louvain (1994)
44. C. De Coster, P. Habets, *Two-Point Boundary Value Problems: Lower and Upper Solutions*, Mathematics in Science and Engineering, vol. 205. (Elsevier B. V., Amsterdam, 2006)
45. P. De Mottoni, A. Schiaffino, Competition systems with periodic coefficients: a geometric approach. J. Math. Biol. **11**(3), 319–335 (1981)
46. K. Deimling, *Nonlinear Functional Analysis* (Springer, Berlin, 1985)
47. M.A. del Pino, R.F. Manásevich, A. Murúa, On the number of 2π periodic solutions for $u'' + g(u) = s(1 + h(t))$ using the Poincaré-Birkhoff theorem. J. Differ. Equ. **95**(2), 240–258 (1992)
48. R. Dieckerhoff, E. Zehnder, Boundedness of solutions via the twist-theorem. Annali della Scuola Normale Superiore di Pisa-Classe di Scienze **14**(1), 79–95 (1987)
49. G. Dinca, J. Mawhin, *Brouwer Degree–the Core of Nonlinear Analysis*, Progress in Nonlinear Differential Equations and their Applications, vol. 95 (Birkhäuser/Springer, Cham, 2021)
50. T. Ding, Boundedness of solutions of Duffing's equation. J. Differ. Equ. **61**(2), 178–207 (1986)

51. T. Ding, H. Huang, F. Zanolin, A priori bounds and periodic solutions for a class of planar systems with applications to Lotka-Volterra equations. Discrete Contin. Dyn. Syst. A **1**(1), 103 (1995)
52. T. Ding, F. Zanolin, Periodic solutions of Duffing's equations with superquadratic potential. J. Differ. Equ. **97**(2), 328–378 (1992)
53. T. Ding, F. Zanolin, Subharmonic solutions of second order nonlinear equations: a time-map approach. Nonlinear Anal. Theory Methods Appl. **20**(5), 509–532 (1993)
54. T.R. Ding, R. Iannacci, F. Zanolin, On periodic solutions of sublinear Duffing equations. J. Math. Anal. Appl. **158**(2), 316–332 (1991)
55. W.Y. Ding, Fixed points of twist mappings and periodic solutions of ordinary differential equations. Acta Math. Sinica **25**(2), 227–235 (1982)
56. W.Y. Ding, A generalization of the Poincaré-Birkhoff theorem. Proc. Am. Math. Soc. 341–346 (1983)
57. A. Dold, *Lectures on Algebraic Topology* (Springer, Berlin, Heidelberg, New York, 1972)
58. J. Dugundji, *Topology* (Allyn and Bacon Inc., Boston, MA, 1966)
59. J. Dugundji, A. Granas, *Fixed Point Theory I* (Polish Sci. Publ., Warszawa, 1982)
60. H.S. Dumas, *The KAM Story* (World Scientific Publishing Company, 2014). A Friendly Introduction to the Content History, and Significance of Classical Kolmogorov-Arnold-Moser Theory
61. C. Fabry, *Periodic Solutions of the Equation* $X'' + F(t, X) = 0$ (Université catholique de Louvain, Institut de mathématique pure et appliquée, 1987)
62. C. Fabry, Landesman-Lazer conditions for periodic boundary value problems with asymmetric nonlinearities. J. Differ. Equ. **116**(2), 405–418 (1995)
63. C. Fabry, A. Fonda, A systematic approach to nonresonance conditions for periodically forced planar Hamiltonian systems. Ann. Mat. Pura Appl. (4) **201**(3), 1033–1074 (2022)
64. C. Fabry, P. Habets, Periodic solutions of second order differential equations with superlinear asymmetric nonlinearities. Archiv der Mathematik **60**(3), 266–276 (1993)
65. C. Fabry, J. Mawhin, Oscillations of a forced asymmetric oscillator at resonance. Nonlinearity **13**(3), 493 (2000)
66. M.L.C. Fernandes, F. Zanolin, Repelling conditions for boundary sets using Liapunov-like functions. ii. Persistence and periodic solutions. J. Differ. Equ. **86**(1), 33–58 (1990)
67. P. Fitzpatrick, The parity as an invariant for detecting bifurcation of the zeroes of one parameter families of nonlinear Fredholm maps, in *Topological Methods for Ordinary Differential Equations* (Springer, 1993), pp. 1–31
68. A. Fonda, Positively homogeneous Hamiltonian systems in the plane. J. Differ. Equ. **200**(1), 162–184 (2004)
69. A. Fonda, *Playing Around Resonance*. Birkhäuser Advanced Texts: Basler Lehrbücher. [Birkhäuser Advanced Texts: Basel Textbooks] (Birkhäuser/Springer, Cham, 2016). An invitation to the search of periodic solutions for second order ordinary differential equations
70. A. Fonda, L. Ghirardelli, Multiple periodic solutions of scalar second order differential equations. Nonlinear Anal. Theory Methods Appl. **72**(11), 4005–4015 (2010)
71. A. Fonda, L. Ghirardelli et al., Multiple periodic solutions of Hamiltonian systems in the plane. Topol. Methods Nonlinear Anal. **36**(1), 27–38 (2010)
72. A. Fonda, P. Gidoni, An avoiding cones condition for the Poincaré-Birkhoff theorem. J. Differ. Equ. **262**(2), 1064–1084 (2017)
73. A. Fonda, P. Habets, Periodic solutions of asymptotically positively homogeneous differential equations. J. Differ. Equ. **81**(1), 68–97 (1989)
74. A. Fonda, G. Klun, A. Sfecci, Well-ordered and non-well-ordered lower and upper solutions for periodic planar systems. Adv. Nonlinear Stud. **21**(2), 397–419 (2021)
75. A. Fonda, R. Manásevich, F. Zanolin, Subharmonic solutions for some second-order differential equations with singularities. SIAM J. Math. Anal. **24**(5), 1294–1311 (1993)
76. A. Fonda, J. Mawhin, M. Willem, Multiple periodic solutions of infinite-dimensional pendulum-like equations. Pure Appl. Funct. Anal. **5**(4), 951–963 (2020)

77. A. Fonda, M. Sabatini, F. Zanolin, Periodic solutions of perturbed Hamiltonian systems in the plane by the use of the Poincaré-Birkhoff theorem. Topol. Methods Nonlinear Anal. **40**(1), 29–52 (2012)
78. A. Fonda, Z. Schneider, F. Zanolin, Periodic oscillations for a nonlinear suspension bridge model. J. Comput. Appl. Math. **52**(1–3), 113–140 (1994)
79. A. Fonda, A. Sfecci, A general method for the existence of periodic solutions of differential systems in the plane. J. Differ. Equ. **252**(2), 1369–1391 (2012)
80. A. Fonda, A. Sfecci, Periodic solutions of weakly coupled superlinear systems. J. Differ. Equ. **260**(3), 2150–2162 (2016)
81. A. Fonda, R. Toader, Subharmonic solutions of Hamiltonian systems displaying some kind of sublinear growth. Adv. Nonlinear Anal. **8**(1), 583–602 (2019)
82. A. Fonda, A.J. Ureña, A higher-dimensional Poincaré-Birkhoff theorem without monotone twist. Comptes Rendus Mathematique **354**(5), 475–479 (2016)
83. A. Fonda, A.J. Ureña, A higher dimensional Poincaré-Birkhoff theorem for Hamiltonian flows. Ann. Inst. H. Poincaré C Anal. Non Linéaire **34**(3), 679–698 (2017)
84. A. Fonda, F. Zanolin, Periodic oscillations of forced pendulums with very small length. Proc. Roy. Soc. Edinb. Sect. A **127**(1), 67–76 (1997)
85. J. Franks, Generalizations of the Poincaré-Birkhoff theorem. Ann. Math. (2) **128**(1), 139–151 (1988)
86. J. Franks, Erratum to "Generalizations of the Poincaré-Birkhoff Theorem". *arXiv preprint math/0410316* (2004)
87. L. Führer, Ein elementarer analytischer Beweis zur Eindeutigkeit des Abbildungsgrades im R^n. Mathematische Nachrichten **54**(1–6), 259–267 (1972)
88. M. Furi, M. Pera, A continuation principle for forced oscillations on differentiable manifolds. Pac. J. Math. **121**(2), 321–338 (1986)
89. M. Furi, M. Pera, On the notion of winding number for closed curves and applications to forced-oscillations on even dimensional spheres. Bollettino Della Unione Matematica Italiana **7**(3), 397–407 (1993)
90. S. Fučik, *Solvability of Nonlinear Equations and Boundary Value Problems*, vol. 4 (Springer, 1981)
91. R.E. Gaines, J.L. Mawhin, *Coincidence Degree, and Nonlinear Differential Equations*, Lecture Notes in Mathematics, vol. 568. (Springer, Berlin-New York, 1977)
92. M. García-Huidobro, R. Manásevich, F. Zanolin, Strongly nonlinear second-order odes with unilateral conditions. Differ. Integral Equ. **6**(5), 1057–1078 (1993)
93. J.R. Graef, On the generalized Liénard equation with negative damping. J. Differ. Equ. **12**(1), 34–62 (1972)
94. A. Granas, The Leray-Schauder index and the fixed point theory for arbitrary ANRs. Bulletin de la Société Mathématique de France **100**, 209–228 (1972)
95. V. Guillemin, A. Pollack, *Differential Topology* (New Jersey, 1974)
96. J. Hale, *Theory of Functional Differential Equations*. Applied Mathematical Sciences, 2nd edn., vol. 3 (Springer, New York-Heidelberg, 1977)
97. P. Hartman, *Ordinary Differential Equations* (Wiley, New York, 1964)
98. P. Hartman, On invariant sets and on a theorem of Ważewski. Proc. Am. Math. Soc. **32**(2), 511–520 (1972)
99. P. Hartman, On boundary value problems for superlinear second order differential equations. J. Differ. Equ. **26**(1), 37–53 (1977)
100. A.R. Hausrath, R. Manasevich, Periodic solutions of a periodically perturbed Lotka-Volterra equation using the Poincaré-Birkhoff theorem. J. Math. Anal. Appl. **157**(1), 1–9 (1991)
101. M. Hirsch, *Differential Topology*. Graduate Texts in Mathematics, vol. 33 (Springer, New York, 1976)
102. H. Jacobowitz, Periodic solutions of $x" + f(x, t) = 0$ via the Poincaré-Birkhoff theorem. J. Differ. Equ. **20**(1), 37–52 (1976)
103. H. Jacobowitz, The existence of the second fixed point: A correction to "periodic solutions of $x'' + f(x, t) = 0$ via the Poincaré-Birkhoff theorem". J. Differ. Equ. **25**(1), 148–149 (1977)

104. J. Kato, On a boundedness condition for solutions of a generalized Liénard equation. J. Differ. Equ. **65**(2), 269–286 (1986)
105. A.M. Krasnosel'skii, M.A. Krasnosel'skii, J. Mawhin, A. Pokrovskii, Generalized guiding functions in a problem on high frequency forced oscillations. Nonlinear Anal. **22**(11), 1357–1371 (1994)
106. M.A. Krasnosel'skiĭ, *The Operator of Translation Along the Trajectories of Differential Equations*. Translations of Mathematical Monographs, vol. 19 (American Mathematical Society, Providence, R.I., 1968). Translated from the Russian by Scripta Technica
107. M.A. Krasnosel'skiĭ, P.P. Zabreĭko, *Geometrical Methods of Nonlinear Analysis*, volume 263 of *Grundlehren der mathematischen Wissenschaften [Fundamental Principles of Mathematical Sciences]* (Springer, Berlin, 1984). Translated from the Russian by Christian C. Fenske
108. S. Laederich, M. Levi, Invariant curves and time-dependent potentials. Ergodic Theory Dyn. Syst. **11**(2), 365–378 (1991)
109. J.P. Lasalle, S. Lefschetz, *Stability by Lyapunov's Direct Method* (1961)
110. P. Le Calvez, J. Wang, Some remarks on the Poincaré-Birkhoff theorem. Proc. Am. Math. Soc. **138**(2), 703–715 (2010)
111. J. Leray, J. Schauder, Topologie et équations fonctionnelles. Ann. Sci. École Norm. Sup. **3**(51), 45–78 (1934)
112. M. Levi, KAM theory for particles in periodic potentials. Ergodic Theory Dyn. Syst. **10**(4), 777–785 (1990)
113. M. Levi, Quasiperiodic motions in superquadratic time-periodic potentials. Commun. Math. Phys. **143**(1), 43–83 (1991)
114. M. Levi, On Littlewood's counterexample of unbounded motions in superquadratic potentials, in *Dynamics Reported* (Springer, 1992), pp. 113–124
115. M. Levi, J. You, Oscillatory escape in a Duffing equation with a polynomial potential. J. Differ. Equ. **140**(2), 415–426 (1997)
116. B. Li, Periodic orbits of autonomous ordinary differential equations: theory and applications. Nonlinear Anal. Theory Methods Appl. **5**(9), 931–958 (1981)
117. X. Li, Boundedness of solutions for Duffing's equations with semilinear potentials. J. Differ. Equ. **176**(1), 248–268 (2001)
118. J.E. Littlewood, Unbounded solutions of an equation $y'' + g(y) = p(t)$, with $p(t)$ periodic and bounded, and $g(y)/y \to \infty$ as $y \to \pm\infty$. J. Lond. Math. Soc. **1**(1), 497–507 (1966)
119. J.E. Littlewood, Unbounded Solutions of $y'' + g(y) = p(t)$. J. Lond. Math. Soc. **1**(1), 491–496 (1966)
120. N.G. Lloyd, *Degree Theory* (Cambridge University Press, Cambridge, 1978)
121. R. Manásevich, J. Mawhin, Periodic solutions for nonlinear systems with p-Laplacian-like operators. J. Differ. Equ. **145**(2), 367–393 (1998)
122. A. Margheri, C. Rebelo, F. Zanolin, Maslov index, Poincaré-Birkhoff theorem and periodic solutions of asymptotically linear planar Hamiltonian systems. J. Differ. Equ. **183**(2), 342–367 (2002)
123. L. Markus, *Lectures in Differentiable Dynamics*, vol. 3 (American Mathematical Society, 1971)
124. R. Martins, A.J. Ureña, The star-shaped condition on Ding's version of the Poincaré-Birkhoff theorem. Bull. Lond. Math. Soc. **39**(5), 803–810 (2007)
125. J.L. Massera, The existence of periodic solutions of systems of differential equations. Duke Math. J. **17**(4), 457–475 (1950)
126. J. Mawhin, Degré topologique et solutions périodiques des systèmes différentiels non linéaires. Bull. Soc. Roy. Sci. Liège **38**, 308–398 (1969)
127. J. Mawhin, Équations intégrales et solutions périodiques des systèmes différentiels non linéaires. Acad. Roy. Belg. Bull. Cl. Sci. **5**(55), 934–947 (1969)
128. J. Mawhin, Equivalence theorems for nonlinear operator equations and coincidence degree theory for some mappings in locally convex topological vector spaces. J. Differ. Equ. **12**, 610–636 (1972)

129. J. Mawhin, An extension of a theorem of A. C. Lazer on forced nonlinear oscillations. J. Math. Anal. Appl. **40**, 20–29 (1972)
130. J. Mawhin, Recent trends in nonlinear boundary value problems, in *Proceedings of the 7th International Conference Nonlinear Oscillations (Berlin)*, vol. 1 (Akademie Berlin, 1977), pp. 51–70
131. J. Mawhin, *Topological Degree Methods in Nonlinear Boundary Value Problems*, CBMS Regional Conference Series in Mathematics, vol. 40 (American Mathematical Society, Providence, RI, 1979)
132. J. Mawhin, *Topological Degree Methods in Nonlinear Boundary Value Problems, CBMS Regional Conference Series in Mathematics*, vol. 40 (American Mathematical Society, Providence, RI, 1979). Expository lectures from the CBMS Regional Conference held at Harvey Mudd College, Claremont, Calif., June 9–15, 1977
133. J. Mawhin, Qualitative behavior of the solutions of periodic first order scalar differential equations with strictly convex coercive nonlinearity, in *Dynamics of Infinite Dimensional Systems* (Springer, 1987), pp. 151–159
134. J. Mawhin, Topological degree and boundary value problems for nonlinear differential equations, in *Topological Methods for Ordinary Differential Equations (Montecatini Terme, 1991), Lecture Notes in Mathematics*, vol. 1537 (Springer, Berlin, 1993), pp. 74–142
135. J. Mawhin, Continuation theorems and periodic solutions of ordinary differential equations, in *Topological Methods in Differential Equations and Inclusions* (Springer, 1995), pp. 291–375
136. J. Mawhin, Leray-Schauder degree: a half century of extensions and applications. Topol. Methods Nonlinear Anal. **14**(2), 195–228 (1999)
137. J. Mawhin, C. Rebelo, F. Zanolin, Continuation theorems for Ambrosetti-Prodi type periodic problems. Commun. Contemp. Math. **2**(1), 87–126 (2000)
138. J. Mawhin, J.R. Ward Jr., Periodic solutions of some forced Liénard differential equations at resonance. Arch. Math. (Basel) **41**(4), 337–351 (1983)
139. J. Mawhin, M. Willem, *Critical Point Theory and Hamiltonian Systems*, Applied Mathematical Sciences, vol. 74 (Springer, New York, 1989)
140. J. Milnor, *Topology from the Differentiable Viewpoint* (Press of Virginia, Charlottesville, 1965)
141. G.R. Morris, A case of boundedness in Littlewood's problem on oscillatory differential equations. Bull. Aust. Math. Soc. **14**(1), 71–93 (1976)
142. J. Moser, Mimeographed lecture notes on ordinary differential equations (New York University, 1969)
143. J. Moser, *Stable and Random Motions in Dynamical Systems*. Annals of Mathematics Studies, No. 77 (Princeton University Press, Princeton, N. J.; University of Tokyo Press, Tokyo, 1973). With special emphasis on celestial mechanics
144. J. Moser, Is the solar system stable? Math. Intell. **1**(2), 65–71 (1978)
145. J. Moser, Dynamical systems, past and present. Proc. Int. Congr. Math. **1**, 381–402 (1998)
146. M. Mrozek, Periodic and stationary trajectories of flows and ordinary differential equations. Univ. Iagel. Acta Math. **27**, 29–37 (1988)
147. M. Nagumo, Über die lage der integralkurven gewöhnlicher differentialgleichungen. *Proceedings of the Physico-Mathematical Society of Japan. 3rd Series*, vol. 24 (1942), pp. 551–559
148. W.D. Neumann, Generalizations of the Poincaré Birkhoff fixed point theorem. Bull. Austral. Math. Soc. **17**(3), 375–389 (1977)
149. R.D. Nussbaum, The fixed point index and fixed point theorems, in *Topological Methods for Ordinary Differential Equations, Lecture Notes in Mathematics*, vol. 1537 (Springer, 1993), pp. 143–205
150. F. Obersnel, P. Omari, Old and new results for first order periodic ODEs without uniqueness: a comprehensive study by lower and upper solutions. Adv. Nonlinear Stud. **4**(3), 323–376 (2004)
151. F. Obersnel, P. Omari, On the periodic Ambrosetti-Prodi problem for a class of ODEs with nonlinearities indefinite in sign. Appl. Math. Lett. **111**, Paper No. 106622, 8 (2021)

152. P. Omari, G. Villari, On a continuation lemma for the study of a certain planar system with applications to Liénard and Rayleigh equations. Results Math. **14**(1–2), 156–173 (1988)
153. P. Omari, G. Villari, F. Zanolin, Periodic solutions of the Liénard equation with one-sided growth restrictions. J. Differ. Equ. **67**(2), 278–293 (1987)
154. P. Omari, F. Zanolin, Periodic solutions of Liénard equations. Rendiconti del Seminario Matematico della Universita di Padova **72**, 203–230 (1984)
155. R. Ortega, Topological degree and stability of periodic solutions for certain differential equations. J. Lond. Math. Soc. **2**(3), 505–516 (1990)
156. R. Ortega, Boundedness in a piecewise linear oscillator and a variant of the small twist theorem. Proc. Lond. Math. Soc. **79**(2), 381–413 (1999)
157. D. Papini, Periodic solutions for a class of Liénard equations. Funkcial. Ekvac. **43**(2), 303–322 (2000)
158. D. Papini, G. Villari, Periodic solutions of a certain generalized Liénard equation. Funkcial. Ekvac. **47**(1), 41–61 (2004)
159. D. Papini, G. Villari, F. Zanolin, Chaotic dynamics in a periodically perturbed Liénard system. Differ. Integral Equ. **32**(11–12), 595–614 (2019)
160. H. Poincaré, Sur un théoreme de géométrie. Rendiconti del Circolo Matematico di Palermo **33**(1), 375–407 (1912)
161. D. Qian, P.J. Torres, Periodic motions of linear impact oscillators via the successor map. SIAM J. Math. Anal. **36**(6), 1707–1725 (2005)
162. P.H. Rabinowitz, Pairs of positive solutions of nonlinear elliptic partial differential equations. Indiana Univ. Math. J. **23**(2), 173–186 (1973)
163. P.H. Rabinowitz, A note on pairs of solutions of a nonlinear Sturm-Liouville problem. manuscripta mathematica **11**(3), 273–282 (1974)
164. C. Rebelo, A note on the Poincaré–Birkhoff fixed point theorem and periodic solutions of planar systems. Nonlinear Anal. Theory Methods Appl. **29**(3), 291–311 (1997)
165. C. Rebelo, F. Zanolin, Multiplicity results for periodic solutions of second order ODE's with asymmetric nonlinearities. Trans. Am. Math. Soc. **348**(6), 2349–2389 (1996)
166. R.M. Redheffer, The theorems of Bony and Brezis on flow-invariant sets. Am. Math. Mon. **79**(7), 740–747 (1972)
167. R. Reissig, R. Reissig, G. Sansone, R. Conti, *Qualitative theorie nichtlinearer differentialgleichungen* (Cremonese, 1963)
168. G.E.H. Reuter, A boundedness theorem for non-linear differential equations of the second order. Proc. Camb. Philos. Soc. **47**, 49–54 (1951)
169. H.L. Royden, *Real Analysis* (The Mac Millan Company, New York, 1963)
170. G. Sansone, R. Conti, *Non-linear Differential Equations*. International Series of Monographs in Pure and Applied Mathematics, vol. 67, revised edition (The Macmillan Company, New York, 1964)
171. J. Schauder, Zur theorie stetiger abbildungen in funktionalräumen. Mathematische Zeitschrift **26**(1), 47–65 (1927)
172. J. Schauder, Invarianz des gebietes in Funktionalraümen. Studia Mathematica **1**(1), 123–139 (1929)
173. J. Schauder, Der fixpunktsatz in funktionalraümen. Studia Mathematica **2**(1), 171–180 (1930)
174. J. Schauder, Über den Zusammenhang zwischen der eindeutigkeit und lösbarkeit partieller Differentialgleichungen zweiter ordnung vom elliptischen typus. Mathematische Annalen **106**(1), 661–721 (1932)
175. J.T. Schwartz, *Nonlinear Functional Analysis*. Notes on Mathematics and its Applications (Gordon and Breach Science Publishers, New York-London-Paris, 1969)
176. G.R. Sell, *Topological Dynamics and Ordinary Differential Equations* (Van Nostrand Reinhold Co., London, 1971)
177. H.W. Siegberg, Some historical remarks concerning degree theory. Am. Math. Mon. **88**(2), 125–139 (1981)
178. E.E. Slaminka, Removing index 0 fixed points for area preserving maps of two-manifolds. Trans. Am. Math. Soc. **340**(1), 429–445 (1993)

179. H.L. Smith, Cooperative systems of differential equations with concave nonlinearities. Nonlinear Anal. Theory Methods Appl. **10**(10), 1037–1052 (1986)
180. R. Srzednicki, Periodic and constant solutions via topological principle of Wazewski. Univ. Iagel. Acta Math. **26**, 183–190 (1987)
181. R. Srzednicki, *A Geometric Method for the Periodic Problem in Ordinary Differential Equations, Séminaire d'Analyse Moderne [Seminar on Modern Analysis]*, vol. 22 (Université de Sherbrooke, Département de Mathématiques et d'Informatique, Sherbrooke, QC, 1992)
182. A. Tineo, An iterative scheme for the n-competing species problem. J. Differ. Equ. **1**(116), 1–15 (1995)
183. P.J. Torres, Existence of closed solutions for a polynomial first order differential equation. J. Math. Anal. Appl. **328**(2), 1108–1116 (2007)
184. R.R. Vance, E.A. Coddington, A nonautonomous model of population growth. J. Math. Biol. **27**(5), 491–506 (1989)
185. G. Vidossich, Towards a theory for periodic solutions to first order differential equations. (59/83/M)
186. G. Villari, F. Zanolin, A continuation theorem with applications to periodically forced Liénard equations in the presence of a separatrix. Ann. Mat. Pura Appl. (4) **180**(4), 387–411 (2002)
187. G. Villari, F. Zanolin, A geometric approach to periodically forced dynamical systems in presence of a separatrix. J. Differ. Equ. **208**(2), 292–311 (2005)
188. W. Walter, Fixpunkte der Poincaré-Abbildung. Archiv der Mathematik **33**(1), 80–84 (1979)
189. Y. Wang, Unboundedness in a duffing equation with polynomial potentials. J. Differ. Equ. **160**(2), 467–479 (2000)
190. Y. Wang, Boundedness for sublinear Duffing equations with time-dependent potentials. J. Differ. Equ. **247**(1), 104–118 (2009)
191. H. Whitney et al., A function not constant on a connected set of critical points. Duke Math. J. **1**(4), 514–517 (1935)
192. S. Yamaguti, M. Mizohata, On the existence of periodic solutions of the nonlinear differential equation $x'' + f(x)x' + g(x) = e(t)$. Mem. Coll. Sci., Univ. Kyoto, Ser. A **27**, 109–113 (1952)
193. J.A. Yorke, Invariance for ordinary differential equations. Math. Syst. Theory **1**(4), 353–372 (1967)
194. T. Yoshizawa, *Stability Theory and the Existence of Periodic Solutions and Almost Periodic Solutions*, vol. 14 (Springer, New York-Heidelberg, 1975)
195. C. Zanini, F. Zanolin, A multiplicity result of periodic solutions for parameter dependent asymmetric non-autonomous equations. Dyn. Contin. Discrete Impuls. Syst. Ser. A Math. Anal **12**(3-4), 343–361 (2005)
196. F. Zanolin, Permanence and positive periodic solutions for Kolmogorov competing species systems. Results Math. **21**(1–2), 224–250 (1992)
197. F. Zanolin, Continuation theorems for the periodic problem via the translation operator. Rend. Sem. Mat. Univ. Politec. Torino **54**(1), 1–23 (1996)
198. E. Zeidler, *Nonlinear Functional Analysis Vol. 1: Fixed-point Theorems* (Springer, Berlin and Heidelberg, 1986)
199. X. Zhao, The qualitative analysis of n-species Lotka-Volterra periodic competition systems. Math. Comput. Model. **15**(11), 3–8 (1991)
200. V. Zharnitsky, Breakdown of stability of motion in superquadratic potentials. Commun. Math. Phys. **189**(1), 165–204 (1997)

Chapter 3
Chaos in the Duffing Equation: With Some Simulations

In this chapter, we describe some examples which exhibit numerical evidence of the presence of complex dynamics, including chaos, for the periodically forced Duffing equations, with and without damping. At the moment we defer the technical meaning of chaotic dynamics and the theory needed to prove analytically these aspects to the next chapter.

3.1 The Duffing Equation: The Case with the Cubic Stiffness Term

Duffing's equation was introduced by G. Duffing in 1918 as already mentioned, as a model of a nonlinear oscillator with a cubic stiffness term. The cubic stiffness term describes the *hardening spring effect* seen in many mechanical problems. The periodically forced case is an equation of the form:

$$\ddot{x} + f(x) = e(t). \tag{3.1}$$

The cubic term appears in $f(x)$ where $f(x) = \beta x + \alpha x^3$, the function $e(\cdot) : \mathbb{R} \to \mathbb{R}$ represents an external perturbation which we will assume to be a (nonconstant) periodic function of a fixed period $T > 0$. A typical example would be the periodic forcing term $e(t) = A \cos \omega t$. The nonlinear term $f(x)$ represents the spring characteristic (in the undamped case) or in

$$\ddot{x} + c\dot{x} + f(x) = e(t), \quad c > 0, \tag{3.2}$$

when a friction term is included. Therefore the Duffing oscillator would be a forced oscillator with a restoring force $f(x) = \beta x + \alpha x^3$. When $\beta > 0$, the spring is called

L. Burra and F. Zanolin, *The Duffing Equation*, Infosys Science Foundation Series in Mathematical Sciences, https://doi.org/10.1007/978-981-97-8301-4_3

a hardening spring when $\alpha > 0$ and a softening spring when $\alpha < 0$ [26]. The unforced Duffing oscillator without damping is given by $\ddot{x} + \beta x + \alpha x^3 = 0$. Writing it as a Hamiltonian system we have $\dot{x} = y, \quad \dot{y} = -\beta x - \alpha x^3$. The corresponding energy for the level lines in the phase-plane is obtained by integration and they are

$$E(x, y) = \frac{1}{2}y^2 + \frac{1}{2}\beta x^2 + \frac{1}{4}\alpha x^4.$$

Also when $\beta > 0$ the potential function is a single well and when $\beta < 0$ the potential is a double well, α is positive in both the cases (see Fig. 3.1 for the potential function for $\beta = \pm 1$ and the corresponding level lines.)

The equilibria of the unforced Duffing oscillator can be evaluated by setting $\dot{x} = 0$ and $\dot{y} = 0, \quad x(\beta + \alpha x^2) = 0$. One point is at $x = 0$ and the other two are at $x = \pm\sqrt{-\beta/\alpha}$, when $\alpha\beta < 0$.

Thus in the case of the standard Duffing equation with $f(x)$ as in (1.3), this approach is well suited for $\beta < 0 < \alpha$ as well as $\alpha < 0 < \beta$. The potential $V(x) = \int_0^x f(s)\,ds = \frac{\beta}{2}x^2 + \frac{\alpha}{4}x^4$ is a double well potential in the former case and the origin is a saddle point, and also a homoclinic point in the phase-plane for the autonomous system (see Fig. 3.1 left panel).

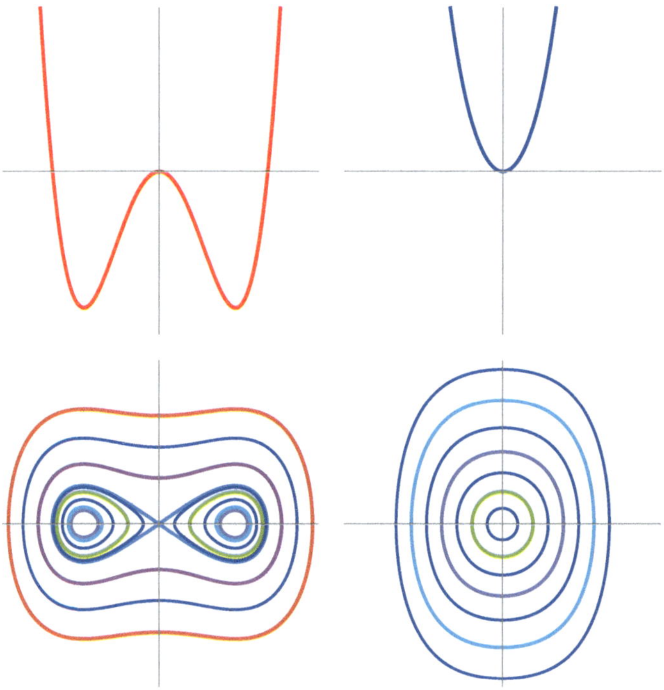

Fig. 3.1 The single well $V(x) = \frac{1}{2}x^2 + \frac{1}{4}x^4$ (right panel) and double well potential $V(x) = -\frac{1}{2}x^2 + \frac{1}{4}x^4$, (the left panel) with their corresponding energy level lines for the Duffing equation

Remark 3.1 In the case of the double well potential, we have a saddle point and, for the Duffing Oscillator, the phase-plane has homoclinic orbits. The Melnikov method, as described in [12], may be used to prove the presence of chaos in this case when a small periodic forcing term is included.

If $\alpha < 0 < \beta$, the potential $V(x) = \frac{\beta}{2}x^2 + \frac{\alpha}{4}x^4$ has a single well located at the origin surrounded on either side by two *humps* (see Fig. 3.2). In the phase-plane, the origin is a local centre bounded by two heteroclinic orbits connecting the equilibrium points (saddle points). The case $\alpha < 0 < \beta$ has been widely studied for the special values of the coefficients $\alpha = -2k^2$ and $\beta = 1 + k^2$ where k is a constant with $0 < k < 1$. In this case, the solutions with $x(0) = 0$ and $\dot{x}(0) = 1$ are denoted by $\mathrm{sn}(t, k)$ and called the Jacobi elliptic sine function (see [21, Sect. 1.6] and [20]) as well as Chap. 1, Sect. 1.3. The effect of the cubic term with the negative coefficient is such that all the solutions outside a neighbourhood of the origin blow up to infinity in finite time. On the other hand, the solutions near the origin typically behave like those of the harmonic oscillator, because the cubic term is quite negligible in a small neighbourhood of the origin compared with the linear term. If one is interested in periodic perturbations of the Duffing equation with these parameters, results can be found in [3, 30] also in connection with the nonlinear Schrödinger equation. We shall not discuss this case and instead, from the dynamics point of view, prefer to analyze the pendulum equation.

The Duffing-Type Equation Explored by Ueda

In this context it is worth mentioning the work of Y. Ueda. He showed evidence of chaotic dynamics in 1961 (using an analog computer) in some ODEs which are

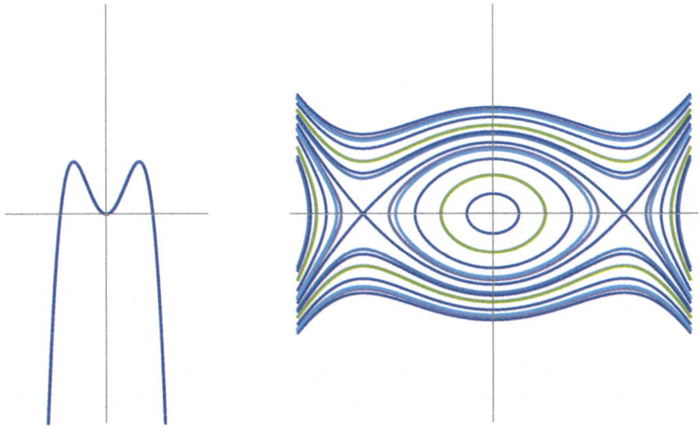

Fig. 3.2 Double hump potential $V(x) = \frac{1}{2}x^2 + \frac{1}{4}x^4$, with their corresponding energy level lines for the Duffing equation

indeed periodically perturbed Duffing equations with a friction term. His work is one of the pioneering examples in this area [1, 27]. When Ueda was a graduate student of Electrical Engineering of Kyoto University, he was involved in using computer simulations to solve some ODEs with an analog computer. In the course of his studies, he noticed unusual behaviour in some output data. He observed non-periodic responses of the system, and peculiar time evolution. This was the characteristic feature recognized by him which is now considered implicit in a chaotic system. Ueda registered the output data of the ODE on a graph paper, dated November 27, 1961.

A year later, Ueda studied another system of forced oscillation of the Duffing type

$$\ddot{x} + k\dot{x} + x^3 = B\cos t, \tag{3.3}$$

where k and B are constant parameters, and discovered chaotic behaviour of the solution $x(t)$ in certain domains of the parameter space (k, B). The chaotic attractor obtained for the case of $k = 0.10$ and $B = 12.0$ was called as the Japanese Attractor, whilst that of the case of $k = 0.05$ and $B = 7.5$ was called as Ueda's Chaotic Attractor. These attractors were discovered around the same year as the publication of the well-known paper of the Lorenz attractor, the beginning of 1963.

3.2 Simulations of the Duffing Equation

In this section, we include some simulations of Ueda's equation as well as some simulations of the Duffing equation and some modifications of the Duffing equation.

Simulations of the Duffing-Type Equation Used by Ueda

The first figure (Fig. 3.3) is a phase plot of the forced Duffing-type oscillator

$$\ddot{x} + k\dot{x} + x^3 = B\cos t,$$

studied by Ueda.

In the two-dimensional case, the Poincaré map can be "visualized" in the numerical simulations and it is a very useful technique to show the dynamics associated with periodically perturbed planar systems. This consists of representing the sequence of points $z_0, \Phi(z_0), \ldots, \Phi^n(z_0), \ldots$ where z_0 is an initial value and Φ is the Poincaré map at the time T, the time T being the period of the forcing term. Simulations representing a sufficiently large number of such points in the plane are usually referred to as *Poincaré sections* or representations of the *Poincaré return map*. Another term which is equivalently used is that of *stroboscopic map*. The reason behind the usage

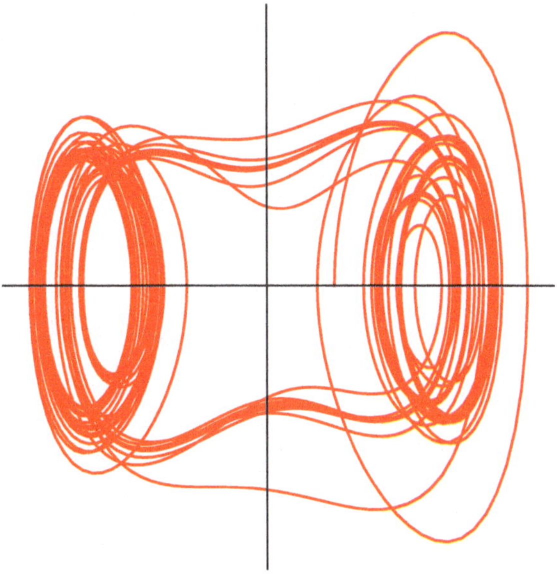

Fig. 3.3 The phase curves of (3.3) with $k = 0.1,\ B = 20,\ x(0) = 1,\ y(0) = 0$

of such terminology is clear from the context and by the geometrical meaning of the Poincaré map. Indeed, given a non-autonomous planar system

$$\begin{cases} \dot{x} = X(t, x, y) \\ \dot{y} = Y(t, x, y) \end{cases}$$

with X and Y functions which are T-periodic in the t-variable, the points $z_0, \Phi(z_0), \ldots, \Phi^n(z_0), \ldots$ can be seen as the projection on the (x, y)-plane of the various intersections of the solution curve $(t, x(t; z_0), y(t; z_0))$ in the *extended phase space* (t, x, y) with the planes $t = 0, t = T, \ldots, t = nT, \ldots$ Setting the problem in the extended phase space corresponds to considering the t-variable as another dimension and consider the system

$$\begin{cases} \dot{t} = 1 \\ \dot{x} = X(t, x, y) \\ \dot{y} = Y(t, x, y). \end{cases}$$

From this point of view, it is natural to adopt the term Poincaré section. Equivalently, we can look at whole trajectory $\{(x(t; z_0), y(t; z_0)) : t \geq 0\}$ in the (x, y)-plane and highlight the points at the times $t = 0, t = T, \ldots, t = nT, \ldots$ From this point of view, it is natural to adopt the term stroboscopic map. Clearly, from the point of view of the resulting diagrams, there is no difference, so that we will adopt interchangeably any of these terms (Figs. 3.4, 3.5 and 3.6).

Next, we have a Poincaré return map of Eq. (3.3). Some authors use a bifurcation diagram to explain chaotic dynamics. To visualize clearly the range of the parameters and the bifurcation the next figure, Fig. 3.7 is a bifurcation diagram which scans 200 values of B, in Eq. (3.3) starting from 3 to 11, for a value of $k = .1$. The equations are integrated for 1500 periods of the driving force $\cos \omega t$. The bifurcation diagram below is similar to the Poincaré map, except that it scans a range of values of B, and gives the x-values on the attractor for each B. We start from the initial point (1, 0) and consider the 1500 iterates of the Poincaré map and we plot on the y-axis the value of

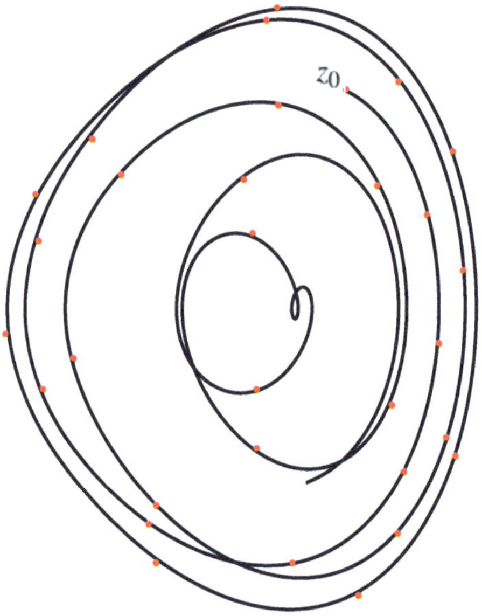

Fig. 3.4 The *stroboscopic* view of the Poincaré map, with the sequence of points $z_0, \Phi(z_0), \ldots, \Phi^n(z_0), \ldots$ shown as points on the solution curve for times $t = 0,\ t = T, \cdots nT$

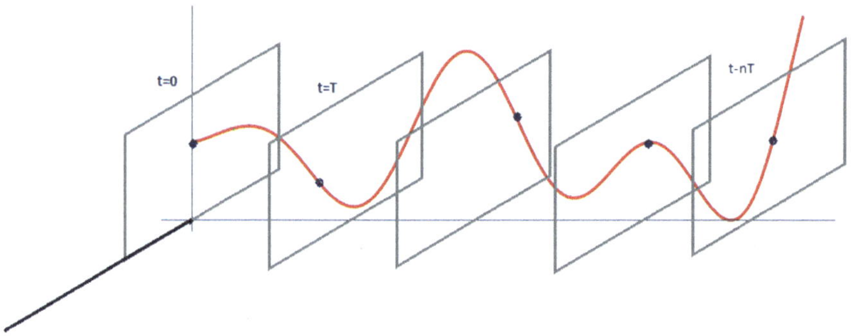

Fig. 3.5 In this figure the solution in the extended phase-plane is seen going through different planes placed at the time intervals $t = 0,\ t = T, \cdots nT$, then these points of intersection are projected onto one plane to get the *Poincaré return map*

Fig. 3.6 The Poincaré return map of Eq. (3.3) with $k = 0.05,\ B = 7.5$, (left) and $k = 0.1,\ B = 11$ (right), $x(0) = 1,\ y(0) = 0$

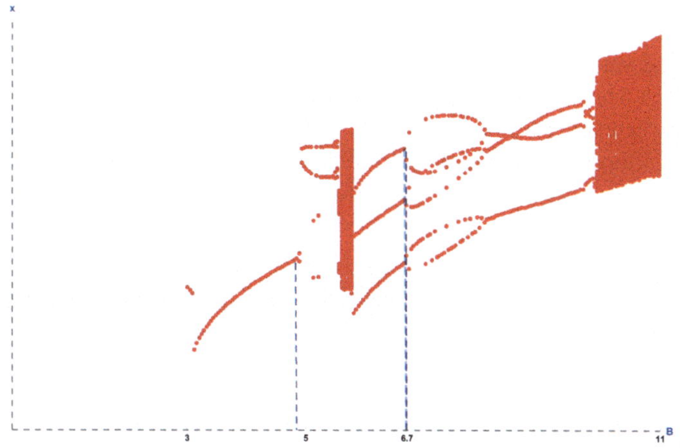

Fig. 3.7 A bifurcation diagram of equation $\ddot{x} + k\dot{x} + x^3 = B\cos t$ moving the parameter $B \in [3, 11]$ (in the abscissa) and $x(t)$ in the $Y-$ axis

x for the last 500 iterates. When we have a unique point or a few scattered points we can interpret it by saying that there is a periodic solution or a subharmonic solution of a small period. When for some B we have a large set of values like a vertical line of points we can argue that the simulation shows subharmonic solutions with a large period or chaotic dynamics. For the simulation we have chosen $k = .1$, $\omega = 1$ ($T = 2\pi/\omega$) and initial conditions $x(0) = 1$, $x'(0) = 0$. We scan for 200 values of B in the given interval. The complexity of the dynamics as B changes from 3 to 11 can be seen clearly. Period doubling is evident in this bifurcation diagram. The period doubling can be seen around the value of the parameter $B = 5$ as well as several period doubling instances around the value of the parameter $B = 6.7$. For other values of the parameter, the behaviour can be seen to be more complicated.

We now go back to the Duffing equation with the cubic term mentioned earlier, that is,

$$\ddot{x} + c\dot{x} + \beta x + \alpha x^3 = B\cos\omega t. \tag{3.4}$$

We include some simulations of the solutions given in Eq. (3.4), considering different values of the parameters. Expressing (3.4) without the damping term $c\dot{x}$ as a system

$$\begin{cases} \dot{x} = y \\ \dot{y} = -\beta x - \alpha x^3 + B\cos\omega t \end{cases} \tag{3.5}$$

and its damped counterpart which is equivalent to Eq. (3.4) is then

$$\begin{cases} \dot{x} = y \\ \dot{y} = -cy - \beta x - \alpha x^3 + B\cos\omega t. \end{cases} \tag{3.6}$$

Choosing the values of $\beta = -1$ and $\alpha = 1$ the potential function is a double well (see Fig. 3.1). That is, we consider a perturbation of the autonomous equation, $\ddot{x} + g(x) = 0$, where $g(x) = -x + x^3$. At $x = 1$ $g(1) = 0$, $g'(1) = 2$. Consider the linearized Duffing equation about the equilibrium point $(1, 0)$, $\ddot{u} + g'(1)u = 0$ which is $\ddot{u} + 2u = 0$. The solutions for this equation are linear combinations of $\sin(\sqrt{g'(1)}t)$ and $\cos(\sqrt{g'(1)}t)$ functions, whose period is $\frac{2\pi}{\sqrt{g'(1)}}$. So the period map has values ranging from $2\pi/\sqrt{2}$ to $+\infty$, and this is monotone increasing.

Still in the case $\alpha > 0 > \beta$, we now consider Eq. (3.5), which has the driving force included in the term $B\cos\omega t$. Figure 3.8 shows the phase curves of Eq. (3.5) with $\alpha = 1$, $\beta = -1$, $B = 0.1$, $\omega = 1.1$, $x(0) = 0$, $y(0) = 0$. It is interesting to compare the outcome with the level lines in Fig. 3.1. It is apparent that the trajectories wind a certain number of times around one of the stable equilibrium points and then switch to the other corresponding to the double well potential.

We now consider the autonomous Duffing equation with a damping term, that is, $\ddot{x} + c\dot{x} + \beta x + \alpha x^3 = 0$. The dynamics is illustrated in Fig. 3.9. In this figure, it

Fig. 3.8 The solution curves of (3.5) with $\alpha = 1$, $\beta = -1$, $B = 0.1$, $\omega = 1.1$, $x(0) = 0$, $y(0) = 0$

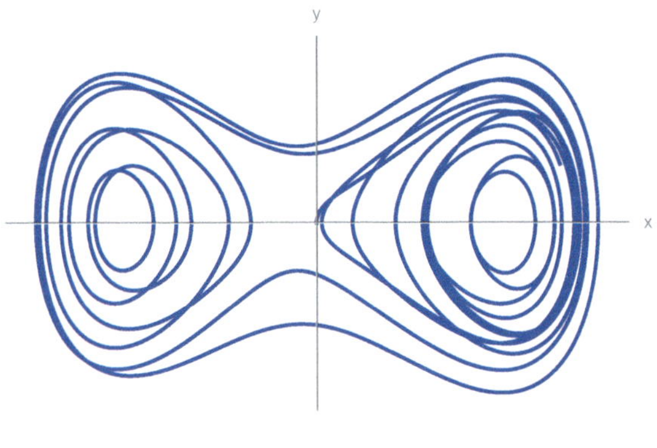

can be seen that the solutions either tend to one equilibrium point $(1, 0)$ or $(-1, 0)$, depending on the initial point. There is a third kind of solution which tends to the third equilibrium point (saddle point), $(0, 0)$.

It is worth noticing that, if we put any friction term to an autonomous Duffing equation, then the only periodic solutions are the constant solutions. Indeed, consider any second-order equation

$$\ddot{x} + c\dot{x} + f(x) = 0 \tag{3.7}$$

with $f : J \to \mathbb{R}$ a continuous function defined on an open interval $J \subset \mathbb{R}$. Then we have

Claim f $c \neq 0$, then the only periodic solutions of (3.7) are the constants z such that $f(z) = 0$.

Indeed, suppose that $x(t)$ is a periodic solution of (3.7) with some period $\tau > 0$. We multiply the equation by $\dot{x}(t)$ and integrate on $[0, \tau]$, to have

$$\int_0^\tau \ddot{x}(t)\dot{x}(t)\,dt + c\int_0^\tau \dot{x}(t)^2\,dt + \int_0^\tau f(x(t))\dot{x}(t)\,dt = 0$$

and hence

$$\int_0^\tau \dot{x}(t)^2\,dt = \frac{1}{c}\left[-\frac{1}{2}(\dot{x}(t))^2 - F(x(t))\right]_0^\tau,$$

where F is a primitive on the interval J of the continuous function f. Hence, by the τ-periodicity of the solution $x(t)$, the right-hand side of the above relation vanishes and we conclude that $\int_0^\tau \dot{x}(t)^2\,dt = 0$ and hence $\dot{x}(t) = 0$ for all $t \in [0, \tau]$ and hence also for all $t \in \mathbb{R}$, by the hypothesis that the solution is τ-periodic. We have thus proved that $x(t) = \text{constant} = z \in J$ and then, replacing the constant value z in (3.7), we find $f(z) = 0$. Figure 3.9 provides a simulation for (3.5) with a small friction. Since the friction is very small, the solutions move very slowly towards the equilibrium points and the diagram gives the "illusion" that they move asymptotically towards a periodic orbit.

If we increase the friction coefficient even marginally the resulting figure is a pronounced spiral, as can be seen in Fig. 3.10. The increase in the friction coefficient from 0.015 to 0.075 makes the solution curves tend to the equilibrium point faster.

Let us now consider both a driving force and damping term in order to remove energy conservation as well as introduce dissipation into the Hamiltonian system. Here c is the friction coefficient as mentioned before and B is the strength of the driving force which oscillates at a frequency ω. We will fix coefficients as $c = 0.05$, $\omega = 1.1$, and initially set $B = 0.1$ This is shown in Fig. 3.11.

Figure 3.11 appears complex, but in fact it is not as complicated as it looks. For the most part, the plot shows the initial time period during which the solutions seem to approach a final behaviour. This early or initial behaviour is called an "initial transient". What seems to appear is an orbit going around the minimum of the potential at $x = 1$.

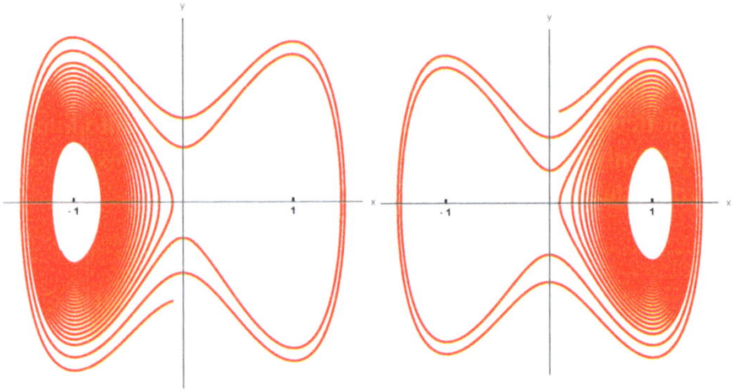

Fig. 3.9 The solution curves of (3.5) with $\alpha = 1$, $\beta = -1$, $c = 0.015$, $B = 0$, $x(0) = -0.1$, $y(0) = -0.5$ (left) and $x(0) = 0.1$, $y(0) = 0.5$ (right)

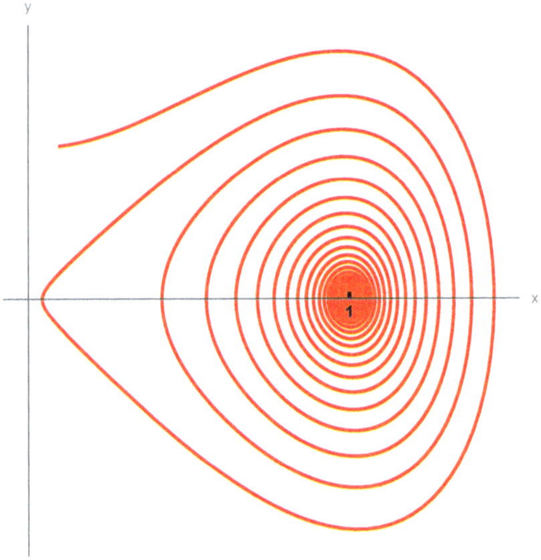

Fig. 3.10 The solution curves of Eq. (3.5) with $\alpha = 1$, $\beta = -1$, $c = 0.075$, $B = 0$, $x(0) = 0.1$, $y(0) = 0.5$

Let us look into how the phase space plots alter when the intensity of the driving force is changed. We choose $B = 0.2$ and increase the time up to 800.

From Fig. 3.12, it can be seen that the particle has moved through both of the "potential wells" at $(-1, 0)$ and $(1, 0)$. The system wanders around between the fixed points at $(-1, 0)$ and $(1, 0)$ in a haphazard or "chaotic" manner. Nonetheless, once again, just about most of this complexity is due to an initial transient.

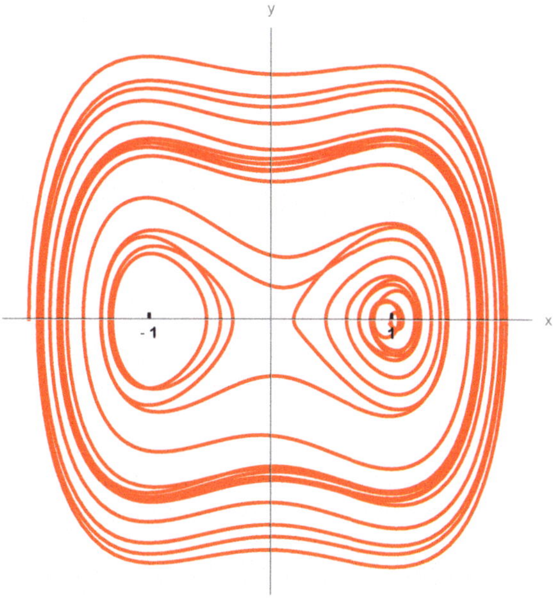

Fig. 3.11 The solution curves of the damped Duffing oscillator equation (3.6) with $\alpha = 1,\ \beta = -1,\ B = 0.1,\ c = 0.05,\ \omega = 1.1,\ x(0) = -2,\ y(0) = 0,\ t$ ranging from 0 to 150

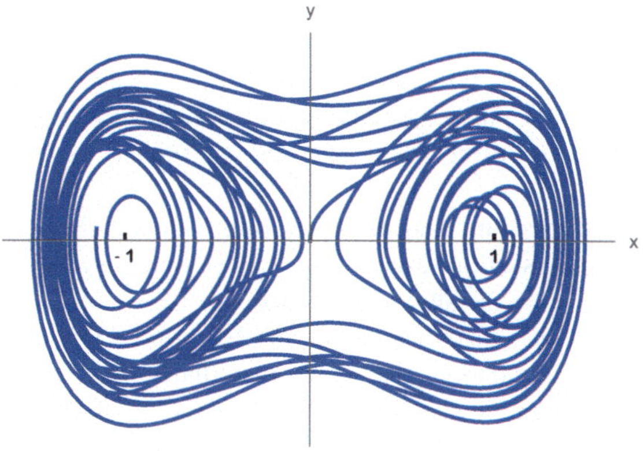

Fig. 3.12 The solution curves of the damped Duffing oscillator (3.6) with $\alpha = 1,\ \beta = -1,\ B = 0.2,\ c = 0.05,\ \omega = 1.1,\ x(0) = 0,\ y(0) = 0,\ t \in [0, 800]$

Poincaré Sections

In order to study chaotic motion, there is a much better method than observing a trajectory move around in phase space. A Poincaré section is constructed on the phase-plane by only plotting points every $T = 2\pi/\omega$ where ω is the angular frequency of the driving force and T is its period.

If the trajectory in phase space is periodic, with this period, then we obtain only one point displayed on the plot. If the orbit has a period equal to twice the period of the driving force, then the Poincaré section will show two points, and so on.

If the system is chaotic, the Poincaré section will consist of a pattern of points called the attractor. The attractor has a structure that is frequently quite beautiful. An

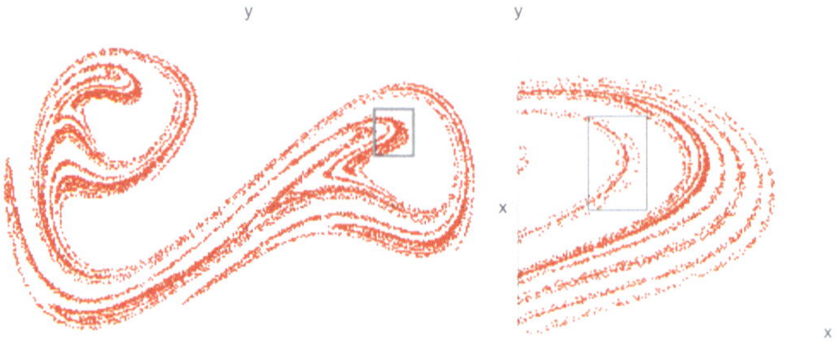

Fig. 3.13 A Poincaré section of the damped Duffing oscillator equation (3.6) with $\alpha = 1,\ \beta = -1,\ B = 0.35,\ c = 0.1,\ \omega = 1.4,\ x(0) = 0,\ y(0) = 0,\ t \in [0,\ 10^5]$ and a blow up of the region in the shaded box

unexpected outcome is that a deterministic system can display unpredictability and apparent chaos and at the same time maintains a consistent global structure. Thus, the Poincaré section is a discrete set of points in the phase space of the particle at every period of the driving force.

The next figure, Fig. 3.13, is that of a Poincaré section of the Duffing two-well oscillator.

This peculiar diagram is the strange attractor. It is the limiting set of points to which the discretized trajectory tends to repeating after every time interval equal to the period of the driving force. It can be noticed that the structure of the system is complex but not completely random. We can see some kind of a structure which persists at different scales similar to a fractal.

A Bifurcation Diagram The bifurcation diagram in Fig. 3.14 is similar to the Poincaré section, except that it goes over an array of values of B, and gives the x-values on the attractor for each B.

One can easily see a pattern of "period doubling" leading into chaos. With the parameters chosen, the system oscillates from a cycle of two to a cycle of four and then eight, eventually entering a region of chaotic dynamics (depending on the precise value of B.) The procedure scans 200 values of B in the range from 0.3 to 0.35, with $c = 0.1$ and $\omega = 1.4$. It starts with $x = 0$ and $x = 1$, i.e. at rest in the $x = 1$ minimum. The equations are integrated for 1500 periods of the driving force with the first 1000 being ignored and the value of x plotted at the end of each of the remaining 500 periods.

To compare the bifurcation diagram of Fig. 3.14 with the resulting attractor in the phase-plane, we consider the Poincaré section with some specific values of the parameter B. This is illustrated in Fig. 3.15. The picture shows 15000 iterates of the Poincaré map ignoring the first 400 transients, with initial points $x(0) = 0$ and $\dot{x}(0) = 0$, $c = 0.1$ and $\omega = 1.4$. The parameter $B = 0.34$. The attractor shows the presence of 8 single points which means that there is an attractive subharmonic solution of order 8. Comparing with Fig. 3.13 which was the Poincaré section for the same values of all parameters except B, which changed from 0.35 to 0.34 in this

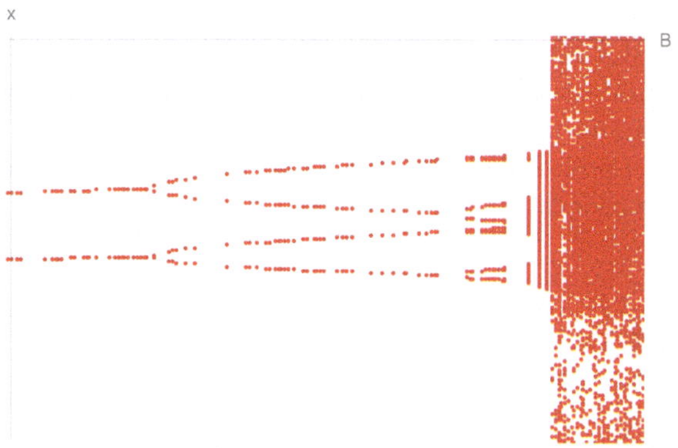

Fig. 3.14 A bifurcation diagram of the damped Duffing oscillator (3.6) with $\alpha = 1,\ \beta = -1$

Fig. 3.15 A Poincaré return map of the damped Duffing oscillator (3.6) with $\alpha = 1,\ \beta = -1$

figure. We come back to this example and prove the presence of chaos rigorously for this equation, with particular values of the frequency ω, using the Melnikov method in Chap. 4, Sect. 4.6.

3.3 Different Variants of the Duffing Equation

As mentioned earlier (see Chap. 2), over a period spanning several years to the present day, a great deal of investigation on the qualitative and numerical aspects of the solutions of the Duffing equations has been done by several authors from different perspectives. For a few classical examples, we recall the works discussed in Chap. 2 on the number of periodic solutions (see [13, 25]), the results connected to the problems of Littlewood and Moser on the boundedness of the solutions (see [19, 24]) and the works on the existence of infinitely many periodic solutions [9, 14, 23]. From a theoretical point of view, rigorous analytical proofs of the existence of chaos for some examples of Duffing equations have been obtained (see [12, 29] and the references therein). A classical and very effective approach to prove the existence of chaotic dynamics rigorously is based on the Melnikov method or its variants. In this setting, a possible way to attack the problem is that of creating a suitable perturbation of an autonomous system possessing homoclinic (or heteroclinic) orbits (see, for instance, [2, 10, 28]). Concrete examples of equations of this form have been proposed in [12, 22] as a magnetoelastic oscillator where the presence of strange attractor can be detected.

3.3.1 A Duffing-Like Differential Equation

In this section, we analyse a different configuration of the Duffing-type equations. We consider the case $\ddot{x} + x^3 + a_2(t)x^2 + a_1(t)x + a_0(t) = 0$ with $a_1(t) = -1$ $a_1(t) = 1$, and $a_0(t) = a\cos\omega t$. Figure 3.16 shows different Poincaré sections constructed in the phase-plane. In these simulations, our control parameter is a, the coefficient of $\cos\omega t$.

In Fig. 3.17, the phase portrait of the equation $\ddot{x} + x^3 + a_2(t)x^2 + a_1(t)x = 0$ with $a_1(t) = -1$ and $a_1(t) = 1$ is shown, drawing one's attention to the homoclinic orbit at the origin and in agreement with the Poincaré section in Fig. 3.16.

3.3.2 The Duffing Equation with Fractional Order

Let us consider a general nonlinearity with fractional order in the Duffing equation

$$\ddot{x} + f(x) = e(t)$$

as

$$f(x) = -ax + bx|x|^{p-1} \tag{3.8}$$

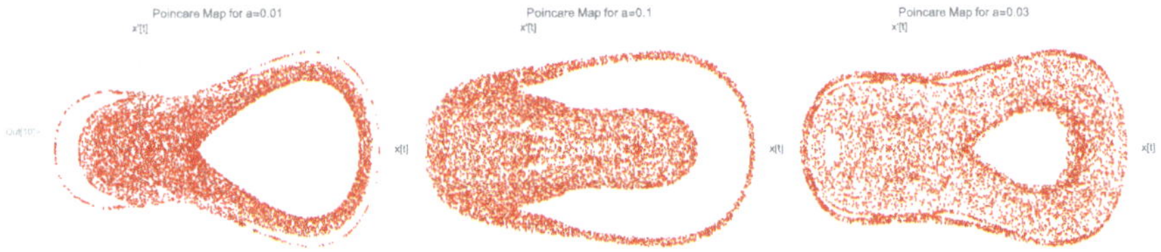

Fig. 3.16 A Poincaré section of a Duffing-like oscillator $\ddot{x} + x^3 + a_2(t)x^2 + a_1(t)x + a_0(t) = 0$ with $a_1(t) = -1$ $a_1(t) = 1$, and $a_0(t) = a\cos\omega t$, $x(0) = .01$, $\dot{x}(0) = .01$, for $a = 0.01,\ 0.1, 0.03$

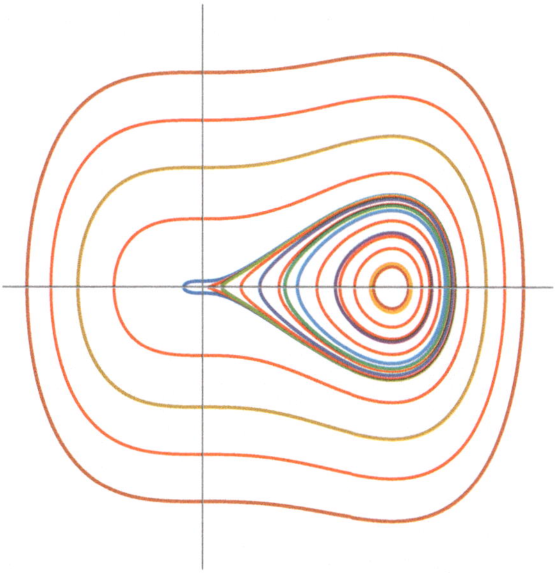

Fig. 3.17 The phase portrait of the unperturbed Duffing-like oscillator $\ddot{x} + x^3 + a_2(t)x^2 + a_1(t)x = 0$ with $a_1(t) = -1$ and $a_1(t) = 1$

with

$$a, b > 0 \quad \text{and} \quad p > 1. \tag{3.9}$$

The Duffing equation with $f(x)$ as in (3.8) is

$$\ddot{x} - ax + bx|x|^{p-1} = e(t), \tag{3.10}$$

which is a generalized Duffing-type oscillator considered by several authors [7, 17] with a fractional order nonlinear term. The constant p is an integer or a fraction. $e(t)$ is a periodic function. Comparing Eq. (3.10) with (3.3), it is apparent that the cubic term is generalized to the fractional order term.

There is physical meaning to Eq. (3.10), it represents the oscillatory movement of a buckled beam with supported or hinged ends. The exponent p in the fraction power is dependent on the material of the beam as well as its bending properties. Many substances have nonlinear elastic properties, for instance, the materials commonly used for building aircraft like aluminium, titanium, etc., where $p > 1$. For other materials like copper and its alloys, the parameter p is a rational number. Materials such as wood and polymers have a value of $p = 3$ and the differential equation for these materials is the Duffing type with the cubic nonlinearity as in (3.3). The Melnikov method (see Chap. 4 for a detailed derivation and explanation of the Melnikov method) has been applied to the equation

$$\ddot{x} - ax + bx|x|^{p-1} = \varepsilon(-c\dot{x} + r\cos(\omega t))$$

in some above-mentioned papers [7, 17]. The constant $p > 1$ in [7] can be an integer or a fraction, hence giving rise to the term *fractional order deflection function* for $f(x)$ in (3.8). As observed in [17], for a general $p > 1$, the computation of integrals in the Melnikov function is a very difficult task.

Other types of restoring forces $f(x)$ including a term of the form $x|x|^{p-1}$ have been widely considered. For instance, in [11], the authors deal with a Duffing-type equation of the form

$$\ddot{x} + x|x|^{p-1} = \sin(\omega t)$$

with $p \neq 1$ and positive. At this juncture we would like to recall from Chap. 2, Sect. 2.12, the work of Littlewood [18, 19] which was mentioned. In this context, we point out that when $p > 1$, the superlinear case is considered and for $0 < p < 1$, the sublinear case is considered by the terminology introduced by Littlewood in [18, 19]. The analysis of the Lyapunov exponents and the Poincaré map in the phase-plane performed in [11], for some values of p (e.g. for $p = 3$), suggests the presence of chaotic dynamics. Quoting the authors, *numerical experiments also show that the form of the periodic drive is not critical. For example, a square wave of similar frequency also produces chaos in these systems* [11, p. 387]. In our work [4] on these kinds of equations, we investigate the undamped scalar equation of the Duffing type

$$(H) \qquad \ddot{x} + f(x) = m_{1,2}(t),$$

as well as its associated planar system

$$(S_{1,2}) \qquad \begin{cases} \dot{x} = y \\ \dot{y} = -f(x) + m_{1,2}(t), \end{cases}$$

where $f : \mathbb{R} \to \mathbb{R}$ is a locally Lipschitz continuous function and $m_{1,2} : \mathbb{R} \to \mathbb{R}$ is a piecewise constant T-periodic forcing term defined on $[0, T[$ by

$$(M) \qquad m_{1,2}(t) := \begin{cases} m_1, & \text{for } t \in [0, \tau_1[\\ m_2, & \text{for } t \in [\tau_1, \tau_1 + \tau_2[, \end{cases} \qquad (m_1 \neq m_2)$$

with

$$\tau_1, \ \tau_2 > 0, \quad \tau_1 + \tau_2 = T.$$

We prove the presence of chaotic dynamics under general conditions for the nonlinear term $f(x)$, including those of (3.8)–(3.9). Our results are *robust*, i.e. stable with respect to small perturbations, hence they are still true for the more general equation

$$(H_{c,e}) \qquad \ddot{x} + c\dot{x} + f(x) = e(t),$$

provided that c is sufficiently small and $e(\cdot) : \mathbb{R} \to \mathbb{R}$ is a T-periodic function with

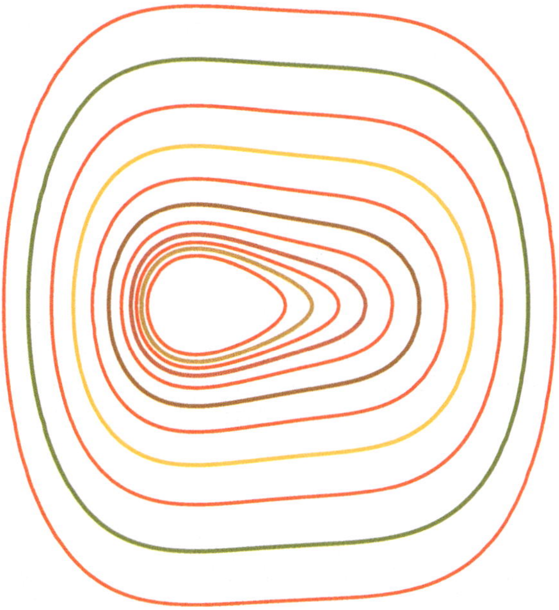

Fig. 3.18 A phase portrait of the energy level lines $E(x, y) = \frac{y^2}{2} + x^2\frac{|x|^2}{4} - \frac{x^2}{2} + 3x$ corresponding to the Duffing equation with $f(x)$ as in (3.8), with $p = 3$, $a = 1$, $b = 1$, $e(t) = 3$

$$\int_0^T |e(t) - m_{1,2}(t)|\, dt < \varepsilon, \tag{3.11}$$

with $\varepsilon > 0$ some small number. In this case, we can take $e(t)$ as smooth as we like (even of class C^∞).

To give us an idea into what the phase portrait of the system (3.10) would look like see Fig. 3.18. This a Hamiltonian system as is also evident from Fig. 3.18.

If we consider Eq. (3.10) with a small friction term and a forcing term $A\cos\omega t$ written in the phase-plane, we have

$$\begin{cases} \dot{x} = & y \\ \dot{y} = & -c\dot{x} + ax - bx|x|^{p-1} + A\cos\omega t. \end{cases} \tag{3.12}$$

A numerical simulation of system (3.12) is shown in Fig. 3.19. Observe that Fig. 3.19 with $p = 5/2$ is similar to Fig. 3.12 with $p = 3$.

Figures 3.19 and 3.20 clearly display the chaotic nature of system (3.12). To get further insight into the chaotic nature of system (3.12), we plot a Poincaré return map for a few values of A, as shown in Fig. 3.21. It can be observed that there is a striking similarity between the Ueda attractor in Figs. 3.6 and 3.21.

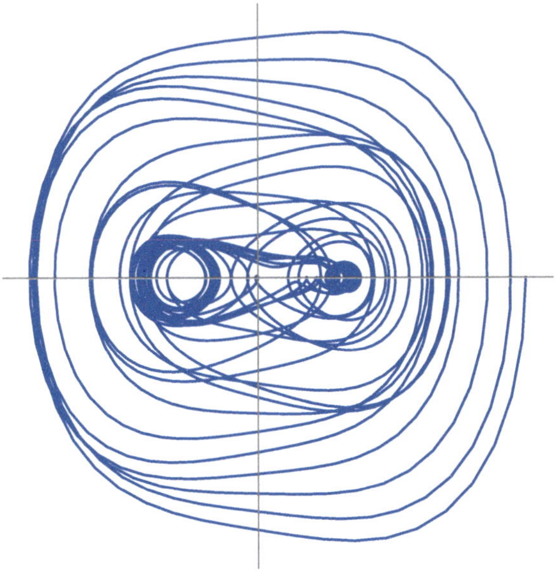

Fig. 3.19 Phase curves of system (3.12) with $p = 5/2,\ a = 1,\ b = 1\ c = 0.1,\ A = 20,\ x(0) = 10,\ y(0) = 0$

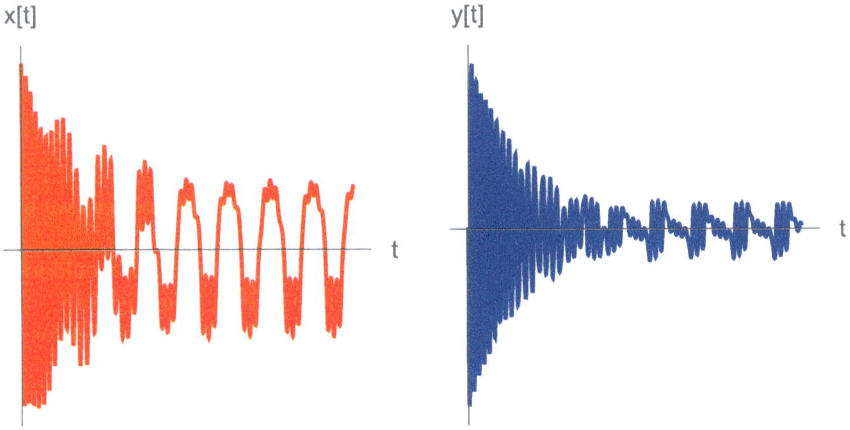

Fig. 3.20 Solution curves of system $x(t)$ vs. t (left) and $y(t)$ vs t (right) (3.12) with $p = 5/2,\ a = 1,\ b = 1\ c = 0.1,\ A = 20,\ x(0) = 10,\ y(0) = 0$

3.3.3 *Duffing Equation with Singularities*

As a further example, we deal with a class of periodically perturbed nonlinear second-order ordinary differential equations of the Duffing type

$$\ddot{x} + g(x) = p(t), \tag{3.13}$$

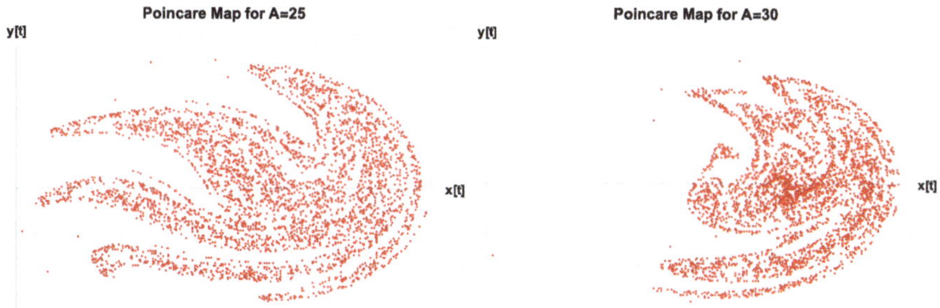

Fig. 3.21 Poincaré return map of system (3.12) with $p = 5/2$, $a = 1$, $b = 1$ $c = 0.01$, $A = 25$, $A = 30$, $x(0) = 1$, $y(0) = 0$

where $g : \mathbb{R}_0^+ \to \mathbb{R}$ is a locally Lipschitz continuous function which has a singularity at the origin of the repulsive type and a superlinear growth at infinity. More precisely, for $\mathbb{R}_0^+ :=]0, +\infty[$, we suppose that the following conditions hold:

$$(g_0) \qquad \lim_{s\to 0^+} g(s) = -\infty \quad \text{and} \quad \lim_{s\to 0^+} G(s) = +\infty \quad \text{for} \quad G' = g \text{ on } \mathbb{R}_0^+;$$

$$(g_\infty) \qquad \lim_{s\to +\infty} \frac{g(s)}{s} = +\infty.$$

The forcing term $p(t)$ is taken as $p : \mathbb{R} \to \mathbb{R}$, a T-periodic function (for some $T > 0$), with $p \in L^1([0, T])$. We suppose that p is nonconstant. To this end, we set

$$\bar{p} := \frac{1}{T}\int_0^T p(t)\,dt, \quad \tilde{p}(t) := p(t) - \bar{p}$$

and denote by $P(t)$ a primitive of $\tilde{p}(t)$ with mean value zero. The function $P(t)$ is an absolutely continuous T-periodic function with $||P||_\infty > 0$ (as, by assumption, $p(t) \neq \bar{p}$ on a set of positive measure). Without loss of generality we suppose, from now on, that $T > 0$ is the minimal period of $P(t)$. Passing to the phase-plane we can equivalently write Eq. (3.13) as a first-order system as

$$\begin{cases} \dot{x} = y \\ \dot{y} = -g(x) + p(t) \end{cases} \tag{3.14}$$

or as

$$\begin{cases} \dot{x} = y + P(t) \\ \dot{y} = -g(x) + \bar{p}. \end{cases} \tag{3.15}$$

We look for *positive* solutions of (3.13), namely, we look for globally defined solutions $x(t)$ of (3.13), in the Carathéodory sense, such that $x(t) > 0$ for all $t \in \mathbb{R}$.

A typical nonlinearity satisfying (g_0) and (g_∞) could be given by

$$g(x) = -\frac{A}{x^\alpha} + Bx^\beta, \quad \text{with } A, B > 0, \ \alpha \geq 1, \beta > 1. \tag{3.16}$$

When we consider the behaviour of $g(x)$ at the origin, expressed by condition (g_0), we are in the case of the so-called *repulsive-type equations*, according to Lazer and Solimini [16]. We recall that in [16], the authors considered the singular equation

$$g(x) = -\frac{1}{x^\alpha} + h(t), \tag{3.17}$$

where $h(t)$ is a T-periodic forcing term and proved that if $\alpha \geq 1$, then (3.17) has a positive T-periodic solution if and only if $h(t)$ has a negative mean value. It was also proved in [16] that this result is optimal in the sense that if $0 < \alpha < 1$, it is possible to choose a function $h(t)$ that has a negative mean value and the equation has no positive T-periodic solutions. Indeed, as an example of such a condition let us consider some simulations of a particular example of Eq. (1.2) from [16] $x''(t) - 1/x(t)^\alpha = h(t)$ written in the phase-plane, with $\alpha = 1$ and $h(t) = -b + \mu \sin \omega t, \quad b > 0$

$$\begin{cases} \dot{x} = y \\ \dot{y} = \frac{1}{x} - b + \mu \sin \omega t. \end{cases} \tag{3.18}$$

The next simulation is that of the Poincaré return map of Eq. (3.18). The picture suggests the presence of chaos.

For a more detailed analysis of the Duffing equation with singularity, in the Lazer-Solimini setting, see our recent paper [6].

With regard to the second condition, that is, (g_∞), our choice is motivated by the analogy with the superlinear Duffing equation on the real line. Another motivation comes from the work of del Pino and Manásevich [8] where a second-order singular equation of the form

$$\ddot{x} + f(t, x) = 0$$

was considered, with $f(t, s)$ satisfying

$$-\infty \leq \limsup_{s \to 0^+} sf(t, s) < 0, \quad \lim_{s \to +\infty} \frac{f(t, s)}{s} = +\infty,$$

uniformly in t. In [8], the authors, using the Poincaré-Birkhoff fixed point problem, proved the existence of infinitely many positive T-periodic solutions of the form $x_n^\pm(t)$ such that $x_n^\pm(t) - 1$ has exactly $2n$ zeros in $[0, T[$ for every integer n sufficiently large (Figs. 3.22, 3.23, and 3.24).

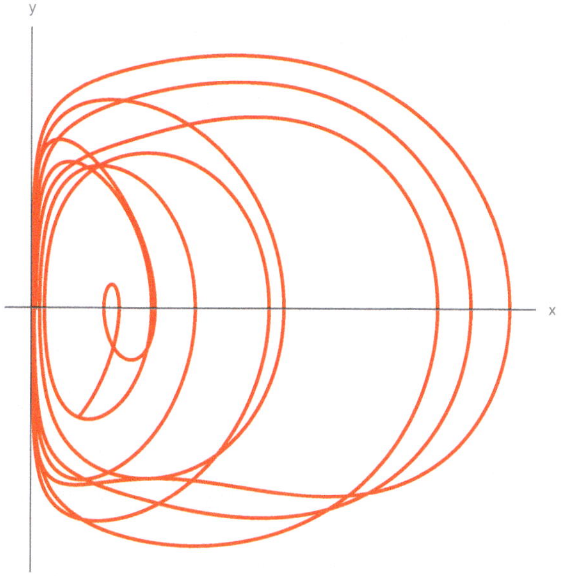

Fig. 3.22 A numerical simulation of (3.18) depicting the solution curve in the phase-plane with $\omega = 1,\ \mu = 1,\ b = 1,\ x(0) = 0.1,\ y(0) = 0$

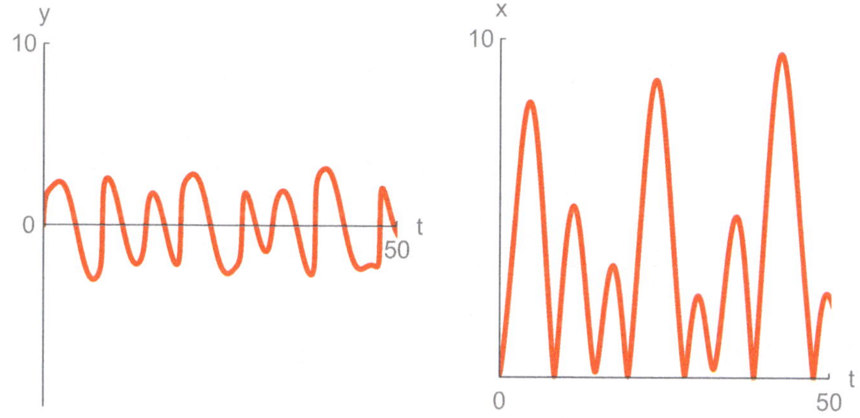

Fig. 3.23 A numerical simulation of (3.18) $y\ vs\ t$ (left) and $x\ vs\ t$ (right) with $\omega = 1,\ \mu = 1,$ $b = 1,\ x(0) = 0.1,\ y(0) = 0$

To get an idea of the dynamics of a system with singularities we consider some simulations of Eq. (3.13), with $g(x)$ as in (3.16) and $p(t) = \mu$. Such an equation written in the phase-plane would be

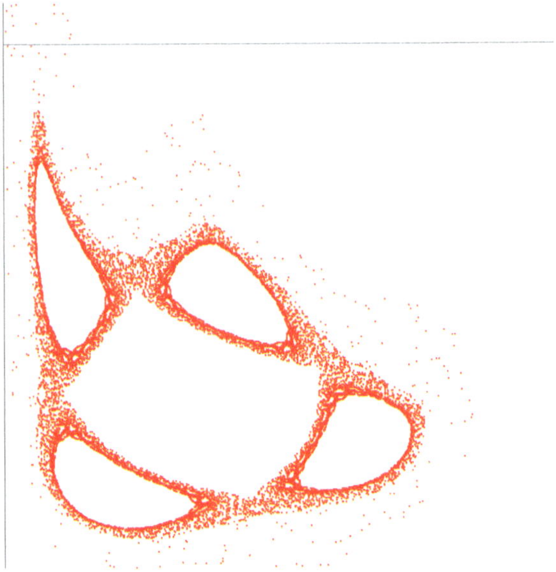

Fig. 3.24 A Poincaré return map of Eq. (3.18) with $\omega = 1,\ b = 1,\ \mu = 0.6,\ x(0) = 0.1,\ y(0) = 0$

$$\begin{cases} \dot{x} = y \\ \dot{y} = \frac{A}{x^{\alpha}} - Bx^{\beta} + \mu, \end{cases} \tag{3.19}$$

where μ is any real parameter. The phase portrait of (3.19) is that of a global centre in the open right half-plane. For a general function $g(x)$ satisfying conditions (g_0) and (g_∞), the phase portrait of Eq. (3.14) is that of a centre outside a compact set in the open right half-plane. That is, there is a closed trajectory Γ_0 in $\mathbb{R}_0^+ \times \mathbb{R}$ including in its interior region all the equilibrium points and such that any trajectory outside Γ_0 is also a closed orbit. These closed orbits are actually energy level lines of the form:

$$\frac{y^2}{2} + G(x) - \mu x = E, \tag{3.20}$$

where $G(x) = -\dfrac{A}{1-\alpha}x^{1-\alpha} + \dfrac{B}{1+\beta}x^{1+\beta}$. Figure 3.25 shows the energy level lines of Eq. (3.20) for different values of E. If we add a small friction term to system (3.19) with the coefficient $c = .01$ and choose the perturbation term $p(t) = \mu \sin \omega t$ that is $\dot{x} = y,\ \dot{y} = -cy + \dfrac{A}{x^{\alpha}} - Bx^{\beta} + \mu \sin \omega t$, the phase curves are as shown in Fig. 3.26.

For the same system, Fig. 3.27 gives the graphs of $x\ vs\ t$ and $y\ vs\ t$.

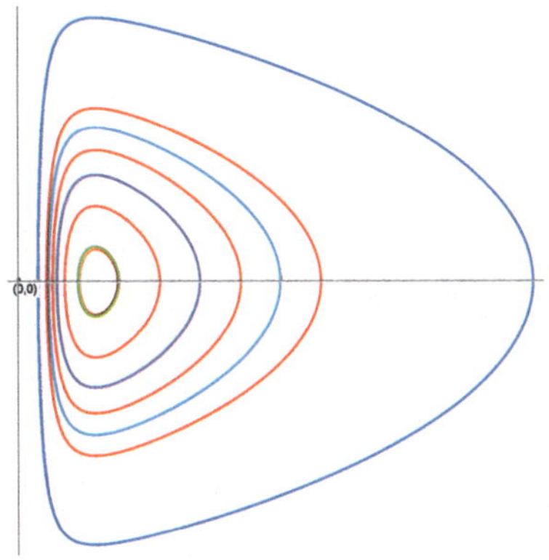

Fig. 3.25 The energy level lines of Eq. (3.20), with $A = 1,\ B = 1\alpha = 3,\ \beta = 2,\ \mu = 18$

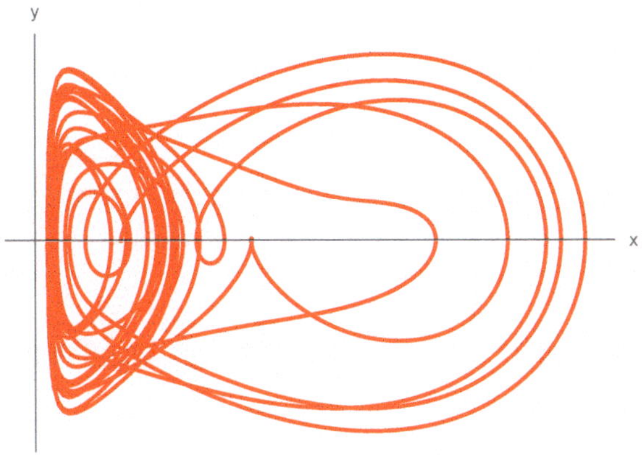

Fig. 3.26 The phase curves of equation $\dot{x} = y,\ \dot{y} = -cy + \dfrac{A}{x^{\alpha}} - Bx^{\beta} + \mu \sin \omega t$ with $A = 1,\ B = 1\alpha = 3,\ \beta = 2,\ \mu = 18,\ \omega = 1.5,\ x(0) = 1,\ y(0) = 0$

As with the previous modifications of the Duffing equation, we also have a Poincaré return map for this type of equation. A look at Fig. 3.28 suggests the presence of structures similar to the so-called "lakes of Wada" [15], which are indicative of the presence of chaos.

For more details on this case of the singular Duffing equation, see our recent paper [5].

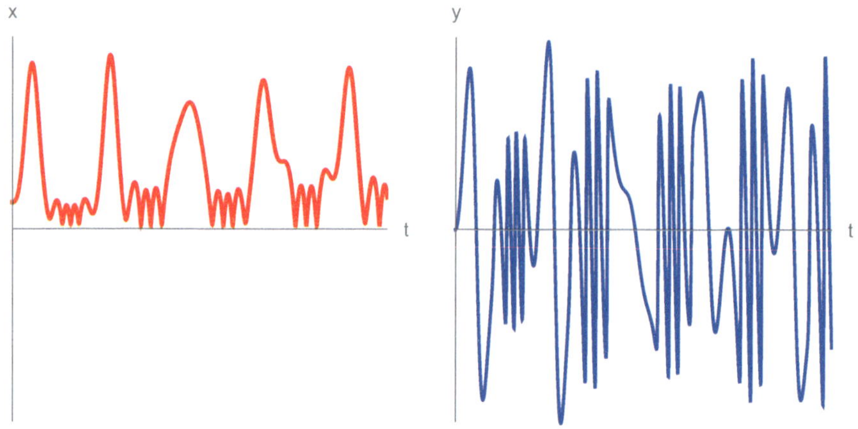

Fig. 3.27 The solution curves of (3.19), x and y with respect to t with $A = 1,\ B = 1\alpha = 3,\ \beta = 2,\ \mu = 18,\ \omega = 1.5,\ \mu$ replaced with $\mu \sin \omega t$ and adding a friction term $(.01\dot{x})$

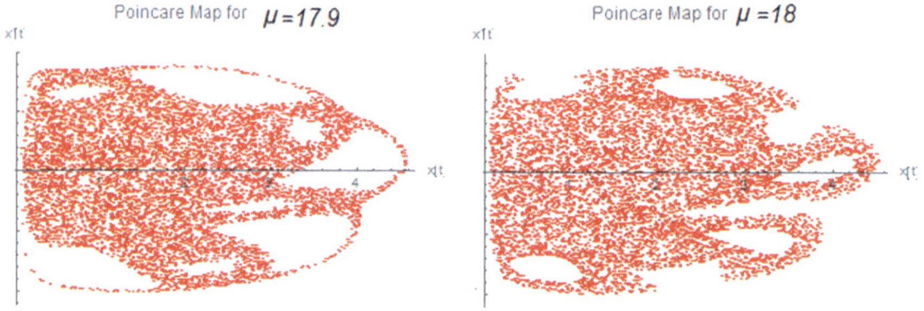

Fig. 3.28 The present figure shows the attractor for the stroboscopic mapping associated with $\ddot{x} - x^{-3} + x^2 = \mu \sin(t)$ for different values of the parameter μ, with $x(0) = 1,\ y(0) = 0$

References

1. R. Abraham, Y. Ueda, *The Chaos Avant-garde: Memories of the Early Days of Chaos Theory*, vol. 39 (World Scientific, 2000)
2. F. Battelli, K.J. Palmer, Chaos in the Duffing equation. J. Differ. Equ. **101**(2), 276–301 (1993)
3. J. Belmonte-Beitia, P.J. Torres, Existence of dark soliton solutions of the cubic nonlinear Schrödinger equation with periodic inhomogeneous nonlinearity. J. Nonlinear Math. Phys. **15**(sup3), 65–72 (2008)
4. L. Burra, F. Zanolin, Complex dynamics in a planar Hamiltonian system for an equation of the duffing type. EPAM J. **2**(1), 3–24 (2016)
5. L. Burra, F. Zanolin, Periodically perturbed superlinear Duffing type equations with singularities: subharmonic solutions and complex dynamics. Dyn. Contin. Discrete Impuls. Syst. Ser. A Math. Anal. **26**(5), 349–364 (2019)
6. L. Burra, F. Zanolin, Monotonicity of the period function and chaotic dynamics in a class of singular ODEs. J. Math. Anal. Appl. **507**(2), Paper No. 125814, 17 (2022)
7. L. Cveticanin, M. Zukovic, Melnikov's criteria and chaos in systems with fractional order deflection. J. Sound Vib. **326**(3–5), 768–779 (2009)

8. M.A. Del Pino, R.F. Manásevich, Infinitely many T-periodic solutions for a problem arising in nonlinear elasticity. J. Differ. Equ. **103**(2), 260–277 (1993)
9. T. Ding, F. Zanolin, Periodic solutions of Duffing's equations with superquadratic potential. J. Differ. Equ. **97**(2), 328–378 (1992)
10. T. Gedeon, H. Kokubu, K. Mischaikow, H. Oka, Chaotic solutions in slowly varying perturbations of Hamiltonian systems with applications to shallow water sloshing. J. Dyn. Differ. Equ. **14**(1), 63–84 (2002)
11. H. Gottlieb, J.C. Sprott, Simplest driven conservative chaotic oscillator. Phys. Lett. A **291**(6), 385–388 (2001)
12. J. Guckenheimer, P. Holmes, *Nonlinear Oscillations, Dynamical Systems, and Bifurcations of Vector Fields* (Pergamon, 1988)
13. J. Hale, *Ordinary Differential Equations*, 2nd edn. (R.E. Krieger Publishing Co., Inc., Huntington, N.Y., 1980)
14. P. Hartman, On boundary value problems for superlinear second order differential equations. J Differ. Equ. **26**(1), 37–53 (1977)
15. J. Kennedy, J. Yorke, Basins of wada. Physica D: Nonlinear Phenom. **51**(1–3), 213–225 (1991)
16. A.C. Lazer, S. Solimini, On periodic solutions of nonlinear differential equations with singularities. Proc. Am. Math. Soc. **99**(1), 109–114 (1987)
17. H. Li, X. Liao, S. Ullah, L. Xiao, Analytical proof on the existence of chaos in a generalized Duffing-type oscillator with fractional-order deflection. Nonlinear Anal. Real World Appl. **13**(6), 2724–2733 (2012)
18. J.E. Littlewood, Unbounded solutions of an equation $y'' + g(y) = p(t)$, with $p(t)$ periodic and bounded, and $g(y)/y \to \infty$ as $y \to \pm\infty$. J. Lond. Math. Soc. **1**(1), 497–507 (1966)
19. J.E. Littlewood, Unbounded solutions of $y'' + g(y) = p(t)$. J. Lond. Math. Soc. **1**(1), 491–496 (1966)
20. K.R. Meyer, Jacobi elliptic functions from a dynamical systems point of view. Am. Math. Mon. **108**(8), 729–737 (2001)
21. K.R. Meyer, G.R. Hall, D. Offin, *Introduction to Hamiltonian Dynamical Systems and the N-Body Problem, Applied Mathematical Sciences*, 2nd edn., vol. 90 (Springer, New York, 2009)
22. F.C. Moon, P.J. Holmes, A magnetoelastic strange attractor. J. Sound Vib. **65**(2), 275–296 (1979)
23. G.R. Morris, An infinite class of periodic solutions of $\ddot{x} + 2x^3 = p(t)$. Proc. Camb. Philos. Soc. **61**, 157–164 (1965)
24. G.R. Morris, A case of boundedness in Littlewood's problem on oscillatory differential equations. Bull. Aust. Math. Soc. **14**(1), 71–93 (1976)
25. F. Nakajima, G. Seifert, The number of periodic solutions of 2-dimensional periodic systems. J. Differ. Equ. **49**(3), 430–440 (1983)
26. J.M.T. Thompson, H. Stewart, *Nonlinear Dynamics and Chaos* (Wiley, 2002)
27. Y. Ueda, Survey of regular and chaotic phenomena in the forced duffing oscillator. Chaos, Solitons Fractals **1**(3), 199–231 (1991)
28. S. Wiggins, On the detection and dynamical consequences of orbits homoclinic to hyperbolic periodic orbits and normally hyperbolic invariant tori in a class of ordinary differential equations. SIAM J. Appl. Math. **48**(2), 262–285 (1988)
29. S. Wiggins, *Introduction to Applied Nonlinear Dynamical Systems and Chaos*, vol. 2 (Springer, 2003)
30. C. Zanini, F. Zanolin, An example of chaos for a cubic nonlinear Schrödinger equation with periodic inhomogeneous nonlinearity. Adv. Nonlinear Stud. **12**(3), 481–499 (2012)

Chapter 4
Topological Methods for the Detection of Chaos

We start this chapter with a quote from Poincaré, the father, so to speak of Chaos Theory

Si la nature n'était pas belle, elle ne vaudrait pas la peine d'être connue, la vie ne vaudrait pas la peine d'être vécue.

If nature were not beautiful, it would not be worth knowing and life would not be worth living.

Jules Henri Poincaré

4.1 Introduction to Some Definitions of Chaotic Dynamics

Chaotic dynamics has proved to be one of the most exciting subjects in dynamical systems theory starting with the last few decades of the twentieth century. Several books and movies on this topic have added to the excitement of this area of mathematics as well as in popular culture. Scientists have found that unusual patterns arise frequently in important deterministic processes of interest to many fields, including chemistry, physics, biology, medicine, engineering, and economics. The fact that a perfectly deterministic system can behave in an apparently unpredictable way was of interest far beyond dynamical systems.

The different definitions of chaos being around at the turn of the century have been configured to meet different purposes and they are based on different levels of mathematical sophistication. Sometimes it is not obvious how these notions of chaos relate to each other.

L. Burra and F. Zanolin, *The Duffing Equation*, Infosys Science Foundation Series in Mathematical Sciences, https://doi.org/10.1007/978-981-97-8301-4_4

Terminology and Concepts: Before we discuss the main ideas of the chapter we would like to explain and define some key concepts which are used here. We also give some relevant definitions of chaos as it is used in the current day [33]. For this, we need some preliminary definitions.

Entropy Entropy is a term one encounters in Thermodynamics and Statistical mechanics. This was a concept introduced by Rudolf Clausius and elaborated by Boltzmann. The concept of information entropy was introduced by Claude Shannon and is sometimes called Shannon entropy in his honour.

In relation to dynamical systems theory we are interested in "topological entropy", a concept invented by Roy Adler and his collaborators. Topological entropy is a non-negative number which measures the complexity of the system. Roughly, it measures the exponential growth rate of the number of distinguishable orbits as time advances. Quoting from Newhouse and Pignataro [43], "The topological entropy" of a system is a quantitative measure of the complexity of its orbits. In a certain sense, it is the maximum amount of information lost per unit of time by the system using measurements with "finite precision". Hence, positive topological entropy indicates orbit complexity in this sense. Positive topological entropy is generally considered to be a hallmark of chaos. Amongst seminal scientific achievements connected with the invention of topological entropy are the first applications of Markov partitions for dynamical systems and the application of symbolic dynamics to coding problems in hyperbolic geometry and information theory.

The original definition of topological entropy was introduced by Adler, Konheim and McAndrew in 1965 [2]. Their idea to assign a number to an open cover to measure its size was inspired by Kolmogorov and Tihomirov [34]. Adler's introduction of topological entropy was an important contribution to ergodic theory and topological dynamics. Topological entropy is the topological analogue of the measure-theoretic dynamical entropy that had been defined by A.N. Kolmogorov just a few years earlier. Then to define topological entropy for continuous maps they strictly imitated the definition of Kolmogorov-Sinai entropy of a measure preserving transformation in ergodic theory.

Topological Entropy

Let X be a compact metric space and $T : X \to X$ be a continuous map. Let $\mathcal{A} = \{A_i\}$ be a finite open cover of X. The topological entropy of the cover $\mathcal{A}$ is defined to be the logarithm, $H(\mathcal{A}) = \log N(\mathcal{A})$, of the smallest number $N(\mathcal{A})$ of sets that can be used in a subcover of $\mathcal{A}$.

To be more precise:

For an open cover $\mathcal{A}$ of X, let $N(\mathcal{A})$ denote the smallest cardinality of a subcover of (i.e. a subfamily of $\mathcal{A}$ whose union still equals X). By compactness, $N(\mathcal{A})$ is always finite. If $\mathcal{A}$ and $\mathcal{B}$ are open covers of X then

$$\mathcal{A} \vee \mathcal{B} = \{A \cap B : A \in \mathcal{A},\ B \in \mathcal{B}\}$$

is called their common refinement. Let $\mathcal{A}^n = \mathcal{A} \vee T^{-1}\mathcal{A} \vee \cdots \vee T^{-n+1}\mathcal{A}$, where $T^{-k}\mathcal{A} = \{T^{-k}A : A \in \mathcal{A}\}$.

Using a subadditivity argument, it can be shown that the limit

$$h(\mathcal{A}, T) = \lim_{n\to\infty} \frac{\log N(\mathcal{A}^n)}{n}$$

exists for any open cover $\mathcal{A}$ and equals $\inf_{n\in\mathbb{N}} \left(\frac{\log N(\mathcal{A}^n)}{n}\right)$. The topological entropy of (X, T) is defined to be the supremum

$$h_s(T) = \sup h(\mathcal{A}, T),$$

where the supremum ranges over all open covers $\mathcal{A}$ of X.

We now give an alternative and equivalent definition due to Bowen and Dinaburg[1] [14, 24], for **topological entropy of mappings.**

Let X be a compact metric space and $T : X \to X$ a continuous map. A set $\mathscr{E} \subset X$ is said to be (n, ε)-separated, if for every $(x, y) \in \mathscr{E}$ with $x \neq y$ there is $i \in \{0, 1, \cdots, n-1\}$ such that $d(T^i x, T^i y) \geq \varepsilon$. Let $s(n, \varepsilon)$ be the maximal cardinality of an (n, ε)-separated set in X. This number is always finite by compactness. Defining

$$\bar{h}(\varepsilon, T) = \limsup_{n\to\infty} \frac{\log s(n, \varepsilon)}{n},$$

the topological entropy is obtained as

$$h_B(T) = \sup_{\varepsilon>0} \bar{h}(\varepsilon, T) = \lim_{\varepsilon\to 0^+} \bar{h}(\varepsilon, T).$$

It should be noted that $s(n, \varepsilon)$ can be substituted in Bowen's definition with a possibly smaller number, $r(n, \varepsilon)$, the minimal cardinality of an (n, ε)-spanning set. A set $\mathscr{E} \subset X$ is (n, ε)-spanning if for every $x \in X$ there is $y \in \mathscr{E}$ such that $d(T^i x, T^i y) < \varepsilon$ for all $i \in \{0, \cdots, n-1\}$. With such a substitution one obtains the same value of $h_B(T)$. For compact metric spaces, the two definitions coincide and we write h_{top} for $h_s = h_B$.

Amongst the several properties of topological entropy, we state the ones that are relevant to our work and subsequent discussion. For simplicity, we denote by $h_{top}(f)$ the topological entropy related to a map f on a compact metric space.

Burns and Weiss [15] established a relation between positive topological entropy and Smale horseshoes: topological entropy generated in a so-called horseshoe: wherein the rectangle is stretched, bent upward, and placed over itself (We describe this horseshoe map a little later, briefly). The number of orbits distinguishable in n

[1] Although the topological entropy can be defined with reference to continuous self-maps of topological spaces (due to Adler), we confine ourselves to the case of metric spaces as given by Bowen and Dinaburg.

steps grows as 2^n, and it can be shown that it generates the topological entropy of $(1/n)\log(2^n) = \log 2$.

For the one-sided or two-sided Bernoulli shift σ on m symbols, (which we shall explain a little later) the following relation holds:

$$h_{top}(\sigma) = \log m.$$

Given a continuous self-map f of a compact topological space X and a compact subset $\mathcal{I} \subset X$ invariant that is $f(\mathcal{I}) = \mathcal{I}$ or positively invariant $f(\mathcal{I}) \subset \mathcal{I}$ then

$$h_{top}(f) \geq h_{top}(f|_{\mathcal{I}})$$

and

$$h_{top}(f^n) = nh_{top}(f), \quad \forall n \geq 1$$

where f^n is the n-th iterate of a continuous self-map f. Also given two continuous maps $f : X \to X$ and $g : Y \to Y$ of two compact topological spaces X and Y and a continuous, onto map $\phi : X \to Y$ such that the diagram

$$\begin{array}{ccc} X & \xrightarrow{f} & X \\ {\scriptstyle\phi}\downarrow & & \downarrow{\scriptstyle\phi} \\ Y & \xrightarrow[g]{} & Y \end{array} \tag{4.1}$$

commutes. i.e. such that $\phi \circ f = g \circ \phi$, then it is true that

$$h_{top}(f) \geq h_{top}(g).$$

If ϕ is also injective, the above inequality is indeed an equality. When the diagram in (4.1) commutes, we say that f and g are topologically semi-conjugate and that ϕ is a semi-conjugacy between them. If ϕ is also one-to-one, then f and g are called topologically conjugate and ϕ is a conjugacy. The concepts of topological conjugacy and semi-conjugacy are reconsidered later in a more formal manner.

Thus, for a continuous self-map f of a compact topological space X and a (positively) invariant compact subset $\mathcal{I} \subset X$, it holds that if $f|_{\mathcal{I}}$ is semi-conjugate to the (one-sided or two-sided) Bernoulli shift σ on m symbols, then

$$h_{top}(f) \geq h_{top}(f|_{\mathcal{I}}) \geq h_{top}(\sigma) = \log(m). \tag{4.2}$$

If $f|_{\mathcal{I}}$ is conjugate to σ, then the second inequality is indeed an equality. In particular, when dealing with chaotic dynamics, we consider discrete dynamical systems, i.e. couples (X, f), where X is a compact metric space and $f : X \to X$ is continuous. Some authors also require that f and surjective.

- Consider a discrete dynamical system or a pair (X, f), X being a metric space[2] with metric d, f a continuous map of X to itself. Sometimes f may be taken to be a homeomorphism.
- The object of the theory of dynamical systems is to determine the asymptotic behaviour of an iterative process. In the study of differential equations dealing with time as the independent variable, the goal is to understand the behaviour of solutions as $t \to \pm\infty$ [22]. Usually, it is necessary to define terms like the *orbit* or the *semiorbit* of the variable x. This is the set of points $x,\ f(x),\ f^2(x),\ \ldots,\ f^n(x)$. The objective in this instance is to understand the behaviour of these points as n becomes larger and larger.
  Thus, an orbit/semiorbit is a sequence $x_n,\ \ n \in \mathbb{Z},$ or $n \in \mathbb{N}$, with $x_n \in X$, such that $x_{n+1} = f(x_n)$ holds good for all n. The sequence $(x_n)_n$ is called the orbit through x_0.

Definition 4.1 Bernoulli system
Symbolic Dynamics We now define the concept of the *sequence space* [22]. We denote by Σ_2, the sequence space given by

$$\Sigma_2 = \{s = (\ldots s_{-2}s_{-1}s_0s_1s_2\ldots)|s_j = 0 \text{ or } 1,\ j \in \mathbb{Z}\}.$$

Σ_2 is a two-sided sequence space on two symbols 0 and 1. By Σ_2^+ we mean one-sided infinite strings of integers like, for example, (111111...) or (101010...) or (001010011...). Namely,

$$\Sigma_2^+ = \{s = (s_0s_1s_2\ldots)|s_j = 0 \text{ or } 1,\ j \in \mathbb{N}\}.$$

These are one-sided sequences consisting of two symbols 0 or 1. Σ_2^+ (as well as Σ_2) can be made into a metric space by defining the distance between $s = (s_0s_1s_2s_3\ldots)$ and $t = (t_0t_1t_2t_3\ldots)$. (see Definition 4.2). In a similar manner we can also define the two-sided sequence space on m symbols as Σ_m, the sequence space given by

$$\Sigma_m = \{s = (\ldots s_{-2}s_{-1}s_0s_1s_2\ldots)|s_j \in \{0,\ldots,m-1\}\text{with } j \in \mathbb{Z}\}.$$

Similarly, the one-sided sequence space on m symbols is Σ_m^+, the sequence space given by

$$\Sigma_m^+ = \{s = (s_0s_1s_2\ldots)|s_j \in \{0,\ldots,m-1\} \text{ with } j \in \mathbb{N}\}.$$

Definition 4.2 The distance functions for the sequence spaces Σ_2 or Σ_2^+ may be defined as

[2] Some authors suppose X to be a complete or a compact metric space. We will explicitly assume further hypotheses on X when required.

$$d(s,\, t) = \sum_{i \in \mathbb{Z}} \frac{|s_i - t_i|}{2^{|i|}} \tag{4.3}$$

$$d'(s,\, t) = \sum_{0}^{\infty} \frac{|s_i - t_i|}{2^i}, \tag{4.4}$$

respectively. For the sequence space Σ_m^+ on m symbols, the distance function may be defined as

$$\hat{d}(\mathbf{s}', \mathbf{s}'') = \sum_{i \in \mathbb{N}} \frac{|s_i' - s_i''|}{m^{i+1}}, \tag{4.5}$$

The distance function for the two-sided sequence space Σ_m may be defined in a similar fashion as the Σ_2 space.

The Shift Map The shift map may be defined on any of the sequence spaces defined so far. For example, the shift map $\sigma : \Sigma_2^+ \to \Sigma_2^+$ is given by

$$\sigma(s_0 s_1 s_2 \ldots) = (s_1 s_2 s_3 \ldots)$$

that is, the shift map shifts the first index to the next index, or shifts the indices one place to the left. In the same manner, we define the shift map in Σ_2 as well as in Σ_m^+ and Σ_m.

The **itinerary** of x is a sequence $S(x) = s_0 s_1 s_2 \ldots$ where $s_j = 0$ or 1, thus the itinerary is a sequence of $0's$ and $1's$.

As defined earlier, we now formally re-write topological conjugacy and semi-conjugacy for clearer understanding.

Topological conjugacy Let $f : A \to A$ and $g : B \to B$ be two maps. f and g are said to be *topologically conjugate* if there exists a **homeomorphism** $h : A \to B$ such that $h \circ f = g \circ h$, h is called a topological conjugacy.

$$\begin{array}{ccc} A & \xrightarrow{f} & A \\ {\scriptstyle h}\downarrow & & \downarrow{\scriptstyle h} \\ B & \xrightarrow[g]{} & B \end{array} \tag{4.6}$$

Semi-conjugacy Let $f : A \to A$ and $g : B \to B$ be two maps. f and g are said to be *topologically semi-conjugate* if there exists a **continuous**, surjective map $h : A \to B$ such that $h \circ f = g \circ h$.

Chaos, in much of literature, is more of a descriptive term than a quantitative term. Even though one can find the term “chaos” defined in textbooks, one could say, looking at the various forms of the definitions of chaos, that there is still no universally accepted single definition. The Bernoulli property is probably what some might prefer as “chaos”, whilst others may prefer the more often used “ sensitive dependence on initial conditions”. This is characteristic of the general problems involved in giving a universally accepted definition of the term "chaos". We present some of the classical definitions of chaos here and highlight the sense in which we use the term "chaos".

Prevalent Definitions of Chaos

The concept of "chaos" has many definitions as is evident from the huge body of research that is available. One can however find some underlying features which are similar. One such fact which almost all the existing definitions of chaos share is the unpredictability of the behaviour of a chaotic system.

- A continuous map $f : X \to X$ on a compact metric space (X, d) is called chaotic in the sense of Li and Yorke if there exists an uncountable subset S called a scrambled set [13, 37]:

◆ $\limsup_{n\to\infty} d(f^n(x),\ f^n(y)) > 0$ for all $x,\ y \in S,\ x \neq y$.
◆ $\liminf_{n\to\infty} d(f^n(x),\ f^n(y)) = 0$ for all $x,\ y \in S,\ x \neq y$.
◆ $\limsup_{n\to\infty} d(f^n(x),\ f^n(p)) > 0$ for all $x, \in S,\ p \in X,\ p$ periodic.

• A continuous map $f : X \to X$ on a compact metric space X is called chaotic in the sense of Block and Coppel [6, 13], if there exists an $m \in \mathbb{N}$ and a compact f^m-invariant subset Y of X such that $f^m \restriction_Y$ is semi-conjugate to the shift on Σ, i.e. if there exists a continuous surjection $h : Y \to \Sigma$ satisfying $h \circ f^m = \sigma \circ h$ on Y. Here σ is the shift map. By Σ we denote here Σ_m or Σ_m^+ for some integer $m \geq 2$.

If $m = 2$ then this notion of chaos is also known as *chaos in the sense of coin-tossing* [33].

As we already mentioned there is no universally accepted definition for a chaotic system, but Devaney's Definition [22] is one that is frequently used.

• A continuous map $f : X \to X$ on a metric space X is called chaotic in the sense of Devaney if the following properties hold:

(i) f is *topologically transitive*, i.e. for any couple of nonempty open subsets $U,\ V \subset X$ there exists an integer $n \geq 1$ such that $U \cap f^n(V) \neq \emptyset$;
(ii) The set of the periodic points for f is dense in X;
(iii) f *shows sensitive dependence on initial conditions on* X, i.e. there exists $\delta > 0$ such that for any $x \in X$ there is a sequence $(x_i)_{i\in\mathbb{N}}$ of points in X such that $x_i \to x$ when $i \to \infty$ and for each $i \in \mathbb{N}$ there exists a positive integer m_i with $d(f^{m_i}(x_i),\ f^{m_i}(x)) \geq \delta$.

The concept of chaos due to Devaney identifies three features as being the essential components of chaos. The final condition of Devaney is *sensitive dependence on initial conditions*. This condition embodies the idea that in a chaotic system very small errors in experimental data ultimately lead to large-scale divergence in the observations. To most people, sensitive dependence on initial conditions is widely thought to be the central concept of chaos. In [8], the authors show that if the first two conditions of Devaney are fulfilled, that is, if $f : X \to X$ is transitive and has dense periodic points then f has sensitive dependence on initial conditions (to avoid the degenerate case it is assumed that X is not a finite set).

The Bernoulli Shift Map

Let us recall that the shift map $\sigma : \Sigma_2^+ \to \Sigma_2^+$ is the map defined by $\sigma(x_0x_1x_2x_3\cdots) = (x_1x_2x_3\cdots)$. In other words, it is the map that deletes the first term in the sequence, thereby shifting the remaining terms one slot to the left. The $k - th$ iterate of the shift map is as simple: it deletes the first k terms.

For example:

- If $\mathbf{x} = (101010\cdots)$ then $\sigma(\mathbf{x}) = (010101\cdots)$ and $\sigma^2(\mathbf{x}) = (101010\cdots) = \mathbf{x}$, so $\mathbf{x}$ is a periodic point of period 2 for σ.
- If $\mathbf{y} = (011111\cdots)$ then $\sigma(\mathbf{y}) = (111111\cdots)$ and $\sigma^2(\mathbf{y}) = (111111\cdots) = \sigma(\mathbf{y})$, so $\mathbf{y}$ is an eventually fixed point for σ.
  It is easy to write down all of the periodic points for σ: they are the periodic sequences $\mathbf{s} = (s_0s_1\cdots s_{n--1}s_0s_1\cdots s_{n--1}\cdots) = (\overline{s_0s_1\cdots s_{n--1}})$.
- It may be observed that σ is a surjective, two-to-one map on Σ_2^+: for any given $\mathbf{y} \in \Sigma_2^+$, there are two $\mathbf{x}$'s satisfying $\sigma(\mathbf{x}) = \mathbf{y} = (y_0y_1y_2\cdots)$, namely, $\mathbf{x} = (0y_0y_1y_2\cdots)$ and $(1y_0y_1y_2\cdots)$.
- We state the following proposition without proof, which is an easy consequence of the definition of distance between two sequences (for details see [22]).

  Proposition (Nearby Sequences): If $\mathbf{x} = (x_0x_1x_2\cdots)$ and $\mathbf{y} = (y_0y_1y_2\cdots)$ are elements of Σ_2 such that $x_i = y_i$ for each $i \leq n$, then $d(\mathbf{x}, \mathbf{y}) \leq 2^{-n}$. Conversely, if $d(\mathbf{x}, \mathbf{y}) < 2^{-n}$, then $x_i = y_i$ for each $i \leq n$.
  In other words: if the early terms of two elements of Σ_2 agree, then the distance between them must be small. Conversely, if two elements of Σ_2 are close together, then their early terms must agree.
- A central property of the shift map is that it is continuous:

  Proposition: The shift map $\sigma : \Sigma_2^+ \to \Sigma_2^+$ is continuous at every point of Σ_2^+.
  Proof: Let $\epsilon > 0$ and a point $\mathbf{a} = (a_0a_1a_2\cdots)$ be given. Since $\epsilon > 0$, there exists a positive integer n such that $2^{-n} < \epsilon$ we then claim that $\delta = 2^{-(n+1)}$ will satisfy the definition of continuity. By the " nearby sequences" proposition, $x = (x_0x_1x_2\cdots)$ and $d(\mathbf{x}, \mathbf{a}) < 2^{-(n+1)}$, then $x_i = a_i$ for $i \leq n + 1$.
  But then $\sigma(\mathbf{x}) = (x_1x_2x_3\cdots)$ and $\sigma(\mathbf{a}) = (a_1a_2a_3\cdots)$, so these sequences agree from the 0th term up through the nth term, so again by the nearby sequences proposition, we see that $d(\sigma(\mathbf{x}), \sigma(\mathbf{a})) \leq 2^{-n} < \epsilon$, as required.
  Since $\mathbf{a}$ was arbitrary, σ is continuous everywhere.

Properties of the Shift Map on the Sequence Space

To motivate the definition of chaos, we will study some of the properties of the shift map on the sequence space Σ_2^+, as we shall see, the Bernoulli shift is a good example of a system which satisfies all the different aspects of chaos, namely, transitivity, sensitivity to initial conditions as well as having a dense set of periodic points and show that it is an ideal candidate which satisfies all the properties of a chaotic map.

• **Proposition**: The set of periodic points for σ is a dense subset of Σ_2^+.

Proof: We first prove that for any sequence $\mathbf{x} \in \Sigma_2^+$, there are periodic points of σ that are arbitrarily close to $\mathbf{x}$. We need to prove that for any $\mathbf{x} \in \Sigma_2^+$ there is a sequence $\mathbf{x}_i$ of periodic points for σ such that $\lim\limits_{i\to\infty} \mathbf{x}_i = \mathbf{x}$

We construct the points explicitly and suppose $\mathbf{x} = (d_0 d_1 d_2 \cdots)$.

Let $\mathbf{x}_i = \overline{(d_0 d_1 d_2 \cdots d_i)}$. Then since the 0th through i^{th} terms of $\mathbf{x}$ and $\mathbf{x}_i$ agree, by our proposition on nearby sequences we immediately have $d(\mathbf{x}, \mathbf{x}_i) \le 2^{-i}$. Also, $\lim\limits_{i\to\infty} \mathbf{x}_i = \mathbf{x}$ since the first n digits of $\mathbf{x}_n$ do not change in any subsequent $\mathbf{x}_i$. □

• **Proposition**:The shift map is transitive.

Proof: Pick any $\mathbf{x}$ and $\mathbf{y}$ in Σ_2^+ and an $\epsilon > 0$. We will construct a $\mathbf{z}$ whose orbit comes within ϵ of both $\mathbf{x}$ and $\mathbf{y}$. Pick n such that $1/2^n < \epsilon$. This means that if a new point has its first $n+1$ terms the same as those of $\mathbf{x}$ then by the nearby sequences theorem it is within ϵ of $\mathbf{x}$ (similarly for $\mathbf{y}$). Construct the following sequence: $\mathbf{z} = (\underbrace{01}_{\text{length 1}}\ \underbrace{0001\ 1011}_{\text{length 2}}\ \underbrace{000110\ 111110\ \cdots}_{\text{length 3}})$ obtained by listing all sequences of length 1, then all sequences of length 2, then all sequences of length 3, and so on.

That is, we glue together all sequences of length 1 and then all sequences of length 2, etc. Hence any sequence of length n will appear in $\mathbf{z}$ at some point.

Now consider the first $n+1$ terms of $\mathbf{x}$. These are a sequence of length $n+1$. Hence they must (by construction) appear somewhere in $\mathbf{z}$. Thus there is a k_1 such that the first $n+1$ terms of $\sigma^{k_1}(\mathbf{z})$ are the same as those of $\mathbf{x}$. So the orbit of $\mathbf{z}$ comes within ϵ of $\mathbf{x}$. Similarly there exists k_2 such that the first $n+1$ terms of $\sigma^{k_2}(\mathbf{z})$ are the same as those of $\mathbf{y}$. Hence the orbit of $\mathbf{z}$ comes within ϵ of both $\mathbf{x}$ and $\mathbf{y}$. □

• **Proposition** The dynamical system defined by the shift map depends sensitively on initial conditions.

Proof: Let us recall the property of sensitive dependence on initial conditions. Simply put, this says that on X, there exists $\beta > 0$ such that for any $x \in X$ and $\epsilon > 0$ there exists a $y \in X$ within ϵ of x and a $k > 0$ such that $f^k(x)$ and $f^k(y)$ are at least β apart. This definition says that there is some "error bar" = β, and no matter which x we pick and which little region around it, we can find a new point y in this region so that the orbits of x and y will begin to diverge, and at some point they will be at least β apart.

Set $\beta = 1$. Let us choose any $\mathbf{x} = (x_i) \in \Sigma_2^+$ and any $\epsilon > 0$. Then let n be such that $1/2^n < \epsilon$. We now pick a point close, (within ϵ) of $\mathbf{x}$ and show that its orbit will diverge from that of $\mathbf{x}$ by at least β. Let $\mathbf{y} = (y_i) \neq \mathbf{x}$, such that $d[\mathbf{x}, \mathbf{y}] < 1/2^n$. This means that the first $n + 1$ terms of $\mathbf{x}$ and $\mathbf{y}$ are the same. Because $\mathbf{x} \neq \mathbf{y}$ there exists $k > n$ such that $x_k \neq y_k$. Consider now the points $\sigma^k(\mathbf{x})$ and $\sigma^k(\mathbf{y})$. The first term $\sigma^k(\mathbf{x})$ is x_k and the first term of $\sigma^k(\mathbf{y})$ is $y_k \neq x_k$. Hence the distance between these two points is:

$$d[\sigma^k(\mathbf{x}), \sigma^k(\mathbf{y})] = \sum_{n \geq 0} \frac{|s_{n+k} - t_{n+k}|}{2^n} \geq \frac{|s_k - t_k|}{2^0} = 1.$$

Thus the orbits of $\mathbf{x}$ and $\mathbf{y}$ diverge by at least β.

From the properties proved here we can say that they are in agreement with the earlier remark that for the one-sided Bernoulli shift σ on 2 symbols, the following relation holds:

$$h_{top}(\sigma) = \log 2.$$

Combining the previous three propositions gives the important result on the shift map: *The shift map defines a chaotic dynamical system* according to the definition due to Devaney. □

Positive topological entropy and sensitivity to initial conditions are related, since both indicate instability in the system. Topological entropy is a local feature, in the sense that it is sufficient to find a (positively) invariant subset of the domain where it is positive in order to infer that it is positive on the whole domain.

It now remains to link the above definitions of chaos with Li-Yorke definition. In this respect we observe that in [12] the authors prove that any continuous and surjective map with positive topological entropy displays sensitivity to initial conditions. Taking sensitivity on a subset of the domain, some authors have obtained results by considering, the stronger property of chaos in the sense of coin-tossing in place of positive entropy.

4.2 The Horseshoe Map of Smale

On November 30, 1889 the French mathematician Henri Poincaré was faced with solving a problem. He had been awarded a prize, offered by King Oscar II of Sweden, for his work on a long standing question in mathematics: The three-body problem. Poincaré's work had a mistake. He owned up to his mistake and later wrote more extensively about what the mistake had revealed to him. Indeed it was only as a result of correcting the error that he discovered the existence of what today are known as transversal homoclinic points. It is interesting that this error is now regarded as marking the birth of chaos theory (See [9, 10] for more details on this story).

Smale thought about the complexity of the problem and offered "chaos" as a solution to it, this was off the beaten track especially since he had previously conjectured that there was no such thing as chaos. Norman Levinson at MIT had pointed out that some existing work with interesting results contradicted what Smale had predicted. Smale wanted to understand what these results meant. He then put his results in a very geometric context so that it could become clearer, see [59].

The result is now known as Smale's horseshoe map. It beautifully encapsulates how chaotic dynamics can arise in mathematics. This study actually went straight to the heart of the issues Poincaré had encountered when dealing with the three-body problem.

The name and the idea of the horseshoe are derived from the celebrated article of Smale [57], that provided a mathematically rigorous along with a geometric proof of the existence of chaos in a planar map (See also [58]).

In the horseshoe map a square S (in our case S will be the unit square) is first shrunk uniformly in one direction and expanded in the other one. Subsequently, the elongated rectangle obtained in the previous step is bent along the original square in order to cross it twice. Let ℓ and m be two constants with $0 < \ell < 1/2$ and $m > 2$. (Fig. 4.1)

Define two sets H_0 and H_1 as follows:

$$
\begin{aligned}
H_0 &= \{(x,\ y) \in \mathbb{R}^2,\ 0 \le x \le 1,\ 0 \le y \le 1/m\} \\
H_1 &= \{(x,\ y) \in \mathbb{R}^2,\ 0 \le x \le 1,\ 1 - 1/m \le y \le 1\}.
\end{aligned}
$$

Define a homeomorphism $M : \mathbb{R}^2 \to \mathbb{R}^2$ by the geometrical construction in Fig. 4.2. It acts on the unit square S as a strong contraction in the horizontal direction, followed by a strong expansion in the vertical direction, folding, and placing back over S. The map may be defined as follows:

$$
M : \begin{bmatrix} x \\ y \end{bmatrix} \mapsto \begin{bmatrix} \ell & 0 \\ 0 & m \end{bmatrix} \begin{bmatrix} x \\ y \end{bmatrix} \qquad \text{for } (x,\ y) \in H_0
$$

$$
M : \begin{bmatrix} x \\ y \end{bmatrix} \mapsto \begin{bmatrix} -\ell & 0 \\ 0 & -m \end{bmatrix} + \begin{bmatrix} 1 \\ m \end{bmatrix} \qquad \text{for } (x,\ y) \in H_1
$$

That is $M(S) \cap S = V_0 \cup V_1$. where $V_{0,1}$ are two disjoint vertical rectangles. Also the two disjoint horizontal rectangles H_i are such that $H_0 \cup H_1 = (M^{-1}(S)) \cap S$. Figure 4.2 shows the typical construction of the horseshoe map in which a square is

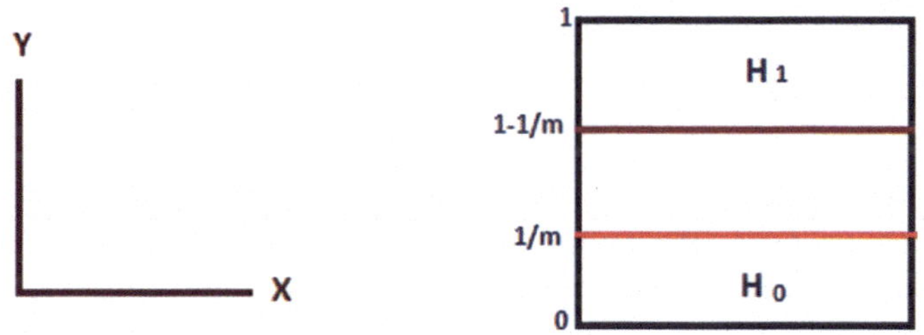

Fig. 4.1 The square S divided into the regions H_0 and H_1 with $m = 3$

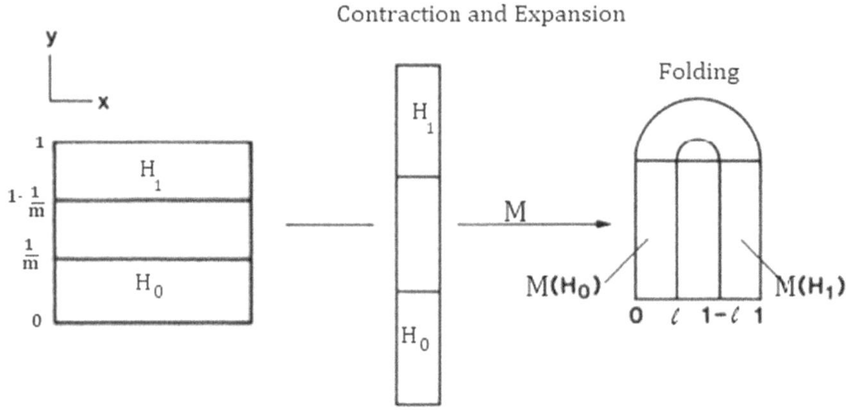

Fig. 4.2 Construction of the horseshoe map: $M(H_i) = V_i,\ i = 0,\ 1,$ with $m = 3$ and $\ell = 1/3$

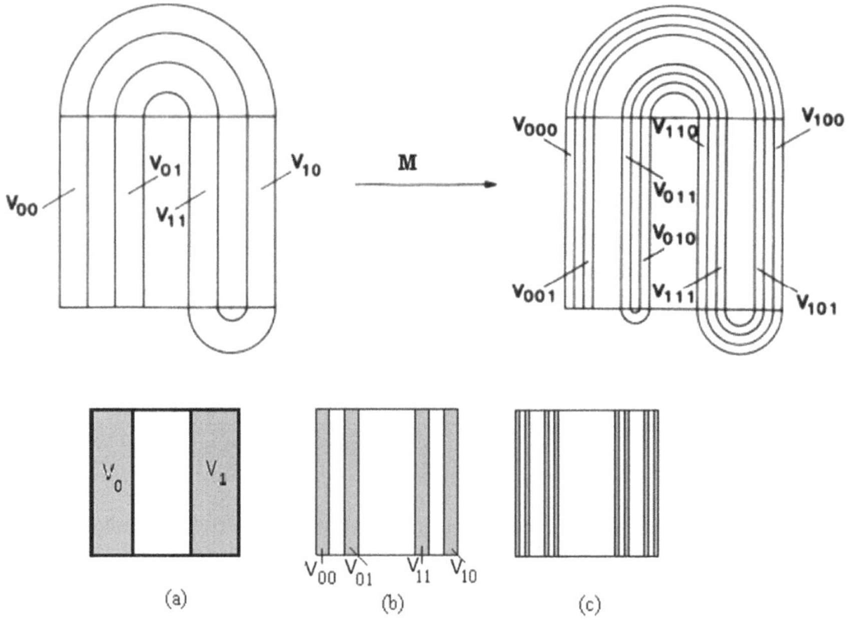

Fig. 4.3 The Smale Horseshoe Map M after more iterations on the square. The coding "01" meaning that it was right at the first iteration, left at the second

first shrunk uniformly in one direction and expanded in the other one. Subsequently, the elongated rectangle obtained in the previous step is bent along the original square in order to cross it twice. The resulting map M is a diffeomorphism with M and M^{-1} transforming the horizontal and vertical lines of the square into similar lines crossing the domain (Fig. 4.3).

In Smale's construction, the set $I_s := \{q \in S : M^k(q) \in S, \ \forall k \in \mathbb{Z}\}$, consists of the points of S which remain in S under the action of M in both forward and backward time. These remain in the square under all the iterates of M, and form a compact invariant set for M. This also contains the periodic points of M as a dense subset; moreover M is topologically transitive on I_S .

Thus M behaves on I_S like the shift map σ on two symbols, as $M|_{I_S}$ and σ are conjugate. This conjugacy, which is given by a homeomorphism $\pi : I_s \to \{0, \ 1\}^{\mathbb{Z}}$ with $\pi \circ M = \sigma \circ \pi$, permits the transfer of the dynamical properties of the Bernoulli shift to $M|_{I_s}$.

In other words, the Smale horseshoe map is topologically conjugate to the two-sided Bernoulli shift on two symbols. Symbolically, it can be represented as: *homeomorphism* π of I_s onto $\Sigma_2 := \{0, 1\}^{\mathbb{Z}}$.

One can use this new tool of "horseshoe dynamics" to prove complex behaviour in different maps. To show that some given map F exhibits complex behaviour we just prove the existence of a horseshoe structure either for the map itself or for an iterate of it. Several authors like in [15] define a horseshoe like set Λ that is invariant under F^{n_0} , for some $n_0 \geq 1$, with the requirement that $F^{n_0}|_\Lambda$ is topologically conjugate to σ, the shift map. As pointed out by Burns and Weiss in [15], the difficulty in finding a horseshoe lies in showing that the map $\pi : \Lambda \to \{0, \ 1\}^{\mathbb{Z}}$ is injective. That is the system is conjugated to the 2-sided Bernoulli shift. We will consider a weaker form of chaos where the map is not necessarily injective but semi-conjugate that is continuous and surjective.

To produce a concrete geometric construction isomorphic to the Smale horseshoe requires the assumption that F be a diffeomorphism,[3] and some hyperbolicity conditions which, roughly speaking, require that the map F under consideration is a diffeomorphism such that both dF and dF^{-1} are contractive along some complementary directions[4,5] need to be fulfilled. The result of Smale is extremely useful in numerical studies of systems with positive entropy. But there are unavoidable errors in the computation of the function F acting on a set Q, $F(Q)$ (Q being the set under iteration) even though these can be made smaller by careful computation. These errors in the computation of the iterates of F, $F^n(Q)$ increase exponentially fast as n increases, because of the positive entropy. Additionally, Smale's assumption of hyperbolicity of F on Q is difficult to verify analytically as well as numerically. The question asked by some authors (see for example [32]) is whether the hyperbolicity

---

[3] [67] A map $\mathbf{g} : U \to V$ is a diffeomorphism if it is a homeomorphism and if $\mathbf{g}$ and the inverse map $\mathbf{g}^{-1}$ are continuously differentiable. A diffeomorphism is a homeomorphism, but the converse is false.

[4] [67] A linear mapping $A : \mathbb{R}^n \to \mathbb{R}^n$ is hyperbolic if none of its eigenvalues has modulus 1. This means that $\mathbb{R}^n$ can be written as a direct sum of two $A-$invariant subspaces E^s and E^u (where s stands for stable and u for unstable)

[5] Let U be an open subset of $\mathbb{R}^n$ and $\mathbf{g} : U \to \mathbb{R}^n$ be a *nonlinear diffeomorphism*.The linearization of $\mathbf{g}$ at x^* is given by the well-known formula (the Jacobean): $\mathbf{D}\,\mathbf{g}(\mathbf{x^*}) = \left[\frac{\partial g_i}{\partial x_j}\right]^n_{i,\,j=1|_{\mathbf{x}=\mathbf{x}^*}}$. The diffeomorphism $\mathbf{g}$ is hyperbolic if $\mathbf{D}\,\mathbf{g}(\mathbf{x^*})$ is a hyperbolic linear diffeomorphism (isomorphism), for any point of U.

assumption may be used in a weakened form of "topological crossing". In answer to this question it has been found that indeed, the dynamics remain just as rich as before. This has caused mathematicians to seek a horseshoe condition with less stringent assumptions, so that geometrical features typical of the Smale horseshoe were intact, without imposing the condition of hyperbolicity. This lead to the study of the ***topological horseshoes*** [15, 32, 75].

If it is assumed that map F (or one of its iterates), satisfying certain geometrical conditions, is semi-conjugate to a full shift on $m \geq 2$ symbols, one could conclude that F displays chaotic dynamics in the sense that F has, for instance, positive topological entropy. So, in this indirect manner, one could conclude that a function possesses all the properties of the shift map that are preserved by the conjugacy or semi-conjugacy relation.

The definition of chaos relevant for the applications in this work

As mentioned earlier, there are several prevailing definitions of chaos. We choose the definition of chaos which is appropriate to the method used to prove the presence of complex dynamics: the "stretching along paths" method, a concept which is explained in detail in the course of the chapter. In the sequel, by a map we mean a continuous map.

Definition 4.3 Let X be a metric space, $\psi : X \supset D_\psi \to X$ be a map and let $D \subset D_\psi$, where D_ψ denotes the domain of ψ. We say that ψ induces chaotic dynamics on two symbols on the set D if there exist two nonempty disjoint compact sets

$$\mathcal{K}_0, \mathcal{K}_1 \subset D,$$

such that, for each two-sided sequence $(s_i)_{i\in\mathbb{Z}} \in \{0, 1\}^{\mathbb{Z}}$ there exists a corresponding sequence $(w_i)_{i\in\mathbb{Z}} \in D^{\mathbb{Z}}$ such that

$$w_i \in K_{s_i} \text{ and } w_{i+1} = \psi(w_i), \forall i \in \mathbb{Z} \tag{4.7}$$

and, whenever $(s_i)_{i\in\mathbb{Z}}$ is a k-periodic sequence (that is, $s_{i+k} = s_i, \forall i \in \mathbb{Z}$) for some $k \geq 1$, there exists a k-periodic sequence $(w_i)_{i\in\mathbb{Z}} \in D^{\mathbb{Z}}$ satisfying (4.7). To put the emphasis on the sets $\mathcal{K}_j' s$, we may also say that ψ induces chaotic dynamics on two symbols on the set D with respect to $\mathcal{K}_0$ and $\mathcal{K}_1$.

This definition is in line with the concept of chaos in the coin-tossing sense explained earlier [33]. With the addition that the definition in [33] is enhanced with the condition on periodic sequences. To get a better handle on Definition 4.3, following the example of Smale in his expository paper [59], let us associate with the set $\mathcal{K}_0$ the name "head" and with the set $\mathcal{K}_1$ the name "tail". If we consider any sequence of symbols $(s_i)_{i\in\mathbb{Z}}$ where, for each i, s_i is either "head" or "tail", then there exists the same itinerary of heads and tails executed by the map ψ. So, we have a sequence $(w_i)_{i\in\mathbb{Z}}$ of points of the metric space X, with $w_{i+1} = \psi(w_i), \forall i \in \mathbb{Z}$, such that $w_i \in \mathcal{K}_0$ or $w_i \in \mathcal{K}_1$ according as the i-th term of the sequence $(s_i)_i$ is "head" or "tail". Our definition extends that in [33], in the sense that any periodic sequence of heads and tails can be obtained by suitable periodic points of the map ψ. For instance consider the following example: There exists a fixed point of ψ in the set $\mathcal{K}_0$ corresponding to the constant sequence of symbols $s_i =$ "head", $\forall\, i \in \mathbb{Z}$. There is also a point of period three $w \in \mathcal{K}_1$ with $\psi(w) \in \mathcal{K}_1$ and $\psi^2(w) \in \mathcal{K}_0$, corresponding to the periodic sequence $\dots TTH\ TTH\ TTH \dots$, and so on.

Using this approach, we can also obtain periodic points for ψ with a minimal period. For example, consider the sequence $\dots HHHT\ HHHT\ HHHT \dots$, obtained by some point $w \in \mathcal{K}_0$ with $\psi(w) \in \mathcal{K}_0$, $\psi^2(w) \in \mathcal{K}_0$, $\psi^3(w) \in \mathcal{K}_1$, $\psi^4(w) = w \in \mathcal{K}_0$, which is a point of minimal period four.

Definition 4.3 is in agreement with other such definitions of chaotic dynamics for ODEs with periodic coefficients (see [19, 63, 73]). The definition of chaos on two symbols may be extended to m symbols. We now proceed to do so.

Definition 4.4 Let Z be a metric space, $\psi : Z \supset D_\psi \to Z$ be a map and let $D \subset D_\psi$. Assume also that $m \geq 2$ is an integer. We say that ψ *induces chaotic dynamics on m symbols in the set D* if there exist m nonempty pairwise disjoint compact sets

$$\mathcal{K}_0\,,\ \mathcal{K}_1\,,\dots,\mathcal{K}_{m-1}\ \subset D,$$

such that, for each two–sided sequence $(s_i)_{i\in\mathbb{Z}} \in \{0,\dots,m-1\}^{\mathbb{Z}}$, there exists a corresponding sequence $(w_i)_{i\in\mathbb{Z}} \in D^{\mathbb{Z}}$ such that

$$w_i \ \in \mathcal{K}_{s_i} \quad \text{and} \quad w_{i+1} = \psi(w_i), \quad \forall\, i \in \mathbb{Z} \tag{4.8}$$

and, whenever $(s_i)_{i\in\mathbb{Z}}$ is a k–periodic sequence (that is, $s_{i+k} = s_i$, $\forall i \in \mathbb{Z}$) for some $k \geq 1$, there exists a k–periodic sequence $(w_i)_{i\in\mathbb{Z}} \in D^{\mathbb{Z}}$ satisfying (4.8). When we want to stress the role of the $\mathcal{K}_j$'s, we also say that ψ *induces chaotic dynamics on m symbols in the set D relatively to* $(\mathcal{K}_0,\dots,\mathcal{K}_{m-1})$.

The fact that the iterates of a map mimic the sequences of coin-tossing has long since been regarded as a key feature of complex dynamics of deterministic systems (see, for instance, Smale's remarks in [59]). This characterization of chaos is similar to that of "topological horseshoes", (see [30, 32]). A specific advantage of the version adopted here is the possibility of obtaining periodic sequences of m symbols by the periodic itineraries generated by periodic points of ψ.

The following Theorem 4.1 gives the relation between the dynamical properties of a map satisfying Definitions 4.3 or 4.4 and the Bernoulli shift. Let Λ be a compact metric space and let $\psi : \Lambda \to \Lambda$ be a continuous map. Let us recall that ψ is *semi-conjugate to the two–sided m–shift* if there exists a continuous surjective mapping $g : \Lambda \to \Sigma_m$ such that

$$g \circ \psi = \sigma \circ g. \tag{4.9}$$

In a similar manner, if we denote by

$$\Sigma_m^+ = \{0,\dots,m-1\}^{\mathbb{N}}$$

the set of the one–sided sequences of m symbols, recall that ψ is *semi-conjugate to the one–sided m–shift* if there exists a continuous surjective mapping $g : \Lambda \to \Sigma_m^+$ such that (4.9) holds.

As stated earlier, the following result connects the concept of semi-conjugation with the Bernoulli shift to the one of chaotic dynamics expressed in Definition 4.4. Its proof may also be adapted from similar arguments in works by authors such as [30, 32] for semidynamical systems induced by continuous maps of metric spaces. We give an outline of this concept.

Theorem 4.1 *[40] Let Z be a metric space, $\psi : Z \supset D_\psi \to Z$ be a map which is continuous and one-to-one on a set $\mathcal{D} \subset D_\psi$ and induces therein chaotic dynamics on $m \geq 2$ symbols (relatively to $(\mathcal{K}_0, \ldots, \mathcal{K}_{m-1})$). Then, there exists a nonempty compact set*

$$\Lambda \subset \bigcup_{j=0}^{m-1} \mathcal{K}_j ,$$

which is invariant for ψ and such that $\psi|_\Lambda$ is semi-conjugate to the two–sided m–shift, so that the topological entropy $h_{top}(\psi)$ satisfies

$$h_{top}(\psi) \geq \log(m).$$

Moreover, the subset $\mathcal{P}$ of Λ made by the periodic points of ψ is dense in Λ and if we denote by $g : \Lambda \to \Sigma_m$ the continuous surjection in Eq.(4.9), it holds also that the counterimage through g of any k–periodic sequence in Σ_m contains at least one k–periodic point.

Proof Setting $\mathcal{K} := \bigcup_{j=0}^{m-1} \mathcal{K}_j$, we define

$$\Lambda_0 := \{w \in \mathcal{K} : \psi^i(w) \in \mathcal{K}, \ \forall i \in \mathbb{Z}\} = \bigcap_{i=-\infty}^{+\infty} \psi^{-i}(\mathcal{K})$$

and

$$\mathcal{P} := \{x \in \Lambda_0 \ : \ \exists k \geq 1 \text{ with } \psi^k(x) = x\}.$$

Since $\mathcal{K}$ is compact and ψ is continuous on $\mathcal{K}$, it follows immediately that also Λ_0 is compact and that $\psi(\Lambda_0) \subset \Lambda_0$ (that is, Λ_0 is invariant for ψ). Let us now define $g_0 : \Lambda_0 \to \Sigma_m$, as

$$g_0(w) := (s_i)_{i \in \mathbb{N}} \Leftrightarrow \psi^i(w) \in \mathcal{K}_{s_i}, \ \forall i \in \mathbb{N}.$$

By Definition 4.4, the map g_0 is surjective and the counterimage through g_0 of any k–periodic sequence in Σ_m contains at least one k–periodic point (belonging to $\mathcal{P}$). The continuity of g_0 comes from the continuity of ψ on Λ_0 , the choice of the distance d and the fact that the sets $\mathcal{K}_j$ are compact and pairwise disjoint. In fact g_0 is uniformly continuous as it is defined on a compact metric space. A direct scrutiny shows that the relation in the equation (4.9) is satisfied and therefore the map g_0 induces a semi-conjugation between $\psi|_{\Lambda_0}$ and the two$-$sided $m-$shift.

Let

$$\Sigma_m^{per} \subset \Sigma_m$$

be the set of the periodic two$-$sided sequences of m symbols.

Now,

$$g_0 : \mathcal{P} \to \Sigma_m^{per},$$

by setting, for each $w \in \mathcal{P}$:

$$g_0(w) := (s_i)_{i\in\mathbb{Z}} \in \Sigma_m^{per} \Leftrightarrow \psi^i(w) \in \mathcal{K}_{s_i}, \ \forall i \in \mathbb{Z}. \tag{4.10}$$

Notice that

$$g_0 \circ \psi(w) = \sigma \circ g_0(w), \quad \forall\, w \in \mathcal{P},$$

where σ is the two$-$sided Bernoulli shift on m symbols.

Now, setting

$$\Lambda := \overline{\mathcal{P}} \subset \Lambda_0\,,$$

it holds that $\psi(\Lambda) \subset \Lambda$, so that Λ is compact and invariant for ψ. At last, we extend the uniformly continuous surjective mapping

$$g_0 : \mathcal{P} \to \Sigma_m^{per} \subset \Sigma_m$$

to a continuous surjective function

$$g : \Lambda \to \Sigma_m\,,$$

such that

$$g \circ \psi(x) = \sigma \circ g(x), \quad \forall\, x \in \Lambda.$$

From the above proved semi-conjugacy condition and by [68, Theorem 7.2] it follows that

$$h_{\text{top}}(\psi) \geq h_{\text{top}}(\sigma) = \log(m).$$

All this was explained earlier in formula (4.2). Hence we see that all the properties listed in the statement of the theorem are satisfied. The proof is complete. □

4.3 Topological Lemmas and Definitions

We now state and prove another result which is more general than the previous Theorem 4.1. In Theorem 4.1 a one-to-one condition is assumed on the map ψ. In the following theorem no such one-to-one condition is assumed. This result is an important and interesting result for the concept of chaotic dynamics as we define it. For simplicity we state the following theorem for $m = 2$ symbols. But the same result may be applied to any arbitrary m.

Theorem 4.2 *[40] Let ψ be a map inducing chaotic dynamics on two symbols on a set $\mathcal{D}$ and which is continuous on*

$$\mathcal{K} := \mathcal{K}_0 \cup \mathcal{K}_1 \subset \mathcal{D},$$

where $\mathcal{K}_0$,$\mathcal{K}_1$ and $\mathcal{D}$ are as in Def.4.3. Defining the nonempty compact set

$$\mathcal{J}_\infty := \bigcap_{n=0}^{\infty} \psi^{-n}(\mathcal{K}) \tag{4.11}$$

then there exists a nonempty compact set

$$\mathcal{J} \subset \mathcal{J}_\infty \subset \mathcal{K},$$

on which the following are fulfilled:

(i) $\mathcal{J}$ is invariant for ψ (i.e. $\psi(\mathcal{J}) = \mathcal{J}$).
(ii) $\psi|\mathcal{J}$ is semi-conjugate to the Bernoulli shift on two symbols, that is, there exists a continuous map π of $\mathcal{J}$ onto $\Sigma_2^+ := \{0, 1\}^{\mathbb{N}}$ endowed with the distance

$$d(\mathbf{s}', \mathbf{s}'') := \Sigma_{j=0}^{\infty} \frac{|s_i' - s_i''|}{2^{i+1}}, \text{ for } \mathbf{s}' = (s_i')_{i \in \mathbb{N}},\ \mathbf{s}'' = (s_i'')_{i \in \mathbb{N}} \in \Sigma_2^+$$

such that the diagram

$$\begin{array}{ccc} \mathcal{J} & \xrightarrow{\psi} & \mathcal{J} \\ {\scriptstyle\pi}\downarrow & & \downarrow{\scriptstyle\pi} \\ \Sigma_2^+ & \xrightarrow[\sigma]{} & \Sigma_2^+ \end{array} \tag{4.12}$$

commutes, where $\sigma : \Sigma_2^+ \to \Sigma_2^+$ is the Bernoulli shift defined by $\sigma((s_i)_i) := (s_{i+1})_i, \forall i \in \mathbb{N}$.
(iii) The set P of the periodic points of $\psi\,|_{\mathcal{J}_\infty}$ is dense in $\mathcal{J}$ and the pre-image $\pi^{-1}(\mathbf{s}) \subset \mathcal{J}$ of every k-periodic sequence $\mathbf{s} = (s_i)_{i \in \mathbb{N}} \in \Sigma_2^+$ contains at least one k-periodic point.
Furthermore, from property (ii) it follows that:

(iv) $h_{top}(\psi) \geq h_{top}(\psi\mid_{\mathcal{J}}) \geq h_{top}(\sigma) = log(2)$, *where* h_{top} *is the topological entropy.*
(v) There exists a compact positively invariant set $\Lambda \subset \mathcal{J}$ *such that* $\psi\mid_{\Lambda}$ *is semi-conjugate to the Bernoulli shift on two symbols, topologically transitive and has sensitive dependence on initial conditions.*

Proof The proof follows the same lines as that of Theorem 4.1. For the reader's convenience, we present all the details. We verify first that the set, $\mathcal{J}_\infty$ in (4.11) is compact and nonempty. As the map ψ is continuous on $\mathcal{K}$, it immediately follows that $\mathcal{J}_\infty$ is closed and, being contained in the compact set $\mathcal{K}$, it is compact, too. The fact that $\mathcal{J}_\infty$ is nonempty follows from Definition 4.3. And this is so if we observe that$z \in \mathcal{J}_\infty \Leftrightarrow \psi^n(z) \in \mathcal{K}, \forall n \geq 0$. This also implies that $\psi(\mathcal{J}_\infty) \subset \mathcal{J}_\infty$: it can be seen that if $z \in \mathcal{J}_\infty$, then also $\psi(z) \in \mathcal{J}_\infty$ Calling $\mathcal{P}$ the subset of $\mathcal{J}_\infty$ consisting of the periodic points of $\psi\mid_{\mathcal{J}_\infty}$, that is,

$$\mathcal{P} := \{w \in \mathcal{J}_\infty : \exists k \in \mathbb{N}\backslash\{0\}, \quad \psi^k(w) = w\},$$

we claim that $\psi(\mathcal{P}) = \mathcal{P}$. In fact, if $z \in \mathcal{P}$, then there exists $l \in \mathbb{N}\backslash\{0\}$ such that $\psi^l(z) = z$. So, on the one hand, $\psi(z) = \psi(\psi^l(z)) = \psi^{l+1}(z) = \psi^l(\psi(z))$ and therefore $\psi(z) \in \mathcal{P}$, too. This implies that $\psi(\mathcal{P}) \subset \mathcal{P}$. By a repetition of the same argument, it is also possible to prove that if $z \in \mathcal{P}$, then $\psi^h(z) \in \mathcal{P}$, for any $h \geq 1$. If $\psi^l(z) = z$, for some $l \in \mathbb{N}\backslash\{0\}$, then there are two possibilities for l, that is, $l = 1$ or $l \geq 2$. In the former case $\psi(z) = z$ and so $z \in \psi(\mathcal{P})$, whilst in the latter $z = \psi^l(z) = \psi(\psi^{l-1}(z))$. Hence, since $\psi^{l-1}(z) \in \mathcal{P}$ whenever $z \in \mathcal{P}$, we have again $z \in \psi(\mathcal{P})$. Therefore it is proved that, if $z \in \mathcal{P}$, then $z \in \psi(\mathcal{P})$, i.e. $\mathcal{P} \subset \psi(\mathcal{P})$. The claim is verified.

At this point it can be observed that, since $\mathcal{P}$ is contained in the compact set $\mathcal{J}_\infty$, also

$$\mathcal{J} := \overline{\mathcal{P}} \subset \mathcal{J}_\infty, \tag{4.13}$$

and moreover $\mathcal{J}$ is compact, being closed in a compact set. From $\psi(\mathcal{P}) = \mathcal{P}$, it follows that

$$\psi(\mathcal{J}) = \psi\overline{(\mathcal{P})} \supset \psi(\mathcal{P}) = \mathcal{P}.$$

But again, by the compactness of $\psi(\mathcal{J})$, it holds that

$$\psi(\mathcal{J}) \supset \overline{\mathcal{P}} = \mathcal{J}.$$

Let us show that also the reverse inclusion is satisfied for $\mathcal{J}$, that is,$\psi(\mathcal{J}) \subset \mathcal{J}$. Since ψ is continuous, we have

$$\psi(\mathcal{P}) = \psi\overline{(\mathcal{P})} \subset \overline{\psi(\mathcal{P})} = \overline{\mathcal{P}} = \mathcal{J}.$$

Hence, the invariance of $\mathcal{J}$ is proved, in agreement with conclusion (i). Let us consider now the diagram

$$\begin{array}{ccc} \mathcal{J}_\infty & \xrightarrow{\psi} & \mathcal{J}_\infty \\ \pi \big\downarrow & & \big\downarrow \pi \\ \Sigma_2^+ & \xrightarrow[\sigma]{} & \Sigma_2^+ \end{array}$$

and define the map $\pi : \mathcal{J}_\infty \to \Sigma_2^+$ by associating with $w \in \mathcal{J}_\infty$ the sequence $(s_n)_{n\in\mathbb{N}} \in \Sigma_2^+$ such that $s_n = j$ if $\psi^n(w) \in \mathcal{K}_j$, for $j = 0, 1$. More formally, it may be noted that for any $w \in \mathcal{J}_\infty$, there exists a unique forward itinerary $(w_i)_{i\in\mathbb{N}}$ such that $w_0 = w$ and $\psi(w_i) = w_{i+1} \in \mathcal{K}$, for every $i \in \mathbb{N}$. Hence the function $g_1 : \mathcal{J}_\infty \to, \mathcal{J}_\infty^{\mathbb{N}}$ which maps any $w \in \mathcal{J}_\infty$ into the one-sided sequence of points from the set $\mathcal{J}_\infty$

$$\mathbf{s}_w := (w_i)_{i\in\mathbb{N}} \ \text{ where } \ w_i := \psi^i(w), \forall i \in \mathbb{N},$$

with the usual convention $\psi^0 = Id_{\mathcal{J}_\infty}$ and $\psi^1 = \psi$, is well defined. Since the sets $\mathcal{K}_0$ and $\mathcal{K}_1$ are disjoint, for every term w_i of $\mathbf{s}_w$ there exists a unique index

$$s_i = s_i(w_i), \ \text{ with } \ s_i \in \{0, 1\},$$

such that $w_i \in \mathcal{K}_{s_i}$. Therefore the map $g_2 : \mathcal{J}_\infty^{\mathbb{N}} \to \Sigma_2^+$,

$$g_2 : \mathbf{s}_w \mapsto (s_i)_{i\in\mathbb{N}} \in \Sigma_2^+ \tag{4.14}$$

is also well defined. Thus, by Definition 4.3 the map

$$\pi := g_2 \circ g_1 : \mathcal{J}_\infty \to \Sigma_2^+ \tag{4.15}$$

is a surjection (onto) that makes the diagram (4.12) commute and the pre-image through π of any k-periodic sequence in Σ_2^+ contains at least one k-periodic point of $\mathcal{J}_\infty$. To verify that π is continuous, we prove continuity in a generic $\bar{z} \in \mathcal{J}_\infty$ as follows: we show that for any $\epsilon > 0$, there exists $\delta > 0$ such that $\forall z \in \mathcal{J}_\infty$ with $d(z, \bar{z}) < \delta$, $d(\pi(z), \pi(\bar{z})) < \epsilon$, with $\hat{d}$ as in the statement of the theorem. Fix $\epsilon > 0$ and let $n \in \mathbb{N}$ such that $0 < 1/2^n < \epsilon$. It is sufficient to prove that $(\pi(z))_i = (\pi(\bar{z}))_i$, for any $i = 0, \cdots, n$. Then, if this is the case, by the definition of $\hat{d}$, it follows that $\hat{d}(\pi(z), \pi(\bar{z})) \leq 1/2^n < \epsilon$ Since $\bar{z} \in \mathcal{J}_\infty$, there exists a sequence $(s_0, \cdots, s_n) \in \{0, 1\}^{n+1}$ such that

$$\bar{z} \in \mathcal{K}_{s_0}, \psi(\bar{z}) \in \mathcal{K}_{s_1}, \cdots, \psi^n(\bar{z}) \in \mathcal{K}_{s_n}.$$

Recalling that $\mathcal{K}_0$ and $\mathcal{K}_1$ are compact and disjoint, it holds that

$$\eta := d(\mathcal{K}_0, \mathcal{K}_1) > 0.$$

Hence, for any $z \in \mathcal{J}_\infty$ with $d(z, \bar{z}) < \eta/2$, it follows that $z \in \mathcal{K}_{s_0}$, too. By the continuity of ψ in $\bar{z}$, there exists $\delta_1 > 0$ such that $\forall z \in \mathcal{J}_\infty$ with $d(z, \bar{z}) < \delta_1$, then $d(\psi(z), \psi(\bar{z})) < \eta/2$. But this means that $\psi(z) \in \mathcal{K}_{s_1}$. Similarly, by the continuity of ψ^2 in $\bar{z}$, there exists $\delta_2 > 0$ such that $\forall z \in \mathcal{J}_\infty$ with $d(z, \bar{z}) < \delta_2$, $d(\psi^2(z); \psi^2(\bar{z})) < \eta/2$ and thus $\psi^2(z) \in \mathcal{K}_{s_2}$, for any such z. Proceeding in this way till the nth iterate of ψ and setting

$$\delta := min\{\frac{\eta}{2}, \delta_1, \cdots, \delta_n\},$$

we find that, for any $z \in \mathcal{J}_\infty$ with $d(z, \bar{z}) < \delta$, it holds that

$$z \in \mathcal{K}_{s_0}, \psi(z) \in \mathcal{K}_{s_1}, \cdots, \psi^n(z) \in \mathcal{K}_{s_n},$$

exactly as for $\bar{z}$. But this means that $(\pi(z))_i = (\pi(\bar{z}))_i$, for any $i = 0, \cdots, n$, and hence $\hat{d}(\pi(z), \pi(\bar{z})) \leq 1/2^n < \epsilon$. The continuity of π is thus proved. Considering in diagram (4.12) the restriction of ψ to $\mathcal{P} \subset \mathcal{J}_\infty$, we find the commutative diagram

$$\begin{array}{ccc} \mathcal{P} & \xrightarrow{\psi} & \mathcal{P} \\ \pi \downarrow & & \downarrow \pi \\ \mathcal{P}_2^+ & \xrightarrow[\sigma]{} & \mathcal{P}_2^+ \end{array} \tag{4.16}$$

where $\mathcal{P}_2^+ \subset \Sigma_2^+$ is the set of the periodic sequences of two symbols. Note that $\pi(\mathcal{P}) = \mathcal{P}_2^+$ from definition 4.3. Recalling the fact that $\mathcal{P}_2^+$ is dense in Σ_2^+ by the continuity of π, it follows that

$$\pi(\mathcal{J}) = \pi(\bar{\mathcal{P}}) \subset \overline{\mathcal{P}_2^+} = \Sigma_2^+.$$

On the other hand $\pi(\mathcal{J})$ is a compact set containing $\pi(\mathcal{P}) = \mathcal{P}_2^+$ and hence

$$\pi(\mathcal{J}) \supset \overline{\mathcal{P}_2^+} = \Sigma_2^+.$$

Therefore, we can conclude that $\pi(\mathcal{J}) = \Sigma_2^+$ and the diagram

$$\begin{array}{ccc} \mathcal{J} & \xrightarrow{\psi} & \mathcal{J} \\ \pi \downarrow & & \downarrow \pi \\ \Sigma_2^+ & \xrightarrow[\sigma]{} & \Sigma_2^+ \end{array}$$

Also the pre-image through π of any k-periodic sequence in Σ_2^+ contains at least one k-periodic point of $\mathcal{J}$ as $P \subset I$. Conclusions (ii) and (iii) are thus proved.

Assertion (iv) regarding the positive topological entropy, comes from property (ii) about the semi-conjugacy to the Bernoulli shift,according formula (4.2). For proof see also [68]. Finally, conclusion (v) regarding the existence of a compact set $\Lambda \subset \mathcal{J}$ which is positively invariant for ψ and such that $\psi\mid_{\Lambda}$ is semi-conjugate to the Bernoulli shift on two symbols, topologically transitive, and has sensitive dependence on initial conditions, follows by utilizing a result by Auslander and Yorke [7]. □

4.4 The Concept of "Stretching" as Related to Chaos

In the present section, we provide an introduction to our approach for detecting the presence of chaos based on the concept of *stretching along the paths* [40].

"Stretching Along the Paths" (SAP) is a method introduced by the authors in [40] to prove the existence of periodic points and chaotic dynamics in discrete-time dynamical systems. Previous versions of this method have been obtained in [46–48]. With this method, as compared with the work of Kennedy and Yorke in [32], allows one to obtain better results with respect to the existence of periodic points because of its specialized framework. It is at the same time mathematically rigorous and based on some elementary geometrical considerations. Thus, it avoids the use of more advanced topological theories and it is relatively easy to apply to specific models arising in applications.

We present some basic definitions needed for this method. Let X be a metric space. A *path* in space X is a continuous map $\gamma : [t_0, t_1] \to X$. Its range is denoted by $\bar{\gamma}$ that is, $\bar{\gamma} := \gamma[t_0, t_1]$. A *sub-path* ω of γ is the restriction of γ to a closed sub-interval of its domain and hence it is defined as $\omega := \gamma|_{[t_0', t_1']}$, for some $[t_0', t_1'] \subset [t_0, t_1]$.

Next, if $M,\ N$ are metric spaces and $\psi : M \supset D_\psi \to N$ is a map which is continuous on a set $\mathcal{M} \subset D_\psi$, then for any path γ in M with $\bar{\gamma} \subset \mathcal{M}$, it follows that $\psi \circ \gamma$ is a path in N with range $\psi(\bar{\gamma})$. In assuming the paths to be defined on [0,1] there is no loss of generality. If $\theta_1 : [a_1, b_1] \to M$ and $\theta_2 : [a_2, b_2] \to M$ with $a_i < b_i,\ \ i = 1, 2$, are two paths in M, we define the equivalence relation $\sim$ between θ_1 and θ_2 by setting $\theta_1 \sim \theta_2$ if there exists a homeomorphism h of $[a_1, b_1]$ onto $[a_2, b_2]$ such that $\theta_2(h(t)) = \theta_1(t), \forall t \in [a_1, b_1]$. It can be proved that if $\theta_1 \sim \theta_2$, then the ranges of θ_1 and θ_2 are the same. Hence, there corresponds an equivalent path defined on $[0, 1]$ for any path γ.

Next we define the concept of an *arc*. which is similar to a path. An arc is the homeomorphic image of the compact interval $[0, 1]$, and an open arc is an arc without its end-points.

A continuum of X is a compact connected subset of X.
By a *generalized rectangle* we mean a set $\mathcal{R} \subset X$ which is homeomorphic to the unit square $[0, 1]^2 \subset \mathbb{R}^2$.

Given a generalized rectangle $\mathcal{R}$ and $\mathcal{S} := [0, 1]^2 \subset \mathbb{R}^2$, consider the associated homeomorphism $h : \mathcal{S} \to h(\mathcal{S}) = \mathcal{R}$. The set $\vartheta R := h(\partial([0, 1]^2))$, where $\partial([0, 1]^2)$ is the boundary of the unit square, is called the *contour* of $\mathcal{R}$. The contour ϑR is well defined as it is independent of the choice of the homeomorphism h. In fact, ϑR is also a homeomorphic image of a simple closed curve,

We call as an *oriented rectangle*, the pair

$$\widetilde{\mathcal{R}} := (\mathcal{R}, \mathcal{R}^-)$$

where $\mathcal{R} \subset X$ is a generalized rectangle and

$$\mathcal{R}^- := \mathcal{R}_l^- \cup \mathcal{R}_r^-$$

is the union of two disjoint compact arcs $\mathcal{R}_l^-$ and $\mathcal{R}_r^- \subset \vartheta\mathcal{R}$ that are designated as *left* and *right* sides of $\mathcal{R}^-$. Since $\vartheta\mathcal{R}$ is a Jordan curve it follows that $\vartheta\mathcal{R}\backslash(\mathcal{R}_l^- \cup \mathcal{R}_r^-)$ consists of two open arcs. Hence the set

$$\mathcal{R}^+ := \overline{\vartheta R \setminus \mathcal{R}^-}$$

is also the union of two disjoint compact arcs. We write

$$\mathcal{R}^+ = \mathcal{R}_b^+ \cup \mathcal{R}_t^+,$$

representing the "base" and "top" components of $\mathcal{R}^+$ as $\mathcal{R}_b^+$ and $\mathcal{R}_t^+$ (sometimes we use the terms "down" and "up" for these components and denote them as $\mathcal{R}_d^+$ and $\mathcal{R}_u^+$, respectively). We would like to mention here that the order in which we label "left" and "right" the two components of $\mathcal{R}^-$ (and, respectively, "base" and "top" the components of $\mathcal{R}^+$) is not important for what follows. This can be seen in Fig. 4.4.

Both the term generalized rectangle for $\mathcal{R}$ and the decomposition of the contour $\vartheta\mathcal{R}$ into $\mathcal{R}^-$ and $\mathcal{R}^+$ are inspired by the construction of rectangular domains around hyperbolic sets arising in the theory of Markov partitions ([29, p.291], [1]), as well as by the Conley-Ważewski theory [21, 62]. Broadly speaking, the sets labelled as $[.]^-$, or as $[.]^+$, in the setting of continuous dynamical systems, would be composed by those points which are moved by the flow outward, or inward, respectively, with respect to $\mathcal{R}$. Here we follow the same analogy for the discrete dynamical systems as well.

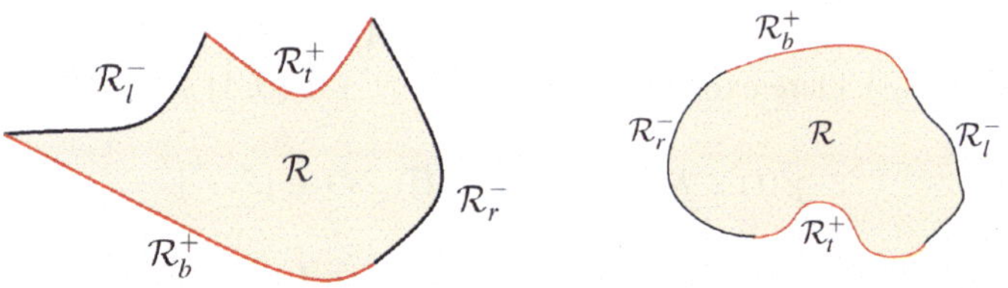

Fig. 4.4 The figure shows examples of oriented rectangles, where "base" and "top" and "left" and "right" are labelled, indicating that the order in which these are chosen is not important

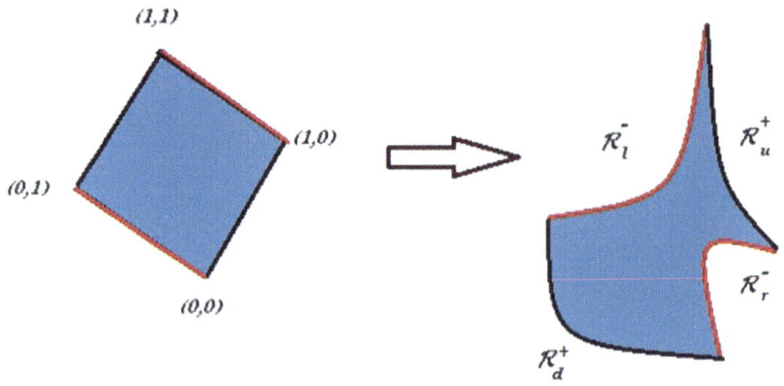

Fig. 4.5 The generalized rectangle with the oriented sides, homeomorphic to the unit square under the homeomorphism g

In the applications discussed here, the space X is the Euclidean plane. The generalized rectangles are compact regions bounded by graphs of some functions. However, the definitions being broad in nature can be applied to more general settings (Fig. 4.5).

Given an oriented rectangle in a metric space X, we can always find a homeomorphism $g : [0, 1]^2 \to g([0, 1]^2) = \mathcal{R}$ (with g possibly different from the map h defined above) such that

$$g(\{0\} \times [0, 1]) = \mathcal{R}_l^-, \quad g(\{1\} \times [0, 1]) = \mathcal{R}_r^-. \tag{4.17}$$

(See [47]).

In light of this we can say that some properties related to $\mathcal{R}$ can be transferred to the unit square $[0, 1]^2 \subset \mathbb{R}^2$ oriented in its natural (left side-right side) manner.

Definition 4.5 Suppose that $\psi : X \supset D_\psi \to X$ is a map defined on a set D_ψ and let $\widetilde{\mathcal{A}} := (\mathcal{A}, \mathcal{A}^-)$ and $\widetilde{\mathcal{B}} := (\mathcal{B}, \mathcal{B}^-)$ be oriented rectangles of a metric space X. Let $\mathcal{K} \subset \mathcal{A} \cap D_\psi$ be a compact set. We say that $(\mathcal{K}, \psi)$ *stretches* $\widetilde{\mathcal{A}}$ *to* $\widetilde{\mathcal{B}}$ *along the paths* and write

$$(\mathcal{K}, \psi) : \widetilde{\mathcal{A}} \Leftrightarrow\!\!\!\to \widetilde{\mathcal{B}}, \tag{4.18}$$

if the following conditions hold:

- ψ is continuous on $\mathcal{K}$;
- for every path $\gamma : [0, 1] \to \mathcal{A}$ such that $\gamma(0) \in \mathcal{A}_l^-$ and $\gamma(1) \in \mathcal{A}_r^-$ (or $\gamma(0) \in \mathcal{A}_r^-$ and $\gamma(1) \in \mathcal{A}_l^-$), there exists a sub-interval $[t', t''] \subset [0, 1]$ such that

  $$\gamma(t) \in \mathcal{K}, \quad \psi(\gamma(t)) \in \mathcal{B}, \quad \forall t \in [t', t'']$$

  and, moreover, $\psi(\gamma(t'))$ and $\psi(\gamma(t''))$ belong to different components of $\mathcal{B}^-$.

In the special case in which $\mathcal{K} = \mathcal{A}$, we simply write

$$\psi : \widetilde{\mathcal{A}} \Leftrightarrow\!\!\!\rightarrow \widetilde{\mathcal{B}}.$$

Definition 4.6 Let $\psi : X \supset D_\psi \to X$ be a map defined on a set on a set D_ψ and let $\widetilde{\mathcal{A}} := (\mathcal{A}, \mathcal{A}^-)$ and $\widetilde{\mathcal{B}} := (\mathcal{B}, \mathcal{B}^-)$ be oriented rectangles of a metric space X. Let $\mathcal{D} \subseteq \mathcal{A} \cap D_\psi$ also be a compact set and let $m \geq 2$. We say that $(\mathcal{D}, \psi)$ stretches $\widetilde{\mathcal{A}}$ to $\widetilde{\mathcal{B}}$ along the paths with crossing number m and write

$$(\mathcal{D}, \psi) : \widetilde{\mathcal{A}} \Leftrightarrow\!\!\!\rightarrow^m \widetilde{\mathcal{B}},$$

if there exist m pairwise disjoint compact sets

$$\mathcal{K}_0, \cdots, \mathcal{K}_{m-1} \subset \mathcal{D}$$

such that

$$(\mathcal{K}_i, \psi) : \widetilde{\mathcal{A}} \Leftrightarrow\!\!\!\rightarrow \widetilde{\mathcal{B}}, \quad i = 0, \cdots, m-1.$$

When $\mathcal{D} = \mathcal{A}$, we simply write

$$\psi : \widetilde{\mathcal{A}} \Leftrightarrow\!\!\!\rightarrow^m \widetilde{\mathcal{B}}.$$

An important feature of the "stretching" property is that, when (4.18) is satisfied with $\mathcal{A} = \mathcal{B}$, is that one could prove the existence of a fixed point for the map ψ in $\mathcal{K}$. Thus, not only is it possible to find the existence of such a point in the domain of ψ, but also can it be localized to the compact set $\mathcal{K}$. Thus, with the stretching condition being satisfied with respect to different, disjoint $\mathcal{K}'s$, multiple fixed points are obtained.

The stretching property thus described is more restrictive than the analogous one considered by Kennedy and Yorke in [32], dealing with connections and preconnections. In fact, the authors, in [32] (horseshoe hypotheses), deal with a locally connected and compact subset Q of the separable metric space X, on which they select two disjoint and compact sets end_0, $end_1 \subset Q$ so that any component of Q intersects both of them. On Q a continuous map $f : Q \to X$ is defined in such a way that every compact connected set $\Gamma \subset Q$ that joins end_0 and end_1 (i.e. a connection according to [32]) admits at least $m \geq 2$ disjoint compact and connected subsets, whose images under f are again connections. Such sub-continua are named *preconnections* and m is the so-called *crossing number*. Making use of Kennedy and Yorke's language and notation, and drawing a parallel, we can see that the sets $\mathcal{R}_l^-$ and $\mathcal{R}_r^-$ in our definition of the oriented rectangle can be seen to be the same as a special instance of the sets end_0 and end_1 described by the authors. In addition to this, any path γ with values in $\mathcal{R}$ and joining $\mathcal{R}_l^-$ with $\mathcal{R}_r^-$ determines a connection via its image $\bar{\gamma}$. Similarly, any sub-path ω of γ with $\omega = \gamma\mid_{[t',t'']}$ as in the "stretching" definition, makes $\bar{\omega}$ a preconnection. As a consequence, when there exist $m \geq 2$ pairwise disjoint compact sets $\mathcal{K}_0, \cdots, \mathcal{K}_{m-1} \subset \mathcal{R}$ such that

$$(\mathcal{K}_i, \psi) : \widetilde{\mathcal{R}} \Leftrightarrow\!\!\!\!\rightarrow \widetilde{\mathcal{R}}, \quad \forall i = 0, \cdots, m-1,$$

we are in the same situation discussed by Kennedy and Yorke with m as a crossing number.

Remark 4.1 Broadly speaking, our stretching condition may be looked upon as a special case of [32]. In [32] the authors prove some kind of chaotic dynamics in the "coin-tossing" sense. However due to the general setting which they consider, the existence of fixed points or periodic points is not guaranteed. Indeed in [32] there are some specific examples where periodic points do not exist. On the contrary, our approach although dealing with a simpler geometric configuration, allows us to obtain fixed points and periodic points as well. This will be proved in the sequel. See [20] for a detailed comparison between the two methods.

Linked Twist Maps

In this section we present a geometrical context in which it is possible to apply the method of "stretching along the paths", the theory of the Linked Twist Maps (for short, LTMs). In this framework, there is a geometric configuration where there are two alternating planar homeomorphisms (or diffeomorphisms) which twist two circular annuli (or two families of them [23, 53]). The domain is two intersecting annuli with two distinct regions of intersection, rectangles A and B. The action of a twist map is to take a line connecting the inner and outer boundaries of the annulus and twist it around inside it. Each annulus is turned onto itself by a homeomorphism which leaves the boundaries of the annulus invariant. Both the maps act in their domain so that a twist effect is produced on both annuli. Typically the twist effect is achieved by assuming that the angular speed is monotone with respect to the radius. If we consider the composition of the two movements in the common regions, we obtain a function which is what we call a "linked twist map" (see [65, 66, 70, 72] for a detailed description of the geometry of the domain of an LTM). A typical LTM configuration is illustrated in Fig. 4.6.

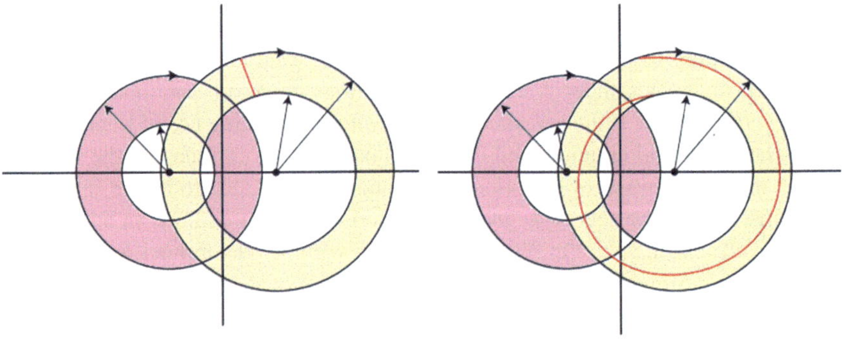

Fig. 4.6 Two linked annuli showing inner and outer circular regions with the regions A and B shown in the intersection

Such maps give a geometric setting for the Smale horseshoes: Under certain conditions, it is possible to prove the existence of a Smale horseshoe inside A and B, as was done in [23]. The requisite assumptions on the twist mappings require, are that they are smooth, monotonicity of the angular speed with respect to the radial coordinate, and preservation of the Lebesgue measure. Since our approach is purely topological, we need just a twist condition on the boundary.

In the past few decades, there has been a growing interest in LTMs. In the early 80s they were studied from a theoretical point of view by Devaney [23], Burton and Easton [18] and Przytycki [52, 53] (just to cite a few contributions in this field), proving some mathematical properties like ergodicity, hyperbolicity , and conjugacy to the Bernoulli shift. However, as observed in [23], such maps also appear naturally in various different applicative contexts, like for example in mathematical models for particle motions in a magnetic field, in differential geometry, and in the study of diffeomorphisms of surfaces. Special configurations related to LTMs can also be found in the restricted three-body problem [42, 56]. Significant applications of LTMs have been obtained in the area of fluid mixing by the pioneering work of Aref [4, 5] and the investigations of Ottino, Sturman, and Wiggins, (see, for instance, [64, 65, 70, 72]). A chapter on the recent work in this area can be found in the book [66]. Our purpose is to show how the framework of LTMs can be adapted to study some natural geometric configurations associated with the study of the periodically perturbed Duffing equation in the phase-plane.

An example of a Linked Twist Map from the Classical Pendulum

Let us consider an example from pendulum-type equations taken from [16, 17]. This is a simple instance, which shows how our approach is suitable for the study of LTMs. This consists of overlapping annuli subject to twist rotations. A similar geometric setting was already considered by Kennedy and Yorke in [31] in the framework of the theory of fluid mixing, where they study planar functions obtained as compositions of a squeezing map and a stirring rotation.

We stress that such examples are only of "pedagogical" nature and, in fact, the chaotic-like dynamics that we obtain could be also proved using different approaches already developed in various papers (like, for instance, [41, 61, 74, 75].)

Consider two systems of equations representing the pendulum equation which we label as I_b and I_a (the system of equations in (I_a) represents the *inverted pendulum*):

$$(I_b): \quad \begin{cases} \dot{x} = \quad y \\ \dot{y} = -\, b\, \sin(\pi x) \end{cases} \qquad \text{for } t \in [0, \tau_b[\tag{4.19}$$

$$(I_a): \quad \begin{cases} \dot{x} = y \\ \dot{y} = a\, \sin(\pi x) \end{cases} \qquad \text{for } t \in [\tau_b, \tau_a + \tau_b[\,. \tag{4.20}$$

where $a, b > 0$ and τ_a , $\tau_b > 0$, with

$$\tau_a + \tau_b = T.$$

With this choice, the weight function is a positive constant equal to b on the time interval $[0, \tau_b[$. At the time $t = \tau_b$, it switches to $-a$ and remains at this value for a length of time τ_a . At the time T the weight function switches again to b, and so on, since a T-periodic function is considered.

The difference between the systems (I_a) and (I_b) is as follows: (I_a) is obtained from the system (I_b) by changing the constant b to $-a$ (where both b and a are positive constants). Figure 4.7 shows the phase portraits of (I_a) on the right and (I_b) on the left, in the strip $-1 \leq x \leq 2$.

Thus for the system (I_a) there are fixed points at $(0, 0)$, $(1, 0)$, $(2, 0)$, (within the domain considered) the point $(1, 0)$ being a centre and the points $(0, 0)$ and $(2, 0)$ saddle points. The system (I_b) has fixed points at $(0, 0)$, $(\pm 1, 0)$, (within the domain considered) the point $(0, 0)$ being a centre and the points $(\pm 1, 0)$ are saddle points.

These two systems of the "classical pendulum" can be easily proved that the angular speed is monotonic decreasing with respect to the energy level lines in the phase-plane, in the sense that points to the inner level lines move faster than points towards the heteroclinic level lines. This follows from the fact that the period map is strictly monotone increasing [28, Sec.V.I. Conservative Systems].

In order to apply the method of the linked twist maps we choose amongst level lines of the phase curves of (I_b) and (I_a) an inner periodic orbit as the inner curve in the annulus of the LTM and two heteroclinic orbits as the outer curve in the annulus.

The application of the stretching along the paths method to the geometry configuration of the linked twist maps is described in detail in [49] and further developed in [38] to the case of planar Hamiltonian systems. In the more recent paper [45] a further extension to reversible systems has been derived. We do not delve into these technical details at this moment because our applications in the next sections are all related to some linked twist maps geometry and then the method will be evident.

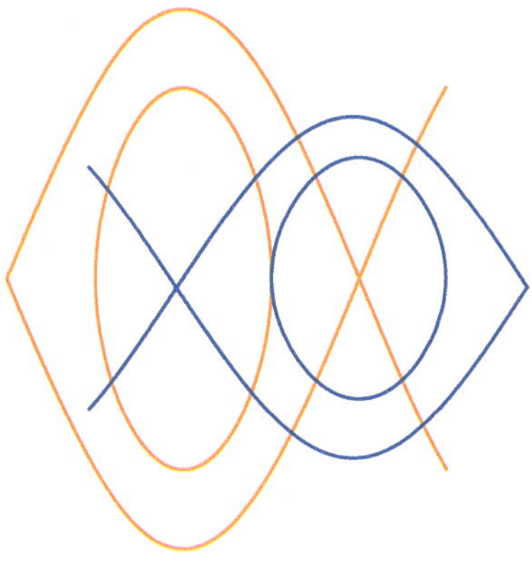

Fig. 4.7 Two linked annuli showing inner (faster) and outer (slower) energy level lines with the intersecting regions

4.5 Theorems and Lemmas

We next establish a few important theorems and lemmas integral to our approach:

Theorem 4.3 *Let $\psi : X \supset D_\psi \to X$ be a map defined on a set D_ψ and let $\widetilde{\mathcal{R}} := (\mathcal{R}, \mathcal{R}^-)$ be an oriented rectangle of a metric space X: If $\mathcal{K} \subset \mathcal{R} \cap D_\psi$ is a compact set for which*

$$(\mathcal{K}, \psi) : \widetilde{\mathcal{R}} \stackrel{\leftrightarrow}{\Leftrightarrow} \widetilde{\mathcal{R}},$$

then there is at least one point $z \in \mathcal{K}$ such that $\psi(z) = z$.

Proof By the definition of the oriented rectangle, one can find a homeomorphism g of the unit square onto $\mathcal{R}$, such that $g(\mathcal{G}) = \mathcal{R}$ and satisfying (4.17), where $\mathcal{G} := [0, 1]^2$.

The presence of a fixed point may be investigated by taking the inverse of g, and confining our search to the compact set $\mathcal{H} := g^{-1}(\mathcal{K})$ for the continuous planar map $\phi := g^{-1} \circ \psi \circ g : \mathcal{G} \supset D_\phi \to \mathbb{R}^2$, defined on $D_\phi := g^{-1}(D_\psi)$. The stretching assumption $(\mathcal{K}, \psi) : \widetilde{\mathcal{R}} \stackrel{\leftrightarrow}{\Leftrightarrow} \widetilde{\mathcal{R}}$ can now be written as

$$(\mathcal{H}, \phi) : \widetilde{\mathcal{G}} \stackrel{\leftrightarrow}{\Leftrightarrow} \widetilde{\mathcal{G}}, \tag{4.21}$$

where the unit square $\mathcal{G}$ is oriented as is done in the usual left-right manner. If w is a fixed point for ϕ in $\mathcal{H}$, then $z := g(w)$ is a fixed point for ψ in $\mathcal{K}$. In order to obtain a fixed point for $\phi = (\phi_1, \phi_2)$ in $\mathcal{G}$, the system of equations (4.22) need to be solved.

$$x_1 - \phi_1(x) = 0, \quad x_2 - \phi_2(x) = 0, \quad \text{for } x = (x_1, x_2) \in \mathcal{H}. \tag{4.22}$$

The first equation of (4.22) gives a suggestion that we introduce the compact set

$$W := \{x \in \mathcal{H} : 0 \le \phi_2(x) \le 1, \ x_1 - \phi_1(x) = 0\}. \tag{4.23}$$

The next step in the proof is to verify the following claim:

Claim: *W contains a continuum (i.e. a compact connected set) $\mathcal{C}$ which joins the lower side $[0, 1] \times \{0\}$ to the upper side $[0, 1] \times \{1\}$ of $\mathcal{G}$.*

Once the existence of such a set $\mathcal{C}$ is proved, the argument is complete. Indeed, by the definition of W, it follows that $\phi_2(z) \in [0, 1], \forall z \in \mathcal{C}$. Hence, for any $p = (p_1, p_2) \in \mathcal{C} \cap ([0, 1] \times \{0\})$ we have $p_2 - \phi_2(p) = 0 - \phi_2(p) \le 0$ and, similarly, for every $p = (p_1, p_2) \in \mathcal{C} \cap ([0, 1] \times \{1\})$ we have $p_2 - \phi_2(p) = 1 - \phi_2(p) \ge 0$. Since $x_2 - \phi_2(x)$ changes its sign (in a weak sense) along the connected set $\mathcal{C}$, by the Bolzano theorem we are guaranteed of the existence of at least a point $w = (w_1, w_2) \in \mathcal{C} \subset \mathcal{H}$ such that $w_2 - \phi_2(w) = 0$ and thus w is a solution of the second equation in (4.22). But, $w \in \mathcal{C} \subset W$ and therefore it is a solution of the first equation in (4.22) too. Therefore, w is a solution of (4.22) and is hence a fixed point of ϕ in $\mathcal{H}$. To complete the proof, one has to just verify the above-stated claim. For this purpose, we use following known property from plane topology [55] that we call the *Crossing*

Lemma as in [49]. It asserts that a compact subset of the unit square $\mathcal{G}$ which cuts any path in $\mathcal{G}$ connecting two opposite sides of the boundary of $\mathcal{G}$ must contain a continuum joining the other two sides. In other words, we have the following.

Lemma 4.1 *Let $\widetilde{\mathcal{R}} := (\mathcal{R}, \mathcal{R}^-)$ be an oriented rectangle in a metric space X and suppose that $S \subset \mathcal{R}$ is a compact set such that*

$$S \cap \bar{\gamma} \neq \emptyset,$$

for each path $\gamma : [0, 1] \to \mathcal{R}$ satisfying $\gamma(0) \in \mathcal{R}_l^-$ and $\gamma(1) \in \mathcal{R}_r^-$, then there exists a compact connected set $\mathcal{C} \subset S$ such that

$$\mathcal{C} \cap \mathcal{R}_d^+ \neq \emptyset, \quad \mathcal{C} \cap \mathcal{R}_u^+ \neq \emptyset.$$

We give only a sketch of the proof. A more detailed explanation can be found in [54] or in [47, 48].

First, using the inverse of the homeomorphism $h : \mathcal{G} \to \mathcal{R}$, we study the compact set $S \subset \mathcal{G}$ having the property of meeting any path γ contained in $\mathcal{G}$, with γ joining the left and the right sides of $\mathcal{G}$. By contradiction, if we assume that S does not contain any compact connected set $\mathcal{C}$ intersecting both the lower and the upper sides of $\mathcal{G}$, by the Kuratowski-Whyburn Lemma as used in [50, Lemma 2.8] (the details of which we do not give here but which can be seen in [36] and in [69]) we may find a decomposition of S into two disjoint compact subsets A and B such that

$$A \cap ([0, 1] \times \{0\}) \neq \emptyset, \qquad A \cap ([0, 1] \times \{1\}) = \emptyset$$

$$B \cap ([0, 1] \times \{1\}) \neq \emptyset, \qquad B \cap ([0, 1] \times \{0\}) = \emptyset.$$

The contradiction is now obtained by showing that there is a path γ contained in $\mathcal{G} \backslash S$ and joining the left and the right sides of $\mathcal{G}$. The existence of such a special path avoiding $S = A \cup B$ may be proved by different techniques of topological or combinatorial nature by the well-known Alexander lemma. (For a more detailed discussion see [26, 35, 50, 55]).

Now, by the assumption (4.21) it follows immediately that the set W fulfils the property of the set S in Lemma 4.1 and therefore it contains a continuum $\mathcal{C}$ connecting the lower side to the upper side of $\mathcal{G}$ as we claimed before. This concludes the proof. □

Remark 4.2 . The result just proved differs from the classical Brouwer fixed point theorem in the following sense. It is a well-known fact that the fixed point property for continuous maps is preserved by homeomorphisms. Therefore, any continuous self-map of a set homeomorphic to the closed unit ball of $\mathbb{R}^N$ has at least one fixed point. In particular, a continuous map ψ satisfying

$$\psi : \mathcal{R} \to \mathcal{R}, \tag{4.24}$$

with $\mathcal{R}$ a generalized rectangle of a metric space X, has a fixed point in $\mathcal{R}$. The situation depicted in Theorem 4.3 is quite different.

(1) Firstly, the stretching assumption $(\mathcal{K}, \psi) : \widetilde{\mathcal{R}} \Leftrightarrow\!\!\!\rightarrow \widetilde{\mathcal{R}}$ *does not imply* $\psi(\mathcal{R}) \subset \mathcal{R}$ and hence neither equation (4.24).

(2) Secondly, ψ needs to be continuous only on $\mathcal{K}$ and not necessarily on the whole set $\mathcal{R}$.

(3) Finally, this result also localizes the presence of a fixed point in the subset $\mathcal{K}$. This means that we are able to obtain a multiplicity of fixed points, provided the stretching condition is satisfied with respect to different compact subsets of $\mathcal{R}$. A development of this perspective is contained in Theorem 4.3 below where we also consider the iterations of a given map, in order to find periodic points. For the sake of simplicity, we discuss only the case of two disjoint compact subsets of $\mathcal{R}$ for which the stretching hypothesis is satisfied. This scenario can be extended easily to the framework of $m \geq 2$ compact pairwise disjoint subsets [51].

We next state and prove a theorem relevant to our approach:

Theorem 4.4 *Let* $\widetilde{\mathcal{R}} := (\mathcal{R}, \mathcal{R}^-)$ *be an oriented rectangle of a metric space* X *and let* $\mathcal{D} \subset \mathcal{R} \cap D_\psi$, *with* D_ψ *the domain of map* $\psi : D_\psi \to X$. *If* $\mathcal{K}_0$ *and* $\mathcal{K}_1$ *are two disjoint compact sets with* $\mathcal{K}_0 \cup \mathcal{K}_1 \subset \mathcal{D}$ *and*

$$(\mathcal{K}_i, \psi) : \widetilde{\mathcal{R}} \Leftrightarrow\!\!\!\rightarrow \widetilde{\mathcal{R}}, \quad for \;\; i = 0, 1,$$

then ψ *induces chaotic dynamics on two symbols on* $\mathcal{D}$ *relatively to* $\mathcal{K}_0$ *and* $\mathcal{K}_1$. *It follows that the map* ψ *possesses the properties* $(i) - (v)$ *of Theorem 4.2.*

The proof follows from some intermediate results. As a first step, we prove a simple lemma which shows that the "stretching" property is preserved by the composition of maps.

Lemma 4.2 *Let* $\phi : X \supset D_\phi \to X$ *and* $\psi : X \supset D_\psi \to X$ *be maps defined on the sets* D_ϕ *and* D_ψ *respectively, and let* $\widetilde{\mathcal{A}} := (\mathcal{A}, \mathcal{A}^-)$, $\widetilde{\mathcal{B}} := (\mathcal{B}, \mathcal{B}^-)$ *and* $\widetilde{\mathcal{C}} := (\mathcal{C}, \mathcal{C}^-)$ *be oriented rectangles of a metric space* X. *Assume that* $\mathcal{H} \subset \mathcal{A} \cap D_\phi$ *and* $\mathcal{K} \subset \mathcal{B} \cap D_\psi$ *are compact sets such that*

$$(\mathcal{H}, \phi) : \widetilde{\mathcal{A}} \Leftrightarrow\!\!\!\rightarrow \widetilde{\mathcal{B}}, \qquad (\mathcal{K}, \psi) : \widetilde{\mathcal{B}} \Leftrightarrow\!\!\!\rightarrow \widetilde{\mathcal{C}}.$$

Then it follows that

$$(\mathcal{H} \cap \phi^{-1}(\mathcal{K}), \psi \circ \phi) : \widetilde{\mathcal{A}} \Leftrightarrow\!\!\!\rightarrow \widetilde{\mathcal{C}}.$$

Proof Let $\gamma : [0, 1] \to \mathcal{A}$ be a path such that $\gamma(0)$ and $\gamma(1)$ belong to the different sides of $\mathcal{A}^-$. Then, since $(\mathcal{H}, \phi) : \widetilde{\mathcal{A}} \Leftrightarrow\!\!\!\rightarrow \widetilde{\mathcal{B}}$, there exists a sub-interval $[t', t''] \subset [0, 1]$ such that

$$\gamma(t) \in \mathcal{H}, \quad \phi(\gamma(t)) \in \mathcal{B} \ \forall t \in [t', t'']$$

and, moreover, $\phi(\gamma(t'))$ and $\phi(\gamma(t''))$ belong to different components of $\mathcal{B}^-$. Let ω be the restriction of γ to the sub-interval $[t', t'']$ and define $\nu := \phi \circ \omega : [t', t''] \to \mathcal{B}$. Note that $\nu(t')$ and $\nu(t'')$ belong to the different sides of $\mathcal{B}^-$ and so, by the stretching hypothesis $(\mathcal{K}, \psi) : \widetilde{\mathcal{B}} \Leftrightarrow\!\!\!\rightarrow \widetilde{\mathcal{C}}$ there exists a sub-interval $[s', s''] \subset [t', t'']$ such that

$$\nu(t) \in \mathcal{K}, \qquad \psi(\nu(t)) \in \mathcal{C}, \ \forall t \in [s', s''],$$

with $\psi(\nu(s'))$ and $\psi(\nu(s''))$ belonging to different components of $\mathcal{C}^-$. But, re-writing all in terms of γ this means that we have found a sub-interval $[s', s''] \subset [0, 1]$such that

$$\gamma(t) \in \mathcal{H} \cap \phi^{-1}(\mathcal{K}), \qquad \psi(\phi(\gamma(t))) \in \mathcal{C}, \quad \forall t \in [s', s'']$$

and $\psi(\phi(\gamma(s')))$ and $\psi(\phi(\gamma(s'')))$ belong to the different sides of $\mathcal{C}^-$. By the arbitrariness of the path γ, the stretching property

$$(\mathcal{H} \cap \phi^{-1}(\mathcal{K}), \phi \circ \psi) : \widetilde{\mathcal{A}} \Leftrightarrow\!\!\!\rightarrow \widetilde{\mathcal{C}}$$

is thus proved. We just point out that the continuity of the composite mapping $\psi \circ \phi$ on the compact set $\mathcal{H} \cap \phi^{-1}(\mathcal{K})$ follows from the continuity of ϕ on $\mathcal{H}$ and of ψ on $\mathcal{K}$, respectively. □

Lemma 4.3 *Let $\psi : X \supset D_\psi \to X$ be a map defined on a set D_ψ and let $\widetilde{\mathcal{R}} := (\mathcal{R}, \mathcal{R}^-)$ be an oriented rectangle of a metric space X. If $\mathcal{K}_0$ and $\mathcal{K}_1$ are two disjoint compact sets with $\mathcal{K}_0 \cup \mathcal{K}_1 \subset \mathcal{R} \cap D_\psi$ and*

$$(\mathcal{K}_i, \psi) : \widetilde{\mathcal{R}} \Leftrightarrow\!\!\!\rightarrow \widetilde{\mathcal{R}}, \quad for\ i = 0, 1,$$

then the following conclusions hold:

- *The map ψ has at least a fixed point in $\mathcal{K}_i$, $i = 0, 1$;*
- *Given an integer $j \geq 2$ and a $j + 1 - tuple$ $(s_0, \cdots, s_j)$ with each $s_i \in \{0, 1\}, i = 0, \cdots, j$ and $s_0 = s_j$ then there exists a point $w \in K_{s_0}$ such that*

$$\psi^i(w) \in \mathcal{K}_{s_i}, \ \forall i = 0, \cdots, j \ \ and \ \ \psi^j(w) = w.$$

Proof The first conclusion is an immediate consequence of Theorem 4.3. As regards the second assertion, it also follows from Theorem 4.3 applied this time to the composite mapping $\phi := \psi^j$. Indeed let us consider the compact set

$$\mathcal{H} := \{x \in \mathcal{K}_{s_0} : \psi^I(x) \in \mathcal{K}_{s_i}, \ \forall i = 1, \cdots, j\} \subset \mathcal{K}_{s_0}$$

Then, by Lemma 4.2 it can be shown that the stretching relation

$$(\mathcal{H}, \phi) : \widetilde{\mathcal{R}} \Leftrightarrow\!\!\!\rightarrow \widetilde{\mathcal{R}}$$

is fulfilled and now the thesis follows from Theorem 4.3 and by the definition of the compact set $\mathcal{H}$. □

This lemma already gives a partial proof for Theorem 4.4, regarding the existence of periodic points. In order to complete the proof of Theorem 4.4 we need another lemma stated now:

Lemma 4.4 *Let $\psi : X \supset D_\psi \to X$ be a map defined on a set D_ψ and let $\widetilde{\mathcal{R}} := (\mathcal{R}, \mathcal{R}^-)$ be an oriented rectangle in a metric space X. Assume that $(\mathcal{K}_j)_{j\in\mathbb{Z}}$ is a sequence of compact sets with $\mathcal{K}_j \subset \mathcal{R} \cap D_\psi$ and*

$$(\mathcal{K}_j, \psi) : \widetilde{\mathcal{R}} \Leftrightarrow\!\!\!\rightarrow \widetilde{\mathcal{R}}, \qquad \forall j \in \mathbb{Z}.$$

Then the following conclusions hold:

- *There is a sequence of points $(w_j)_{j\in\mathbb{Z}}$ such that $w_j \in \mathcal{K}_j$ and $\psi(w_j) = w_{j+1}$, for all $j \in \mathbb{Z}$;*
- *For any $l < m$ couple of integers such that $\mathcal{K}_l = \mathcal{K}_m$, there exists a finite sequence of points $(z_i)_{l\le i\le m}$, with $z_i \in \mathcal{K}_i$ and $\psi(z_i) = z_{i+1}$ for each $i = l, \cdots, m-1$, such that $z_m = z_l$ that is, z_l is a fixed point of ψ^{m-l}.*

Proof We prove the second point first. By Lemma 4.2 it holds that

$$(\mathcal{H}, \psi^{m-l}) : \widetilde{\mathcal{R}} \Leftrightarrow\!\!\!\rightarrow \widetilde{\mathcal{R}}, \tag{4.25}$$

where

$$\mathcal{H} := \{x \in \mathcal{K}_l : \psi^{i-l}(x) \in \mathcal{K}_i, \ \forall i = l+1, \cdots, m\}.$$

Setting $\phi := \psi^{m-l}$, we can now write condition (4.25) as $(\mathcal{H}, \phi) : \widetilde{\mathcal{R}} \Leftrightarrow\!\!\!\rightarrow \widetilde{\mathcal{R}}$ and therefore the thesis follows by Theorem 4.3 and by the definition of the compact set $\mathcal{H}$. Note that ϕ is continuous on $\mathcal{H} \subset \mathcal{K}_l$ by the continuity of the map ψ on each $\mathcal{K}_i$, $i = l, \cdots, m-1$. Let's turn now to the first assertion. In analogy to what was done before, let us define the closed set

$$\mathcal{S} := \{x \in \mathcal{K}_0 : \psi^i(x) \in \mathcal{K}_i, \ \forall i \ge 1\} \tag{4.26}$$

and fix a path $\gamma : [0, 1] \to \mathcal{R}$ with $\gamma(0)$ and $\gamma(1)$ belonging to different components of $\mathcal{R}^-$. Then, since $(\mathcal{K}_0, \psi) : \widetilde{\mathcal{R}} \Leftrightarrow\!\!\!\rightarrow \widetilde{\mathcal{R}}$, there exists a sub-path $\gamma_0 : [t_0', t_0''] \to \mathcal{K}_0$ of γ such that $\psi(\gamma_0([t_0', t_0''])) \subset \mathcal{R}$, with $\psi(\gamma_0(t_0'))$ and $\psi(\gamma_0(t_0''))$ belonging to different sides of $\mathcal{R}^-$. Similarly, since $(\mathcal{K}_1, \psi) : \widetilde{\mathcal{R}} \Leftrightarrow\!\!\!\rightarrow \widetilde{\mathcal{R}}$ we can find a sub-path $\omega_1 : [t_1', t_1''] \to \mathcal{K}_1$ of $\omega_0 := \psi(\gamma_0)$ such that $\psi(\omega_1([t_1', t_1''])) \subset \mathcal{R}$, with $\psi(\omega_1(t_1'))$ and $\psi(\omega_1(t_1''))$ belonging to different components of $\mathcal{R}^-$. Setting

$$\Gamma_1 := \{x \in \gamma_0([t_0', t_0'']) : \psi(x) \in \omega_1([t_1', t_1''])\} \subset \{x \in \mathcal{K}_0 : \psi(x) \in \mathcal{K}_1\}$$

and proceeding by induction, we obtain a decreasing sequence of nonempty compact sets

$$\Gamma := \gamma([0, 1]) \supset \Gamma_0 := \gamma_0([t_0', t_0'']) \supset \cdots \supset \Gamma_n \supset \Gamma_{n+1} \supset \cdots$$

such that $\psi^{j+1}(\Gamma_j)$ joins the two components of $\mathcal{R}^-$, for $j \geq 0$. Moreover, for every $i \geq 1$, it holds that

$$\Gamma_i \subset \{x \in \mathcal{K}_0 : \psi^j(x) \in \mathcal{K}_j, \quad \forall j \ : 1 \leq j \leq i\}.$$

By the Cantor Lemma we have that $\bigcap_{j=0}^{+\infty} \Gamma_j \neq \emptyset$ and for any $w \in \bigcap_{j=0}^{+\infty} \Gamma_j$ it holds that $\psi^n(w) \in \mathcal{K}_n, \ \forall n \in \mathbb{N}$. Thus the set $\mathcal{S}$ is nonempty. The thesis follows by a standard diagonal argument which allows to extend the result to bi-infinite sequences once it has been proved for one-sided sequences. □

We now state the following theorem which is central to our proof of the presence of chaotic dynamics.

Theorem 4.5 ([50]) *Let* $\psi_r : X \supset D_{\psi_r} \to X$ *and* $\psi_s : X \supset D_{\psi_s} \to X$ *be given maps. Let* $\widetilde{\mathcal{M}} := (\mathcal{M}, \mathcal{M}^-)$ *and* $\widetilde{\mathcal{N}} := (\mathcal{N}, \mathcal{N}^-)$ *be oriented rectangles in* X *with* $\mathcal{M} \subset D_{\psi_r}$ *and* $\mathcal{N} \subset D_{\psi_s}$. *Suppose that the following conditions are satisfied:*

($\mathbf{H_r}$) *There exists* $m \geq 2$ *pairwise disjoint compact sets* $\mathcal{K}_1, \ldots, \mathcal{K}_m \subset \mathcal{M}$ *such that*

$$(\mathcal{K}_i, \psi_r) : \widetilde{\mathcal{M}} \Leftrightarrow\!\!\!\!\rightarrow \widetilde{\mathcal{N}}, \quad \forall i = 1, \ldots, m;$$

($\mathbf{H_s}$) $\psi_s : \widetilde{\mathcal{N}} \Leftrightarrow\!\!\!\!\rightarrow \widetilde{\mathcal{M}}$.

Then the map $\psi := \psi_s \circ \psi_r$ *induces chaotic dynamics on* m *symbols in the set*

$$\mathcal{K} := \bigcup_{i=1}^{m} \mathcal{K}_i .$$

Moreover, for each sequence of m *symbols* $\mathbf{s} = (s_n)_n \in \{1, \ldots, m\}^{\mathbb{N}}$, *there exists a compact connected set* $\mathcal{C}_s \subset \mathcal{K}_{s_0}$ *with*

$$\mathcal{C}_s \cap \mathcal{M}_b^+ \neq \emptyset, \quad \mathcal{C}_s \cap \mathcal{M}_t^+ \neq \emptyset$$

and such that, for every $w \in \mathcal{C}_s$ *there exists a sequence* $(y_n)_n$ *with* $y_0 = w$ *and* $y_n \in \mathcal{K}_{s_n}, \quad \psi(y_n) = y_{n+1}, \ \forall n \geq 0$.

An explanation of the concepts ($\mathbf{H_r}$) and ($\mathbf{H_s}$) can be visualized with the help of Figs. 4.8 and 4.9.

We would like to explain the figure by giving the condition ($\mathbf{H_r}$). In Fig. 4.8 there is a path across two disjoint compact sets $\mathcal{K}_1$ and $\mathcal{K}_2, \subset \widetilde{\mathcal{M}}$ with two sub-paths shown by darkened lines one in each compact set. Following the Definition 4.5 of *stretching along paths* we see that the map ψ_r stretches these two sub-paths across $\widetilde{\mathcal{N}}$, shown as two darkened vertical paths intersecting $\mathcal{N}_l^-$ and$\mathcal{N}_r^-$. In Fig. 4.9, for ($\mathbf{H_s}$), a sub-path shown as a darkened part of a vertical path cutting $\mathcal{N}_l^-$ and $\mathcal{N}_r^-$, is stretched

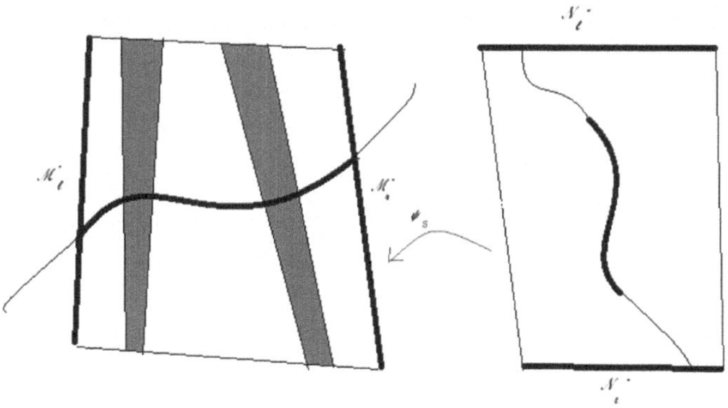

Fig. 4.8 The condition ($\mathbf{H_r}$)

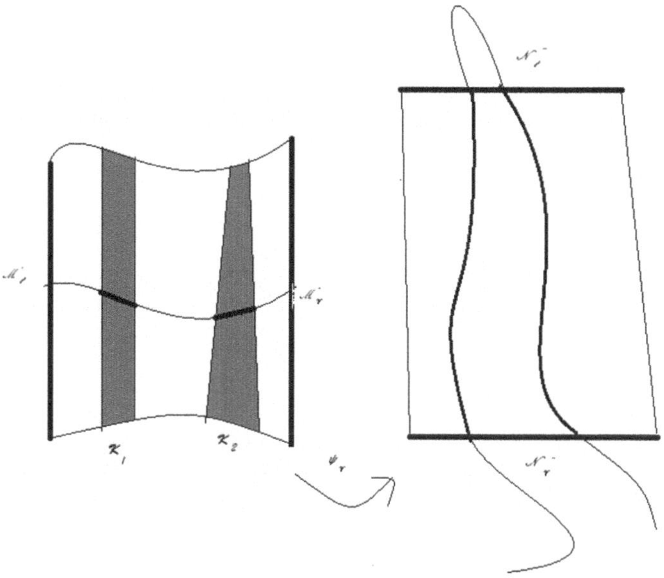

Fig. 4.9 The condition ($\mathbf{H_s}$)

across $\widetilde{\mathcal{M}}$ in the correct direction, under the stretching action of the map ψ_s, which is across the ends $\mathcal{M}_l^-$and$\mathcal{M}_r^-$. The proof follows from Theorem 4.4 applied to the set $\mathcal{R} = \mathcal{M}$. The assertion about the compact connected set $\mathcal{C}_\mathbf{s}$ is a consequence of Lemma 4.1. See [50] for a more detailed proof and [49] for some remarks and extensions. In the applications of Theorem 4.5 to the ODE models considered in the present monograph, we take $X \equiv \mathbb{R}^2$ and the maps ψ_r , ψ_s to be the Poincaré maps associated with some planar systems for different time intervals.

- An application related to this method is presented in detail in Chap. 6, Sect. 6.5. Therein, the topological rectangles are obtained as intersections of annular regions constructed completely with periodic orbits of autonomous systems satisfying a twist condition at the boundary due to the monotonicity of the period map.
- In Chap. 5 the same method is used. But the technical details are only outlined in order to avoid unnecessary repetitions.
- In Chap. 7, the construction is slightly different because the annular regions are bounded by periodic and homoclinic orbits. In this case, the requirement of monotonicity for the period map is not needed for the twist condition. Also in this chapter, we provide the technical details for the implementation of our method.

See also [39] for further reading.

4.6 The Melnikov Method

We give here some information about the Melnikov method which is an other classical analytical tool to prove in a rigorous manner chaotic dynamics in a differential system. For more details see [27, 71].

Stable/Unstable Manifolds of Fixed Points

Let $f\colon \mathbb{R}^2 \to \mathbb{R}^2$ be a sufficiently smooth map and assume that $p \in \mathbb{R}^2$ is a fixed point of f. Let $U \subset \mathbb{R}^2$ be a neighbourhood of p, and define the set

$$W_1^s(p, U) = \{w \in U : f^n(w) \in U \text{ for all } n \geq 0\}.$$

This set is referred to as the local stable set of p relative to U.

Assume that is p is a hyperbolic fixed point, (as discussed in Sect. (4.2)) i.e. if none of the eigenvalues of $Df(p)$ are on the unit circle, which is also a saddle point. Then the stable manifold theorem gives that there exists a neighbourhood $U \subset \mathbb{R}^2$ of p so that $W_1^s(p, U)$ is a 1-dimensional embedded arc tangent to the stable eigenspace at p. If f is analytic then the embedding is analytic.

The invariant set

$$W^s(p) = \left\{w \in \mathbb{R}^2 : \lim_{n\to\infty} f^n(w) = p\right\},$$

is the stable manifold of p, and consists of all points which accumulate under forward iteration of the map to the fixed point p. If f is invertible then we have that

$$W^s(p) = \bigcup_{n=0}^{\infty} f^{-n}\left[W_1^s(p, U)\right],$$

and the stable manifold is obtained as the union of all backwards iterates of a local stable manifold.

We say that the sequence $\{x_j\}_{j=-\infty}^{0}$ is a backward orbit of x_0 if

$$f(x_j) = x_{j+1},$$

for all $j \leq -1$. If $U \subset \mathbb{R}^2$ and $x_j \in U$ for all $j \leq 0$ we say that x_0 has a backward orbit in U. If

$$\lim_{j \to -\infty} x_j = p,$$

we say that x_0 has a backward orbit accumulating at p. Let $U \subset \mathbb{R}^2$ be an open neighbourhood of p and define the set

$$W_1^u(p, U) = \{w \in U : w \text{ has a backward orbit in } U\}.$$

We refer to this as a local unstable manifold of p relative to U. Since we assumed that p is hyperbolic, the unstable manifold theorem gives that there exists an open neighbourhood U of p so that $W_1^u(p, U)$ is a 1- dimensional embedded arc, tangent at p to the unstable eigenspace of $Df(p)$. If f is analytic the embedding is analytic. If f is invertible then the unstable manifold of p under f can be seen as the stable manifold of p under f^{-1}. In the case that f is invertible, the unstable manifold of p is

$$\begin{aligned} W^u(p) &= \left\{w \in \mathbb{R}^2 : \lim_{n \to \infty} f^{-n}(w) = p\right\} \\ &= \bigcup_{n=0}^{\infty} f^n\left[W_1^u(p, U)\right]. \end{aligned}$$

The study of a nonlinear dynamical system begins by considering the fixed/periodic points, their linear stability, and their local stable and unstable manifolds. In order to incorporate this local information into a global picture of the dynamics one then wants to understand the connecting orbits. Questions about connecting orbits are naturally reformulated as questions about where and how the stable and unstable manifolds intersect. Continuous dynamical systems do not have transverse homoclinic orbits, (this is the case in the left most panel in Fig. 4.10) in the sense which will be described below. But the Poincaré map and its iterates of a continuous non-autonomous dynamical system could be considered as an example of this. If we consider the unstable manifold around the point of intersection, and map it forward in time via the Poincaré map the images of this piece under the iteration of the Poincaré map eventually follow the unstable manifold to come back to intersect the stable manifold again. This is illustrated in Fig. 4.10. The figure shows on the left, the stable and unstable manifolds coinciding. In this case, p is a homoclinic point. The figure in the middle shows the stable and the unstable manifolds intersecting transversally at the point q which is not a fixed point. The figure on the right depicts a hypothetical behaviour of the stable and unstable manifolds after the intersection point q with a sequence of other transversal intersection points accumulating back to p.

Likewise, the stable manifold around the point of transversal intersection (see the centre panel in Fig. 4.10) is mapped backward, eventually following the stable manifold to induce new intersections. Consequently, the stable and unstable manifold forms a web, the structure of which is very complicated. A phenomenon first discovered by Poincaré himself albeit accidentally.

A typical situation is that we consider a hyperbolic fixed point, so that the dimension of the stable manifold plus the dimension of the unstable manifold add up to the dimension of the whole space. A transverse intersection of the stable/unstable manifolds of a hyperbolic fixed point is again a single point. Such intersections give rise to homoclinic connecting orbits, and they are of special interest. Smale's *tangle* theorem [44, 58] says that the existence of a transverse homoclinic intersection point implies the existence of hyperbolic chaotic motions.

Melnikov's method is a well-established technique for detecting homoclinic bifurcation of perturbed autonomous or forced systems by providing an analytic criterion for the onset of chaos based on transversal homoclinic intersections. A planar continuous dynamical system generated by an autonomous differential equation with a hyperbolic equilibrium point, connected to itself by a homoclinic orbit, can behave chaotically when perturbed by a time-periodic function. Melnikov's method can be used to check whether the system possesses chaotic dynamics. The method is based on the so-called Melnikov function.

When considering a two-dimensional dynamical system, a time-periodic perturbation of this system may cause chaotic behaviour of the solutions. Usually, transversal intersections of stable and unstable manifolds are the cause of this behaviour. When an unperturbed two-dimensional system contains a hyperbolic fixed point that is connected to itself by a homoclinic orbit, this homoclinic orbit can be interpreted as the union of the unstable and the stable manifold, and the fixed point itself. The perturbation can cause the stable and the unstable manifold to split up and intersect transversely. When we consider a perturbed two-dimensional dynamical system that is periodic in time, the homoclinic Melnikov method can be used to prove that there exist transverse homoclinic orbits to hyperbolic periodic orbits.

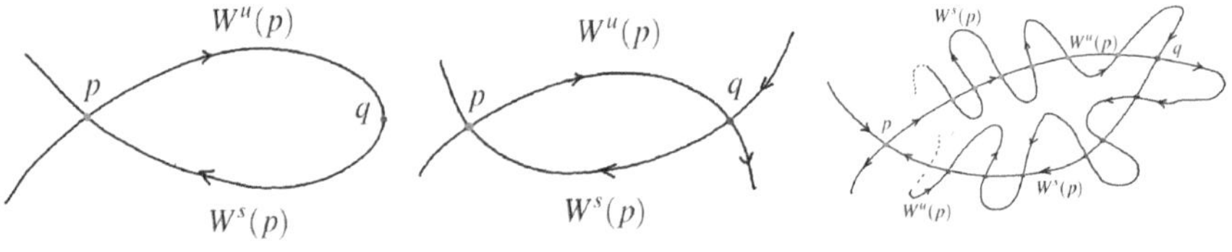

Fig. 4.10 The stable and the unstable manifold at a fixed point p

Using the Melnikov method, firstly, the unperturbed system must be considered. The homoclinic manifold, which could be expressed in terms of the union of the unstable manifold, the stable manifold, and the hyperbolic fixed point is analysed, and then the Melnikov function is derived. In the next section an example of an application of the Melnikov to the Duffing equation is shown.

The Melnikov Method Applied to the Duffing Equation

Let us recall the autonomous Duffing equation

$$\ddot{x} + g(x) = 0 \tag{4.27}$$

and its equivalent planar conservative system is given by the two-dimensional system

$$\begin{cases} \dot{x} = y \\ \dot{y} = -g(x). \end{cases} \tag{4.28}$$

Here, as before, $g : J \to \mathbb{R}$ is a locally Lipschitz continuous function defined on an open interval $J \subset \mathbb{R}$ where g vanishes at least at a point. Without loss of generality (using, if necessary, a simple change of variables) we can suppose that $0 \in J$ and $g(0) = 0$, so that the origin is an equilibrium point of (4.28).

As we saw earlier, system (4.28) is a planar Hamiltonian system with associated Hamiltonian (energy function)

$$H(x, y) = \frac{1}{2}y^2 + G(x), \qquad G(x) = \int_0^x g(s)\, ds.$$

Also as seen earlier, we have that if $(x(t), y(t))$ is any solution of (4.28) then $(x(t), y(t)) \in \Gamma^c$ for all t, where

$$\Gamma^c := \{(x, y) \in \mathbb{R}^2 : H(x, y) = c\}, \quad \text{for } c = H(x(0), y(0)).$$

A large volume of literature has been devoted to the investigation of the perturbations of (4.27) by periodic forcing terms, as well as by damping terms, leading to

$$\ddot{x} + g(t, x) = 0 \tag{4.29}$$

and to

$$\ddot{x} + c\dot{x} + g(t, x) = 0, \tag{4.30}$$

respectively. Just to recall, in (4.29) and (4.30) the nonlinear restoring force $g : \mathbb{R} \times J \to \mathbb{R}$ is a continuous function, locally Lipschitzian in the x-variable and periodic in the t-variable of a fixed period $T > 0$. The domain J for the x-variable is an open interval of the real line. The most common case is given by $J = \mathbb{R}$, however research has been devoted also to the case when $J \neq \mathbb{R}$, in order to describe some physical models where the points at the boundary of J represent possible singularities of the vector field.

To study chaotic-like solutions in [27, Sect. 4.5] the Melnikov method is applied to the perturbed counterpart of equation (4.27). That is, a special form of (4.30),

$$\ddot{x} + \varepsilon c\dot{x} + g(x) = \varepsilon p(\omega t), \tag{4.31}$$

where $c, \omega > 0$ and $p(\cdot)$ is a nonconstant periodic forcing term. According to [27], the functions $g(x)$ and $p(t)$ should be sufficiently smooth (of class C^r with $r \geq 2$). The main assumption in [27] requires the presence of a homoclinic solution $\gamma_0(t) = (q_0(t), v_0(t))$ at the origin for the autonomous system (4.28). Then the Melnikov function associated with $\gamma(t)$, is introduced as

$$M(\alpha) := \int_{-\infty}^{+\infty} \left(v_0(t)p(\omega(t+\alpha)) - cv_0(t)^2\right) dt.$$

Melnikov's theory (cf. [27, Theorem 4.5.3]) guarantees that, if $M(\alpha)$ has a simple zero for some $\alpha = \alpha_0$, then, for $\varepsilon > 0$ sufficiently small, there exists a hyperbolic fixed point $P_\varepsilon^{\alpha_0}$ near the origin for system

$$\begin{cases} \dot{x} = y \\ \dot{y} = -g(x) - \varepsilon p(\omega t) - \varepsilon c y \end{cases} \tag{4.32}$$

such that its stable and unstable manifolds $W^s(P_\varepsilon^{\alpha_0})$ and $W^u(P_\varepsilon^{\alpha_0})$ intersect transversely. This property, in turn, via the Smale-Birkhoff theorem implies that some N-th iterate of the Poincaré map Φ_ε associated with system (4.32), has a Smale horseshoe and therefore it has an invariant set which is conjugate to the Bernoulli shift on a set of symbols. As a consequence, Φ_ε^N presents chaotic dynamics in the sense of Devaney. Further relevant results in this directions can be found in [11] for the equation

$$\ddot{x} + \lambda a(t)g(x) = 0$$

and

$$\ddot{x} + g(x) = \varepsilon p(t).$$

The Melnikov method, although very powerful and relevant for the point of view of the applications is typically applied in situations where the unperturbed Duffing system (4.28) possesses homoclinic or heteroclinic solutions. Moreover, the analysis of the Melnikov function $M(\alpha)$ or its variants can be extremely difficult when an explicit analytic form of the homoclinic solution $\gamma(t)$ is not known. Finally, conclusions may be deduced typically for small perturbations, namely, for $\varepsilon > 0$ sufficiently small.

To illustrate the Melnikov method we consider a classical application of this method to the periodically perturbed Duffing equation (4.31) for the choice of g given by

$$g(x) = -x + x^3. \tag{4.33}$$

Considering the associated autonomous system (4.28), the origin is a saddle point, and the energy level line $H(x, y) = 0$, that is

$$\frac{1}{2}y^2 - \frac{1}{2}x^2 + \frac{1}{4}x^4 = 0$$

consists of two homoclinic orbits

$$\Gamma_+^0 := \{(x, y) : x > 0, H(x, y) = 0\}, \quad \Gamma_-^0 := \{(x, y) : x < 0, H(x, y) = 0\}$$

and the origin. Each of the orbits $\Gamma_\pm^0$ is the locus of a homoclinic solution $\gamma_0^\pm(t) = (q_0^\pm(t), v_0^\pm(t))$, expressed explicitly by the following parametric equations

$$q_0^+(t) = \sqrt{2}\operatorname{sech} t, \quad v_0^+(t) = -\sqrt{2}\operatorname{sech} t \tanh t$$

and $\gamma_0^-(t) = -\gamma_0^+(t)$.

Then, in case of (4.32) with $g(x)$ as in (4.33), we can explicitly compute the Melnikov function as

$$\begin{aligned} M(\alpha) &= \int_{-\infty}^{+\infty} \left(v_0(t)p(\omega(t+\alpha)) - cv_0(t)^2\right) dt \\ &= -\sqrt{2}\int_{-\infty}^{+\infty} \operatorname{sech} t \tanh t\, p(\omega(t+\alpha))\, dt - 2c\int_{-\infty}^{+\infty} \operatorname{sech}^2 t \tanh^2 t\, dt \end{aligned}$$

Using the fact that

$$\int_{-\infty}^{+\infty} \operatorname{sech}^2 t \tanh^2 t\, dt = \frac{2}{3},$$

we obtain the explicit formula

$$M(\alpha) = -\frac{4}{3}c - \sqrt{2}\int_{-\infty}^{+\infty} \operatorname{sech} t \tanh t\, p(\omega(t+\alpha))\, dt.$$

This latter expression can be evaluated whenever a specific analytical expression of $p(t)$ is known. For instance, following [27], for

$$p(t) := E \cos t,$$

the following form of $M(\alpha)$ is obtained.

$$\begin{aligned} M(\alpha) &= -\frac{4}{3}c - \sqrt{2}E \int_{-\infty}^{+\infty} \operatorname{sech} t \tanh t \cos(\omega(t+\alpha))\, dt \\ &= -\frac{4}{3}c - \sqrt{2}E\cos(\omega\alpha) \int_{-\infty}^{+\infty} \operatorname{sech} t \tanh t \cos(\omega t)\, dt \\ &\quad + \sqrt{2}E\sin(\omega\alpha) \int_{-\infty}^{+\infty} \operatorname{sech} t \tanh t \sin(\omega t)\, dt. \end{aligned}$$

Since the function $t \mapsto \operatorname{sech} t \tanh t$ is odd, the first integral vanishes. Concerning the second integral, we observe that the function $t \mapsto \operatorname{sech} t \tanh t \sin(\omega t)$ is even, so that

$$\int_{-\infty}^{+\infty} \operatorname{sech} t \tanh t \sin(\omega t)\, dt = 2\int_{0}^{+\infty} \operatorname{sech} t \tanh t \sin(\omega t)\, dt := K(\omega).$$

The analytical expression of $K(\omega)$ can be found using the method of the residues (cf. [27]) and is given by

$$K(\omega) = \pi\omega \operatorname{sech}\left(\frac{\pi\omega}{2}\right).$$

In conclusion, for (4.32) with $g(x)$ as in (4.33) and in the special case of $p(t) = E\cos(t)$, we obtain

$$M(\alpha) = \sqrt{2}EK(\omega)\sin(\omega\alpha) - \frac{4}{3}c.$$

This function has simple zeros in the variable α and therefore the Melnikov method guarantees the presence of chaotic dynamics for an N-th iterate of the Poincaré map associated with system (4.32) for $\varepsilon > 0$ sufficiently small, if an only if

$$\frac{E}{c} > \frac{4}{3\sqrt{2}K(\omega)}.$$

As one can see from this example taken from [27], the Melnikov method in principle can be applied to a wide class of equations; on the other hand, it may be very difficult to prove the result about the existence of simple zeros for the function $M(\alpha)$ without knowing the specific analytical expression of the homoclinic solution.

A Second Example

As another example of the application of the Melnikov method, we consider a problem analysed in [60].

Consider the equation

$$\ddot{u} + f(u) = p(t), \tag{4.34}$$

where $p(t)$ is a T-periodic forcing term and the nonlinearity $f(u)$ is a positive function, with global minimum at $u = 0$, which is decreasing on the negative reals and increasing on the positive ones. The counterpart of (4.34) with a damping term

$$\ddot{u} + c\dot{u} + f(u) = p(t),$$

may also be studied as well. In this example, the graph of the function f has the shape of a parabola, and our study is related to the so-called Ambrosetti-Prodi problem [3] for the periodic case. In particular we consider the equation (4.34) as a perturbation of $\ddot{u} + f(u) = k$ where $k \in \mathbb{R}$. For $k > 0$ the associated planar phase portrait is that of a local centre enclosed by a homoclinic trajectory of a hyperbolic saddle point, and such saddle-centre geometry gives a hint for the application of a Melnikov-type approach. Let us consider this system in the phase-plane. Equation (4.34) can be written equivalently as a planar system in the phase-plane (x, y):

$$\begin{cases} \dot{x} = y, \\ \dot{y} = -f(x) + k. \end{cases} \tag{4.35}$$

As a possible illustrative example of the function $f(x)$ could be given by $f(u) = \sqrt{1 + u^2} - 1$.

In [11], Battelli and Palmer found a general result for the Duffing equation

$$\ddot{x} + a^2 g(x) = p(t), \tag{4.36}$$

where $p(t)$ is T-periodic and both $g(x)$ and $p(t)$ are sufficiently smooth (at least of class C^{r+3} for $r \geq 5$). Assuming the existence of homoclinic or heteroclinic points $z^{\pm}$ with

$$g(z^{\pm}) = 0, \quad g'(z^{\pm}) < 0$$

for the autonomous equation

$$\ddot{x} + g(x) = 0,$$

the authors in [11] proved the existence of transversal intersection (as in Melnikov theorem) for the Poincaré map associated with (4.36), provided that $a > 0$ is sufficiently large and one of the following conditions holds:

- $z^+ \neq z^-$ and there exists t_0 such that $p(t_0) = 0 \neq \dot{p}(t_0)$: *The case of heteroclinic points*;
- $z^+ = z^-$ and there exists t_0 such that $\dot{p}(t_0) = 0 \neq \ddot{p}(t_0)$: *The case of homoclinic points*.

The advantage of this result with respect to the classical Melnikov formulation (as presented in [27]) depends on the fact that an explicit analytic expression for the homoclinic/heteroclinic solutions is not needed. There are still, however, some assumptions on the forcing term which require simple zeros or simple zeros for the derivative.

We propose a different application by assuming the following conditions. (H_1) $f : \mathbb{R} \to \mathbb{R}$ *is a strictly convex function of class* C^r, *for* $r \geq 2$, *with* $f(0) = 0$, $f(s) > 0$ *for all* $s \neq 0$ *and* $\lim_{|s|\to+\infty} f(s) = +\infty$.

Then, we consider the Melnikov function associated with system (4.34) for

$$p(t) = k + \epsilon p_0(t), \quad k > 0.$$

Without loss of generality, we can also suppose that $p_0(t)$ changes its sign. In particular, by transferring the mean value of p_0 to the constant k, we can assume

$$\int_0^T p_0(t)\, dt = 0. \tag{4.37}$$

Now the Melnikov function is given by

$$\Delta(\alpha) := \int_{-\infty}^{+\infty} \dot{q}(t) p_0(t + \alpha)\, dt. \tag{4.38}$$

Notice that, by the T-periodicity of $p_0(t)$, it turns out that also $\Delta(\alpha)$ is a T-periodic function and $q(t)$ is a homoclinic solution of Equation (4.35). Moreover, from (4.37) we have $\int_0^T \Delta(\alpha)\, d\alpha = 0$, so that either $\Delta \equiv 0$ or $\Delta(\alpha)$ changes its sign. An application of the Melnikov method to system

$$\begin{cases} \dot{x} = y, \\ \dot{y} = -f(x) + k + \varepsilon p_0(t), \end{cases} \tag{4.39}$$

gives the following result (cf. [27, Th. 4.5.3] or [71, Th. 28.1.7]).

Theorem 4.6 *Assume* (H_1) *and let* $(q(t), \dot{q}(t))$ *be the homoclinic solution at the saddle point for the autonomous system* (4.35) *for some* $k > 0$. *Let also* p_0 *be a sufficiently smooth,* C^r *for* $r \geq 2$, T*-periodic function satisfying* (4.37). *If there exists* $\alpha \in [0, T[$ *such that* $\Delta(\alpha) = 0$ *and* $\Delta'(\alpha) \neq 0$, *then there is* $\varepsilon_0 > 0$ *such that for each* ε *with* $0 < |\varepsilon| < \varepsilon_0$ *a Smale horseshoe occurs for some iterate of the Poincaré map associated with system* (4.39).

It may be interesting to compare Theorem 4.6 with [25, Cor:1]. In [25] the authors study a general class of equations including those of the form

$$\ddot{x} + h(x)\dot{x} + f(x) = k + \epsilon p_0(t).$$

Under the assumption $f(x) \to +\infty$ for $x \to \pm\infty$, the following result is proved in [25]: $\exists\ k_0$ such that

$$\begin{cases} k < k_0 & \text{there are no T-periodic solutions} \\ k = k_0 & \text{there is at least one T-periodic solution} \\ k > k_0 & \text{there are at least two T-periodic solutions.} \end{cases}$$

Our contribution shows that for $k > k_0$ when there are two periodic solutions, we can also have chaotic dynamics via the Melnikov method if ϵ is sufficiently small.

The phase-plane of (4.35) shows an equilibrium point which is a centre, and a hyperbolic saddle equilibrium point with a homoclinic orbit enclosing the centre. As mentioned above, this is the classical scheme considered in Melnikov's theory, where system (4.35) can be viewed as a perturbation of the autonomous system (4.34). Melnikov's method is a powerful tool and "one of the few analytical methods available for the detection and the study of chaotic motions" (quoting [27, p. 186]). The result expressed in Theorem 4.6 is robust for small smooth perturbations. More in detail, the presence of a Smale horseshoe is guaranteed also for the system

$$\begin{cases} \dot{x} = y, \\ \dot{y} = -cy - f(x) + k + \varepsilon p_0(t), \end{cases}$$

provided that c is sufficiently small, depending on ε. Hence the result applies to equation

$$\ddot{u} + c\dot{u} + f(u) = k + \varepsilon p_0(t)$$

as well. More precisely, if we write the coefficient c as

$$c := \varepsilon c_0 ,$$

the Melnikov function takes the form

$$\Delta(\alpha) := \int_{-\infty}^{+\infty} \left(\dot{q}(t) p_0(t+\alpha) - c_0 \dot{q}(t)^2\right) dt \tag{4.40}$$

and Theorem 4.6 applies to system

$$\begin{cases} \dot{x} = y, \\ \dot{y} = -\varepsilon c_0 y - f(x) + k + \varepsilon p_0(t). \end{cases}$$

Usually, the test of the existence of a simple zero for the Melnikov function is a difficult task, especially if an explicit analytical expression for $q(t)$ is not given or is sometimes difficult to obtain. The first important and pioneering applications of this method to some second-order nonlinear ODEs, such as the pendulum or the Duffing equation, have taken advantage of the fact that the expression of $q(t)$ was known (see [27, p. 191]).

As we mentioned before, a general result, which does not require any specific assumption on $q(t)$ by involving only a simply verifiable condition on $p_0(t)$, was obtained by Battelli and Palmer in [11]. This result applies to system (4.39) provided that the period of the forcing term is sufficiently large.

We show now an application of Theorem 4.6 to the AP (Ambrosetti-Prodi) periodic problem [3], following [60]. In order to do this, given $\omega > 0$, we suppose that $p_0(t) := \sin(\omega t)$ is the periodic forcing term of period $T := 2\pi/\omega$. Using the properties of $p_0(t)$ we can show the following consequence.

Corollary 4.1 *Assume (H_1) and let $(q(t), q'(t))$ be the homoclinic solution at the saddle point for the autonomous system (4.34) for some $k > 0$. Then, for any $\omega > 0$ there exists $\varepsilon_0 = \varepsilon_0(\omega) > 0$ such that for each ε with $0 < |\varepsilon| < \varepsilon_0$ a Smale horseshoe occurs for some iterate of the Poincaré map associated with system*

$$\begin{cases} \dot{x} = y, \\ \dot{y} = -f(x) + k + \varepsilon \sin(\omega t). \end{cases} \tag{4.41}$$

The same result also holds for the damped system

$$\begin{cases} \dot{x} = y, \\ \dot{y} = -\varepsilon c_0 y - f(x) + k + \varepsilon \sin(\omega t), \end{cases} \tag{4.42}$$

for c_0 sufficiently small.

Proof For simplicity, we investigate only the frictionless case because with a similar argument one can also derive the result when a small friction term c_0 is present.

Recalling that $\dot{q}(t)$ is an odd function, from (4.38) we obtain

$$\Delta(\alpha) = \int_{-\infty}^{+\infty} q'(t) \sin(\omega t + \omega\alpha)\, dt = -2\omega \cos(\omega\alpha)\eta(\omega),$$

for

$$\eta(\omega) := \int_0^{+\infty} \tilde{q}(t) \cos(\omega t)\, dt, \quad \text{with } \tilde{q}(t) := q(t) - x_u,$$

where x_u is the abscissa of the homoclinic saddle point, namely, the negative solution of $f(x) = k$.

In this manner, we have reduced the search of a simple zero for $\Delta(\alpha)$ to the verification that $\eta(\omega) \neq 0$.

Since $\eta'(\omega) = -\omega \int_0^{+\infty} \tilde{q}(t) \sin(\omega t)\, dt = -\int_0^{+\infty} \tilde{q}(\xi/\omega) \sin(\xi)\, d\xi$, we find that

$$-\eta'(\omega) = \sum_{j=0}^{\infty} (-1)^j \int_{j\pi}^{(j+1)\pi} \tilde{q}\left(\frac{\xi}{\omega}\right) |\sin(\xi)|\, d\xi = \sum_{j=0}^{\infty} (-1)^j \Xi_j$$

where we have set

$$\Xi_j := \int_0^{\pi} \tilde{q}\left(\frac{t + j\pi}{\omega}\right) \sin(t)\, dt.$$

By observing that $\tilde{q}(t)$ is positive and decreasing on $[0, +\infty[$, it follows that the sequence $(\Xi_j)_j$ is positive, decreasing and $\Xi_j \to 0$ as $j \to +\infty$. The theory of alternating series guarantees that $\sum(-1)^j \Xi_j > 0$ and hence $\eta'(\omega) < 0$ for each $\omega > 0$. Since $\eta(\omega) \to \int_0^{+\infty} \tilde{q}(t)\, dt > 0$ as $\omega \to 0^+$, we conclude that either $\eta(\omega) > 0$ for each $\omega > 0$ or $\eta(\omega)$ vanishes exactly once. On the other hand, by the Riemann-Lebesgue lemma, it follows that $\eta(\omega) \to 0$ as $\omega \to +\infty$. This implies that the second alternative never occurs because η is strictly decreasing. Hence, in view of Theorem 4.6 the proof is completed. □

For more details and analysis see [60].

References

1. R. Adler, Symbolic dynamics and Markov partitions. Bull. Am. Math. Soc. **35**(1), 1–56 (1998)
2. R. Adler, A.G. Konheim, M.H. McAndrew, Topological entropy. Trans. Am. Math. Soc. **114**(2), 309–319 (1965)
3. A. Ambrosetti, G. Prodi, On the inversion of some differentiable mappings with singularities between Banach spaces. Ann. Mat. Pura Appl. **4**(93), 231–246 (1972)
4. H. Aref, Stirring by chaotic advection. J. Fluid Mech. **143**, 1–21 (1984)
5. H. Aref, The development of chaotic advection. Phys. Fluids **14**(4), 1315–1325 (2002)
6. B. Aulbach, B. Kieninger, On three definitions of chaos. Nonlinear Dyn. Syst. Theory **1**(1), 23–37 (2001)
7. J. Auslander, J.A. Yorke, Interval maps, factors of maps, and chaos. Tohoku Math. J. Second Ser. **32**(2), 177–188 (1980)
8. J. Banks, J. Brooks, G. Cairns, G. Davis, P. Stacey, On Devaney's definition of chaos. Am. Math. Mon. **99**(4), 332–334 (1992)
9. J. Barrow-Green, Oscar II's prize competition and the error in Poincaré's memoir on the three body problem. Arch. Hist. Exact Sci. **48**(2), 107–131 (1994)
10. J. Barrow-Green. *Poincaré and the three body problem*, volume 11 of *History of Mathematics*. American Mathematical Society, Providence, RI; London Mathematical Society, London, 1997
11. F. Battelli, K.J. Palmer, Chaos in the Duffing equation. J. Differ. Equ. **101**(2), 276–301 (1993)
12. F. Blanchard, E. Glasner, S. Kolyada, A. Maass, On Li-yorke pairs. Journal fur die reine und angewandte Mathematik **547**, 51–68 (2002)
13. L. S. Block and W. A. Coppel. *Dynamics in one dimension*. Springer, 2006
14. R. Bowen, Topological entropy for noncompact sets. Trans. Amer. Math. Soc. **184**, 125–136 (1973)
15. K. Burns, H. Weiss, A geometric criterion for positive topological entropy. Commun. Math. Phys. **172**(1), 95–118 (1995)
16. L. Burra, F. Zanolin, Chaotic dynamics in a simple class of Hamiltonian systems with applications to a pendulum with variable length. Differ. Integr. Equ. **22**(9–10), 927–948 (2009)
17. L. Burra, F. Zanolin, Chaotic dynamics in a vertically driven planar pendulum. Nonlinear Anal. **72**(3–4), 1462–1476 (2010)
18. R. Burton and R. W. Easton. Ergodicity of linked twist maps. In *Global Theory of Dynamical Systems*, pages 35–49. Springer, 1980

19. A. Capietto, W. Dambrosio, D. Papini, Superlinear indefinite equations on the real line and chaotic dynamics. J. Differ. Equ. **181**(2), 419–438 (2002)
20. G. Cian, Some remarks on topological horseshoes and applications. Nonlinear Anal. Real World Appl. **16**, 74–89 (2014)
21. Charles Conley, *Isolated invariant sets and the Morse index, CBMS Regional Conference Series in Mathematics*, vol. 38 (American Mathematical Society, Providence, RI, 1978)
22. R. Devaney. *An introduction to chaotic dynamical systems*. CRC Press, 2018
23. R.L. Devaney, Subshifts of finite type in linked twist mappings. Proceed. Am. Math. Soc. **71**(2), 334–338 (1978)
24. E.I. Dinaburg, On the relations among various entropy characteristics of dynamical systems. Math. USSR-Izvestiya **5**(2), 337 (1971)
25. C. Fabry, J. Mawhin, M.N. Nkashama, A multiplicity result for periodic solutions of forced nonlinear second order ordinary differential equations. Bull. London Math. Soc. **18**(2), 173–180 (1986)
26. D. Gale, The game of Hex and the Brouwer fixed-point theorem. Am. Math. Mon. **86**(10), 818–827 (1979)
27. J. Guckenheimer and P. Holmes. *Nonlinear oscillations, dynamical systems, and bifurcations of vector fields*. Pergamon, 1988
28. J. Hale, *Ordinary differential equations*, 2nd edn. (R.E. Krieger Publishing Co., Inc, Huntington, N.Y., 1980)
29. B. Hasselblatt and A. Katok. *A first course in dynamics: with a panorama of recent developments*. Cambridge University Press, 2003
30. J. Kennedy, S. Koçak, J. Yorke, A chaos lemma. Am. Math. Mon. **108**(5), 411–423 (2001)
31. J. Kennedy, J. Yorke, The topology of stirred fluids. Topol. Appl. **80**(3), 201–238 (1997)
32. J. Kennedy, J. Yorke, Topological horseshoes. Trans. Am. Math. Soc. **353**(6), 2513–2530 (2001)
33. U. Kirchgraber, D. Stoffer, On the definition of chaos. ZAMM-Journal of Applied Mathematics and Mechanics/Zeitschrift für Angewandte Mathematik und Mechanik **69**(7), 175–185 (1989)
34. A.N. Kolmogorov, V.M. Tikhomirov, ε-entropy and ε-capacity of sets in function spaces. Uspekhi Matematicheskikh Nauk **14**(2), 3–86 (1959)
35. W. Kulpa, M. Pordzik, L. Socha, M. Turzanski, L1 cheapest paths in Fjord scenery. Eur. J. Oper. Res. **161**(3), 736–753 (2005)
36. K. Kuratowski. Topology, a. vol. ii, pwn, 1968
37. T.Y. Li, J.A. Yorke, Period three implies chaos. Amer. Math. Monthly **82**(10), 985–992 (1975)
38. A. Margheri, C. Rebelo, F. Zanolin, Chaos in periodically perturbed planar Hamiltonian systems using linked twist maps. J. Differ. Equ. **249**(12), 3233–3257 (2010)
39. A. Margheri, C. Rebelo, F. Zanolin, Complex dynamics in pendulum-type equations with variable length. J. Dynam. Differ. Equ. **25**(3), 627–652 (2013)
40. A. Medio, M. Pireddu, F. Zanolin, Chaotic dynamics for maps in one and two dimensions: a geometrical method and applications to economics. Int. J. Bifur. Chaos **19**(10), 3283–3309 (2009)
41. K. Mischaikow, M. Mrozek, Isolating neighborhoods and chaos. Japan J. Ind. Appl. Math. **12**(2), 205–236 (1995)
42. J. Moser. *Stable and random motions in dynamical systems*. Annals of Mathematics Studies, No. 77. Princeton University Press, Princeton, N. J.; University of Tokyo Press, Tokyo, 1973. With special emphasis on celestial mechanics
43. S. Newhouse, T. Pignataro, On the estimation of topological entropy. J. Stat. Phys. **72**(5–6), 1331–1351 (1993)
44. J. Palis and F. Takens. *Hyperbolicity and sensitive chaotic dynamics at homoclinic bifurcations: Fractal dimensions and infinitely many attractors in dynamics*, volume 35. Cambridge University Press, 1995
45. D. Papini, G. Villari, F. Zanolin, Chaotic dynamics in a periodically perturbed Liénard system. Differ. Integr. Equ. **32**(11–12), 595–614 (2019)
46. D. Papini, F. Zanolin, Periodic points and chaotic-like dynamics of planar maps associated to nonlinear Hill's equations with indefinite weight. Georgian Math. J. **9**(2), 339–366 (2002)

47. D. Papini, F. Zanolin, Fixed points, periodic points, and coin-tossing sequences for mappings defined on two-dimensional cells. Fixed Point Theory Appl. **2004**(2), 126568 (2004)
48. D. Papini, F. Zanolin, On the periodic boundary value problem and chaotic-like dynamics for nonlinear Hill's equations. Adv. Nonlinear Stud. **4**(1), 71–91 (2004)
49. A. Pascoletti, M. Pireddu, and F. Zanolin. Multiple periodic solutions and complex dynamics for second order ODEs via linked twist maps. In *The 8th Colloquium on the Qualitative Theory of Differential Equations*, volume 8 of *Proc. Colloq. Qual. Theory Differ. Equ.*, pages No. 14, 32. Electron. J. Qual. Theory Differ. Equ., Szeged, 2008
50. A. Pascoletti, F. Zanolin, Example of a suspension bridge ODE model exhibiting chaotic dynamics: a topological approach. J. Math. Anal. Appl. **339**(2), 1179–1198 (2008)
51. M. Pireddu, F. Zanolin, Cutting surfaces and applications to periodic points and chaotic-like dynamics. Topol. Methods Nonlinear Anal. **30**(2), 279–319 (2007)
52. F. Przytycki. Ergodicity of toral linked twist mappings. *Ann. Sci. École Norm. Sup. (4)*, 16(3):345–354 (1984), 1983
53. F. Przytycki, Periodic points of linked twist mappings. Studia Mathematica **83**(1), 19–24 (1986)
54. C. Rebelo, F. Zanolin, On the existence and multiplicity of branches of nodal solutions for a class of parameter-dependent Sturm-Liouville problems via the shooting map. Differ. Integr. Equ. **13**(10–12), 1473–1502 (2000)
55. D.E. Sanderson, Advanced plane topology from an elementary standpoint. Math. Mag. **53**(2), 81–89 (1980)
56. A. Sivaramakrishnan, Linked twist maps in Hamiltonian systems. Celestial Mech. Dynam. Astronom. **46**(1), 35–48 (1989)
57. S. Smale. Diffeomorphisms with many periodic points. In *Differential and Combinatorial Topology (A Symposium in Honor of Marston Morse)*, pages 63–80. Princeton Univ. Press, Princeton, N.J., 1965
58. S. Smale, Differentiable dynamical systems. Bull. Am. Math. Soc. **73**(6), 747–817 (1967)
59. S. Smale, Finding a horseshoe on the beaches of Rio. Math. Intell. **20**(1), 39–44 (1998)
60. E. Sovrano, F. Zanolin, The Ambrosetti-Prodi periodic problem: different routes to complex dynamics. Dyn. Syst. Appl. **26**(3–4), 589–626 (2017)
61. R. Srzednicki, A generalization of the Lefschetz fixed point theorem and detection of chaos. Proceed. Am. Math. Soc. **128**(4), 1231–1239 (2000)
62. R. Srzednicki, Wazewski method and Conley index. Handbook Differ. Equ. **1**, 591–684 (2004)
63. R. Srzednicki, K. Wójcik, A geometric method for detecting chaotic dynamics. J. Differ. Equ. **135**(1), 66–82 (1997)
64. R. Sturman, S.W. Meier, J.M. Ottino, S. Wiggins, Linked twist map formalism in two and three dimensions applied to mixing in tumbled granular flows. J. Fluid Mech. **602**, 129–174 (2008)
65. R. Sturman, J. M. Ottino, and S. Wiggins. *The mathematical foundations of mixing: the linked twist map as a paradigm in applications: micro to macro, fluids to solids*, volume 22. Cambridge University Press, 2006
66. P.J. Torres, *Mathematical models with singularities: A zoo of singular creatures, Atlantis Briefs in Differential Equations*, vol. 1 (Atlantis Press, Paris, 2015)
67. T. Vialar, *Complex and chaotic nonlinear dynamics* (Advances in Economics and Finance, Mathematics and Statistics, Springer, 2009)
68. P. Walters. *An introduction to ergodic theory*, volume 79. Springer Science & Business Media, 2000
69. G. T. Whyburn. *Topological analysis*. Princeton Mathematical Series, No. 23. Princeton University Press, Princeton, NJ, revised edition, 1964
70. S. Wiggins, Chaos in the dynamics generated by sequences of maps, with applications to chaotic advection in flows with aperiodic time dependence. Zeitschrift für angewandte Mathematik und Physik ZAMP **50**(4), 585–616 (1999)
71. S. Wiggins. *Introduction to applied nonlinear dynamical systems and chaos*, volume 2. Springer Science & Business Media, 2003
72. S. Wiggins and J. M. Ottino. Foundations of chaotic mixing. *Philosophical Transactions of the Royal Society of London. Series A: Mathematical, Physical and Engineering Sciences*, 362(1818):937–970, 2004

73. K. Wójcik, P. Zgliczyński, Isolating segments, fixed point index, and symbolic dynamics. J. Diff. Equ. **161**(2), 245–288 (2000)
74. P. Zgliczyński, Fixed point index for iterations of maps, topological horseshoe and chaos. Topol. Methods Nonlinear Anal. **8**(1), 169–177 (1996)
75. P. Zgliczyński, M. Gidea, Covering relations for multidimensional dynamical systems. J. Differ. Equ. **202**(1), 32–58 (2004)

Chapter 5
Applications to the Superlinear Duffing Equation

5.1 Some Variants of the Duffing Equation

In this chapter we deal with two applications of variants of the Duffing equation. The first is the case of the Duffing-type equation including a fractional order term. Here we prove the existence of complex dynamics in a periodically perturbed Hamiltonian system of the form $\ddot{x} + f(x) = e(t)$, where $e(\cdot) : \mathbb{R} \to \mathbb{R}$ is a T-periodic forcing term which can be approximated by a piecewise constant function $m_{1,2}(t)$ defined on $[0, T[$. To prove the existence of chaotic dynamics we provide a general criterion, based on the monotonicity of the time mapping, which applies when the nonlinear term $f(x)$ has superlinear/sublinear growth at infinity, consistent with the terminology previously introduced in Chap. 2. As an application of the general result we consider the case of a fractional order term $f(x) \sim bx|x|^{p-1}$ with $b > 0$ and $p > 1$. In the second application, we prove the existence of infinitely many subharmonic solutions as well as the presence of a full symbolic dynamics for a class of second-order nonlinear ordinary differential equations with a singularity of repulsive type at the origin and a superlinear growth at infinity. From a dynamical point of view, a singularity of the repulsive type at the origin has the same effect as a superlinear term.

Our results, although dealing with a planar Hamiltonian system, are stable for small perturbations as commented before, and we can allow the presence of a damping coefficient as well.

5.2 An Equation of the Duffing Type with Superlinear Growth

We are interested in detecting the presence of chaotic dynamics in some planar Hamiltonian systems under periodic perturbations. We focus our attention on the periodically perturbed the Duffing equation $\ddot{x} + f(x) = e(t)$.

L. Burra and F. Zanolin, *The Duffing Equation*, Infosys Science Foundation Series in Mathematical Sciences, https://doi.org/10.1007/978-981-97-8301-4_5

[1]Our study deals with a very simplified situation compared to the general theory developed by Meyer and Sell in [30] and applied to perturbations of some Duffing-type equations by means of forcing terms which are not necessarily periodic. On the other hand, our approach, which is based on simple topological arguments, allows to give rigorous proof to the existence of chaotic dynamics under very natural and easily verifiable hypotheses on the nonlinear term $f(x)$. In this application we consider a general nonlinearity including the case

$$f(x) = -ax + bx|x|^{p-1} \tag{5.1}$$

with

$$a, b > 0 \quad \text{and} \quad p > 1. \tag{5.2}$$

The Melnikov method has been applied to the equation

$$\ddot{x} - ax + bx|x|^{p-1} = \varepsilon(-c\dot{x} + r\cos(\omega t))$$

in some papers [11, 27]. The constant $p > 1$ in [11] can be an integer or a fraction, hence giving rise to the term *fractional order deflection function* for $f(x)$ in (5.1). As observed in [27], for a general $p > 1$, the computation of integrals in the Melnikov function is a very difficult task. A preliminary analysis and simulations for this case of study can be seen Chap. 3, Sect. 3.3.2.

Restoring forces $f(x)$ including a term of the form $x|x|^{p-1}$ have been widely considered. For instance, in [22] the authors deal with a Duffing-type equation of the form

$$\ddot{x} + x|x|^{p-1} = \sin(\omega t)$$

with $p \neq 1$ and positive. The analysis of the Lyapunov exponents and the Poincaré map in the phase-plane performed in [22], for some values of p (e.g. for $p = 3$), suggests the presence of chaotic dynamics. Quoting the authors, *numerical experiments also show that the form of the periodic drive is not critical. For example, a square wave of similar frequency also produces chaos in these systems* [22, p. 387]. The method that we use requires that $b > 0$, but it can be also applied to $a \geq 0$ or $a \leq 0$ as well.

Statement of the Main Result

We investigate the undamped scalar equation of the Duffing type

$$(H) \qquad \ddot{x} + f(x) = m_{1,2}(t),$$

[1] We denote here the nonlinear term by $f(x)$ instead of $g(x)$ as was done in the book so far, for the reason that we represent later by $g(x)$ a suitable modification of $f(x)$.

as well as its associated planar system

$$(S)\qquad \begin{cases} \dot{x} = y \\ \dot{y} = -f(x) + m_{1,2}(t) \end{cases}$$

where $f : \mathbb{R} \to \mathbb{R}$ is a locally Lipschitz continuous function and $m_{1,2} : \mathbb{R} \to \mathbb{R}$ is a piecewise constant T-periodic forcing term defined on $[0, T[$ by:

$$(M)\qquad m_{1,2}(t) := \begin{cases} m_1, & \text{for } t \in [0, \tau_1[\\ m_2, & \text{for } t \in [\tau_1, \tau_1 + \tau_2[, \end{cases} \qquad (m_1 \neq m_2)$$

with

$$\tau_1,\ \tau_2 > 0, \quad \tau_1 + \tau_2 = T.$$

We shall prove the presence of chaotic dynamics under general conditions for the nonlinear term $f(x)$, including those of (5.1)–(5.2). As usual, our results are *robust*, i.e. stable with respect to small perturbations, hence they are still true for the more general equation

$$(H_{c,e})\qquad \ddot{x} + c\dot{x} + f(x) = e(t),$$

provided that c is sufficiently small and $e(\cdot) : \mathbb{R} \to \mathbb{R}$ is a T-periodic function with

$$\int_0^T |e(t) - m_{1,2}(t)|\, dt < \varepsilon, \tag{5.3}$$

with $\varepsilon > 0$ some small number. In this case, we can take $e(t)$ as smooth as we like (even of class C^∞). In order to compare our applications with other results in the literature, for the main part of this application, we deal with a function $f(x)$ which has superlinear growth at infinity, namely,

$$\lim_{|x|\to\infty} \frac{f(x)}{x} = +\infty,$$

in order to include the case

$$f(x) = -ax + bx|x|^{p-1} \quad \text{with } b > 0,\ p > 1\ (a \in \mathbb{R},\ \text{arbitrary}).$$

Here p can be any real number. Our approach produces similar results in the sublinear case

$$f(x) = bx|x|^{p-1} \quad \text{with } b > 0,\ 0 < p < 1,$$

but we do not explicitly deal with this case.

The choice of a T-periodic stepwise forcing term $m_{1,2}(t)$ is motivated in order to make the presentation of our method more transparent. Moreover, such a choice allows us to enter into the setting of some previous works where our theory has been already rigorously developed [4, 28]. See Chap. 4 for a detailed development of the theory.

Second-order ODEs subject to a square wave periodic perturbation have been investigated by several authors (see [8, 25, 26, 32], just to quote a sampling). From a general point of view, using such a stepwise perturbation is not more arbitrary than using a special forcing term like $e(t) = \cos(\omega t)$. Mechanical or electrical devices which can be modelled by second-order equations subject to a square wave are easy to explain and have been used as pedagogical examples, too [8].

Because of the form of $m_{1,2}(t)$, the system (S) switches periodically between two different autonomous Hamiltonian systems

$$(S_1) \qquad \begin{cases} \dot{x} = y \\ \dot{y} = -f(x) + m_1 \end{cases}$$

and

$$(S_2) \qquad \begin{cases} \dot{x} = y \\ \dot{y} = -f(x) + m_2 \end{cases}$$

with $(S) = (S_1)$ for $t \in [0, \tau_1[$ and $(S) = (S_2)$ for $t \in [\tau_1, \tau_1 + \tau_2[$. We derive a general result where the only assumption is a monotonicity criterion for the time mapping associated to the autonomous system. Such a criterion can be simply checked by means of a natural growth condition on $f(x)$ (see (f_{sup}) below). As is well known, switched systems play an important role in control theory (see [3]).

We state now our main result for equation (H).

Theorem 5.1 *Let $f : \mathbb{R} \to \mathbb{R}$ be a continuously differentiable function such that*

$$(f_\infty) \qquad \lim_{|x|\to\infty} f(x)\, sgn(x) = +\infty$$

and

$$(f_{\text{sup}}) \qquad \lim_{|x|\to+\infty} (xf'(x) - f(x))\, sgn(x) = +\infty$$

hold. Then for each T-periodic forcing term $m_{1,2}(t)$ as in (M) there exist T_1^* and T_2^* such that equation (H) presents chaotic dynamics provided that $\tau_1 > T_1^*$ and $\tau_2 > T_2^*$. The result is also stable with respect to small perturbations and hence it holds for equation $(H_{c,e})$ provided that $e(t)$ satisfies (5.3) *and ε, c are sufficiently small.*

Here *chaotic dynamics* is intended in the sense of Chap. 4 Definitions 4.3 and 4.4. Notice that (f_∞) and (f_{sup}) imply that $f(x)/x \to +\infty$ as $x \to \pm\infty$, so that $f(x)$ has a superlinear growth at infinity. Moreover, a simple calculation shows that the function

$$f(x) := -ax + b|x|^{p-1}x, \quad \text{for } b > 0,\, p > 1$$

satisfies (f_∞) and (f_{sup}) (independently of $a \in \mathbb{R}$).

We observe that if $f : \mathbb{R} \to \mathbb{R}$ is locally Lipschitz continuous and satisfies (f_∞), then for any T-periodic function $e(\cdot) : \mathbb{R} \to \mathbb{R}$ with $e(\cdot) \in L^1(0, T)$ and any $c \in \mathbb{R}$, there is a unique solution of the differential system

$$\begin{cases} \dot{x} = y \\ \dot{y} = -cy - f(x) + e(t) \end{cases} \tag{5.4}$$

satisfying the initial condition $(x(t_0), y(t_0)) = z_0 \in \mathbb{R}^2$ (here $t_0 \in [0, T]$ and z_0 are arbitrarily chosen initial points). We denote such a solution as $\zeta(t; t_0, z_0)$ and observe that it is continuable to the whole real line. Since $e(t)$ is not necessarily continuous, solutions of (5.4) are meant in the generalized (Carathéodory) sense [23]. Clearly, (5.4) is precisely equation $(H_{c,\varepsilon})$ in the phase-plane. Obviously, system (5.4) includes also (S) and hence equation (H). The presence of chaos according to Definitions 4.3, 4.4, is detected in regions that we call *topological rectangles-generalized rectangles* (see Chap. 4) because they are regions of the plane homeomorphic to the unit square. In order to explain how our approach works, we provide an illustrative example, analysing the argument our proof to the case of the nonlinear function

$$f(x) = -ax + bx|x|^{p-1}, \qquad b > 0,\, p > 1$$

with superlinear growth at infinity.

With this choice of $f(x)$, system (S) can be written in as the superposition of the two autonomous systems

$$(S_1) \qquad \begin{cases} \dot{x} = y \\ \dot{y} = ax - bx|x|^{p-1} + m_1 \end{cases}$$

and

$$(S_2) \qquad \begin{cases} \dot{x} = y \\ \dot{y} = ax - bx|x|^{p-1} + m_2 \end{cases}$$

which switch one to the other in a T-periodic fashion. According to condition (M), the effect of (S_1) lasts for a time interval of length τ_1 whilst (S_2) lasts for a time interval of length τ_2 (with $\tau_1 + \tau_2 = T$). We define the energy associated with (S_1) and (S_2) as $\mathcal{E}_1(x, y)$ and $\mathcal{E}_2(x, y)$ with $\mathcal{E}_1(x, y) := \mathcal{E}(x, y) - m_1 x$ and $\mathcal{E}_2(x, y) := \mathcal{E}(x, y) - m_2 x$, where $\mathcal{E}(x, y) := \frac{y^2}{2} - a\frac{x^2}{2} + b\frac{1}{p+1}|x|^p$. (see Fig. 5.1).

As a next step, we construct two rectangular regions $\mathcal{R}_1$ and $\mathcal{R}_2$ which are obtained as intersections of suitably chosen annuli surrounding the equilibrium points in the phase-plane (see Fig. 5.2).

We can prove the presence of chaotic dynamics both in region $\mathcal{R}_1$ and in region $\mathcal{R}_2$ provided that the period of the orbits defining each of the two annuli is different which

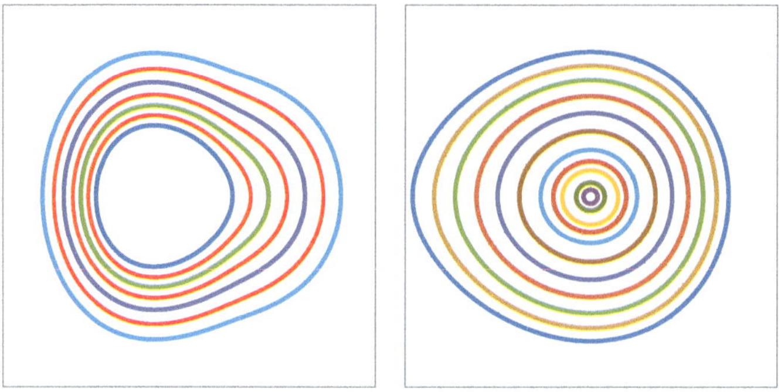

Fig. 5.1 Energy level lines associated to systems (S_1) and (S_2) for $p = 3/2$, $a = 1$ and $b = 1$ $m_1 = 3$ (left) and, $m_2 = -4$ (right)

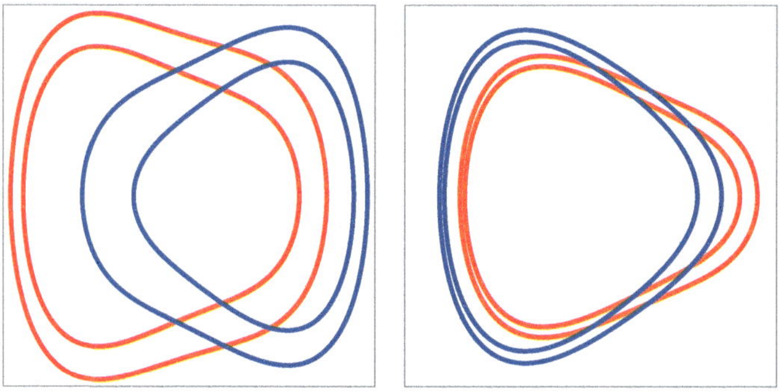

Fig. 5.2 The Linked Rectangles (the regions in the intersections of the pairs of level lines) $\mathcal{R}_1$ and $\mathcal{R}_2$ formed by the overlapping annuli of (S_1), for $m_1 = 3$ and of (S_2), for $m_2 = -4$ (left panel). The same configuration for $m_1 = 4$ and $m_2 = 7$ (right panel)

is clearly satisfied when the period map is monotone. The geometrical configuration is similar to the one considered in [33, 34] for the Volterra prey-predator planar system.

5.3 Technical Estimates and Proof of the Main Result

In this section, we show how to construct two linked annular regions associated to energy level lines of (S_1) and (S_2) in such a manner that

- There is a gap between the periods of the orbits which constitute the boundaries of the annular regions; actually, we prove that the time mapping is strictly monotone.
- The two annular regions cross each other in a proper manner in order to obtain two topological rectangles where our method can be applied.

Both conditions will be proved for an arbitrary superlinear function $f(x)$ satisfying the assumptions of Theorem 5.1.

To fix the ideas, along with the proof, we suppose that

$$m_1 < m_2 .$$

The case in which the reverse inequality holds can be treated in a completely symmetric manner. The geometry considered here is precisely that which is elaborated in detail in Chap. 4.

5.3.1 *Monotonicity of the Time Mappings*

We need to prove the monotonicity of the time mapping associated to a system of the form

$$\begin{cases} \dot{x} = y \\ \dot{y} = -g(x) \end{cases} \tag{5.5}$$

where $g : \mathbb{R} \to \mathbb{R}$ is a continuously differentiable function such that

$$\lim_{|x| \to +\infty} g(x)\text{sign}(x) = +\infty. \tag{5.6}$$

Our result will be applied to the choice $g(x) := f(x) - m$ (for $m \in \mathbb{R}$ a given constant). We also define

$$G(x) := \int_0^x g(s)\, ds.$$

From (5.6) we know that there exists $R > 0$ such that $G(x)$ is strictly decreasing on $(-\infty, -R)$, strictly increasing on $(R, +\infty)$ and, moreover, $G(x) \to +\infty$ as $x \to \pm\infty$. Due to the conservative nature of system (5.5) we know that the associated orbits lie on the energy level lines

$$E(x, y) = \frac{1}{2}y^2 + G(x) = k$$

which are symmetric with respect to the x-axis in the phase-plane. Such level lines are closed curves (hence periodic orbits) provided that

$$k > k^* := \max_{x \in [-R, R]} G(x).$$

For every $k > k^*$ there exists a unique pair (c_k^-, c_k^+) with

$$c_k^- < -R < 0 < R < c_k^+$$

such that

$$G(c_k^-) = G(c_k^+) = k.$$

For any fixed $k > k^*$, the period of the orbit

$$\Gamma^k := \{(x, y) \in \mathbb{R}^2 \,:\, E(x, y) = k\}$$

is given by

$$\tau(k) := \tau^-(k) + \tau^+(k),$$

where

$$\tau^-(k) := 2\int_{c_k^-}^{0} \frac{dx}{\sqrt{2(G(c_k^-) - G(x))}}$$

and

$$\tau^+(k) := 2\int_{0}^{c_k^-} \frac{dx}{\sqrt{2(G(c_k^+) - G(x))}}.$$

Hence, in order to prove the monotonicity of the time mapping, it is sufficient to show that the map

$$\phi(c) := \left| \int_0^c \frac{dx}{\sqrt{(G(c) - G(x))}} \right|$$

is strictly monotone in a neighbourhood of infinity, namely, for $|c|$ sufficiently large.

For simplicity, we discuss only the case in which c is positive and large. Our analysis follows the lines of [31].

Via a rescaling, we write $\phi(c)$ as

$$\phi(c) = \int_0^1 \frac{d\xi}{\sqrt{(G(c) - G(c\xi))/c^2}}.$$

Our goal is to show that for $c < c'$ with $c > R$ and sufficiently large, it follows that

$$(G(c) - G(c\xi))/c^2 < (G(c') - G(c'\xi))/c'^2, \quad \forall 0 \le \xi < 1.$$

Hence, by differentiating with respect to c, we get that the above condition holds true provided that for the auxiliary function

$$h(c) := cg(c) - 2G(c)$$

we have that

$$h(c) > h(\xi c), \quad \forall \xi \in [0, 1[\,.$$

This latter condition holds good for every $c > 0$ sufficiently large, if we suppose that $h(c) \to +\infty$ as $c \to +\infty$ and $h'(c) > 0$ for $c > 0$ large enough (Compare this with Sect. 1.4, Chap. 1).

If we apply the above considerations to the case

$$g(x) := f(x) - m, \quad \text{for } m \in \mathbb{R},$$

we obtain the following result.

Lemma 5.1 *Let $f : \mathbb{R} \to \mathbb{R}$ be a continuously differentiable function such that*

$$\lim_{|x| \to +\infty} f(x)\text{sign}(x) = +\infty \tag{5.7}$$

and

$$\lim_{|x| \to +\infty} (xf'(x) - f(x))\text{sign}(x) = +\infty. \tag{5.8}$$

Then, for every $m \in \mathbb{R}$ there exists k_m^* such that for every $k > k_m^*$, the orbits of (5.5) *(for $g(x) := f(x) - m$) are periodic with a strictly monotonic decreasing period.*

Proof We consider the function $h(c)$ for the special choice of $g(x) = f(x) - m$. Then we have

$$h'(c) = cf'(c) - f(c) + m.$$

Clearly, $h'(c) \to +\infty$ as $c \to +\infty$ if $xf'(x) - f(x) \to +\infty$ as $x \to +\infty$. Similarly, one checks the symmetric condition for $c << 0$. □

5.3.2 *Linking Annular Regions*

Next, we construct linked annuli by means of energy level lines associated with systems (S_1) and (S_2). To this end, we study the system

$$\begin{cases} \frac{1}{2}y^2 + F(x) - m_1 x = \alpha \in [\alpha_0, \alpha_1] \\ \frac{1}{2}y^2 + F(x) - m_2 x = \beta \in [\beta_0, \beta_1]. \end{cases} \tag{5.9}$$

The parameters α_i, β_i for $i = 0, 1$ will be defined in due course. Our goal is to determine two regions (symmetric with respect to the x-axis) which are homeomorphic to rectangles. Hence, we proceed as follows:

Let us fix $\bar{x} \in \mathbb{R}$ such that

$$f(\bar{x}) = \bar{m} := \frac{m_1 + m_2}{2}$$

and choose an interval $[x_0, x_1]$ with $x_0 < \bar{x} < x_1$ such that

$$m_1 < f(x) < m_2, \quad \forall x \in [x_0, x_1].$$

Now, in the (α, β)-plane, we consider the infinite strip

$$\mathscr{S} := \{(\alpha, \beta) : \alpha - x_1(m_2 - m_1) \le \beta \le \alpha - x_0(m_2 - m_1)\}$$

which contains the line

$$\mathscr{L} : \quad \beta = \alpha - \bar{x}(m_2 - m_1).$$

Let us also define

$$m^* := (m_2 - m_1) + \max\{|m_1|, |m_2|\}$$

and

$$c^* := \max_{x \in [x_0, x_1]} F(x) + m^* |x| .$$

Now, for any choice of $(\bar{\alpha}, \bar{\beta}) \in \mathscr{L}$, with $\bar{\alpha}, \bar{\beta} > c^*$, we can select two intervals $[\alpha_0, \alpha_1]$ and $[\beta_0, \beta_1]$, with

$$\alpha_0 < \bar{\alpha} < \alpha_1, \quad \beta_0 < \bar{\beta} < \beta_1$$

such that

$$[\alpha_0, \alpha_1] \times [\beta_0, \beta_1] \subseteq \mathscr{S}. \tag{5.10}$$

Since $f(x) - m_{1,2}$ never vanishes for $x \in [x_0, x_1]$, we have that the energy function is strictly monotone in x. Therefore, we obtain two rectangles with smooth boundaries (except at the vertices). See Fig. 5.2 for the rectangles described here.

5.3.3 *Proof of Theorem 5.1*

Assume conditions (f_∞) and $f_{\text{sup}})$. Let m_1 and m_2 be also given with $m_2 > m_1$. For $i = 1, 2$ let k_i^* be such that for each $k > k_i^*$ the time mapping $k \to \tau^i(k)$ associated to system (S_i) is strictly decreasing (the existence of k_i^* and the strict monotonicity of the corresponding time mapping follow from Lemma 5.1).

Next, we define

$$k^{**} := \max\{k_1^*, k_2^*\}$$

and construct two closed annular regions

$$\mathcal{A} := \left\{ \frac{1}{2} y^2 + F(x) - m_1 x \in [\alpha_0, \alpha_1] \right\}$$

and

$$\mathcal{B} := \left\{ \frac{1}{2}y^2 + F(x) - m_1 x \in [\beta_0, \beta_1] \right\}.$$

The coefficients $\alpha_0, \alpha_1, \beta_0, \beta_1$ are chosen so that (5.10) is satisfied and also

$$\alpha_0 \, , \beta_0 > k^{**}.$$

By the result in Sect. 5.3.2 we find that

$$\mathcal{A} \cap \mathcal{B} = \mathcal{M} \sqcup \mathcal{N},$$

where $\mathcal{M}$ and $\mathcal{N}$ are two topological rectangles symmetric with respect to the x-axis. These are represented as $\mathcal{R}_1$ and $\mathcal{R}_2$ in Fig. 5.2. For instance, we can call $\mathcal{M}$ the set which lies in the upper half-plane and $\mathcal{N}$ the symmetric one. In this manner, we have the *link condition* (LC) of [28, p.3248] satisfied. Accordingly, Theorem 4.1 in [28, p.3249] applies to $\mathcal{R}$ for

$$\mathcal{R} := \mathcal{M} \quad \text{or} \quad \mathcal{R} := \mathcal{N}$$

and therefore we conclude that for any integer $\ell \geq 2$ there exist T_1^* and T_2^* such that if

$$\tau_i > T_i^* \quad \text{for} \quad i = 1, 2,$$

the Poincaré map of system (S) induces chaotic dynamics on ℓ symbols on $\mathcal{R}$. (This is a case of linked twist maps detailed in Chap. 4).

We observe that T_1^* and T_2^* can be computed in terms of ℓ and the periods of the orbits bounding the annuli $\mathcal{A}$ and $\mathcal{B}$. Indeed, T_i^* can be estimated as a multiple of

$$\frac{\tau^i(k_0)\tau^i(k_1)}{\tau^i(k_0) - \tau^i(k_1)},$$

where k_0 is the energy level of the inner orbit and k_1 is the energy level of the outer orbit (see [28]). As mentioned, the result is also stable with respect to small perturbations and therefore it applies to equation $(H_{c,\varepsilon})$, as well. □

5.4 The Duffing Equation with Singularities

From the time of the pioneering work of Yoshisuke Ueda, (see Chap. 3) who discovered the presence of "chaotic attractors" for the periodically perturbed Duffing equation with a small friction

$$\ddot{x} + c\dot{x} + x^3 = B\cos(t),$$

a lot of research has followed in view of proving the existence of chaotic dynamics for such superlinear and periodically perturbed ODEs (See Chap. 3, Sect. 3.1 for an introduction to this). Typical features of these equations are the presence of a small friction coefficient along with a nonlinear term with superlinear growth. It has been observed that a repulsive singularity at the origin with the "strong force condition" mimics, from a dynamical point of view, a superlinear effect. In fact, the energy of the solutions becomes arbitrarily large in a neighbourhood of the origin and the closer the orbits are to the singularity, the faster is the motion (in time). Such a superlinear-type effect, due to the singularity, has been already exploited in [12, 15] concerning the multiplicity of periodic solutions. Thus, keeping in mind these features, it is a natural question to look for chaotic attractors in such equations. (See Chap. 3 for simulations of these attractors) (Fig. 5.3).

In this chapter we study a class of periodically perturbed nonlinear second-order ordinary differential equations of the Duffing type

$$\ddot{x} + g(x) = p(t) \tag{5.11}$$

where $g : \mathbb{R}_0^+ \to \mathbb{R}$ is a locally Lipschitz continuous function which has a singularity at the origin of the repulsive type and a superlinear growth at infinity. More precisely, for $\mathbb{R}_0^+ :=]0, +\infty[$, we suppose that the following conditions hold:

(g_0) $\quad \lim_{s \to 0^+} g(s) = -\infty$ and $\lim_{s \to 0^+} G(s) = +\infty$ for $G' = g$ on $\mathbb{R}_0^+$;

(g_∞) $\quad \lim_{s \to +\infty} \dfrac{g(s)}{s} = +\infty.$

Concerning the forcing term $p(t)$ we suppose that $p : \mathbb{R} \to \mathbb{R}$ is a T-periodic function (for some $T > 0$), with $p \in L^1([0, T])$. More specific assumptions on $p(t)$ will

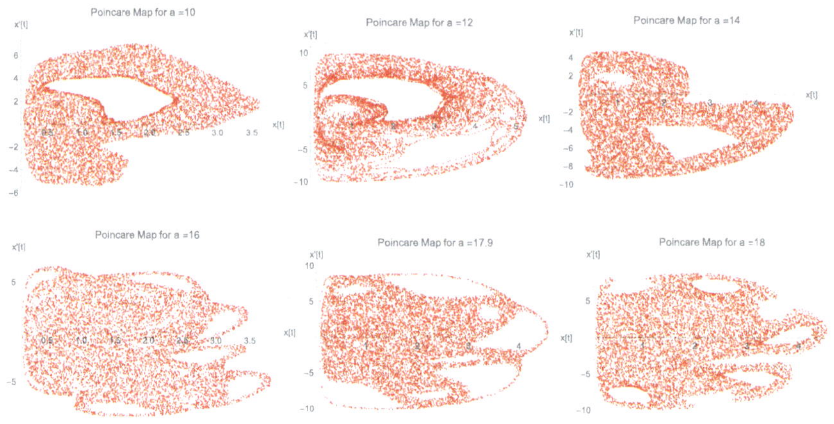

Fig. 5.3 The present figure shows the attractor for the Poincaré mapping (stroboscopic mapping) associated with $\ddot{x} - x^{-3} + x^2 = A\sin(t)$ for different values of the parameter A

be detailed later. In any case, in order to avoid trivialities, we suppose that p is nonconstant. To this end, we set

$$\bar{p} := \frac{1}{T}\int_0^T p(t)\,dt, \quad \tilde{p}(t) := p(t) - \bar{p}$$

and denote by $P(t)$ a primitive of $\tilde{p}(t)$ with mean value zero. The function $P(t)$ is an absolutely continuous T-periodic function with $||P||_\infty > 0$ (as, by assumption, $p(t) \neq \bar{p}$ on a set of positive measure). Without loss of generality, we suppose, from now on, that $T > 0$ is the minimal period of $P(t)$. Passing to the phase-plane we can equivalently write Eq. (5.1) as a first-order system as

$$\begin{cases} \dot{x} = y \\ \dot{y} = -g(x) + p(t) \end{cases} \tag{5.12}$$

or as

$$\begin{cases} \dot{x} = y + P(t) \\ \dot{y} = -g(x) + \bar{p}. \end{cases} \tag{5.13}$$

We look for *positive* solutions of (5.11), namely, we look for globally defined solutions $x(t)$ of (5.11), in the Carathéodory sense, such that $x(t) > 0$ for all $t \in \mathbb{R}$.

A typical nonlinearity satisfying (g_0) and (g_∞) could be given by

$$g(x) = -\frac{A}{x^\alpha} + Bx^\beta, \quad \text{with } A, B > 0, \ \alpha \geq 1, \beta > 1. \tag{5.14}$$

Concerning the behaviour of $g(x)$ at the origin, expressed by condition (g_0), we are in the so-called case of *repulsive type equations*, according to Lazer and Solimini [24]. The divergence of $G(x)$ at the origin is also related to the "strong force" condition introduced by Gordon [21] and referred to by Torres in [36]. We recall that in [24], the authors considered the singular equation with

$$g(x) = -\frac{1}{x^\alpha}, \tag{5.15}$$

and proved that if $\alpha \geq 1$, then (5.11) has a positive T-periodic solution if and only if $p(t)$ has a negative mean value. It was also proved in [24] that this result is optimal in the sense that if $0 < \alpha < 1$, it is possible to choose a function $p(t)$ that has a negative mean value and the equation has no positive T-periodic solutions. With regard to the second condition, that is (g_∞), our choice is motivated by the analogy with the superlinear Duffing equation on the real line (discussed in the previous section). Another motivation comes from the work of del Pino and Manásevich [12] where a second-order singular equation of the form

$$\ddot{x} + f(t, x) = 0,$$

was considered, with $f(t, s)$ satisfying f

$$-\infty \leq \limsup_{s \to 0^+} s f(t, s) < 0, \qquad \lim_{s \to +\infty} \frac{f(t, s)}{s} = +\infty,$$

uniformly in t. In [12], the authors, using the Poincaré-Birkhoff fixed point theorem, proved the existence of infinitely many positive T-periodic solutions of the form $x_n^{\pm}(t)$ such that $x_n^{\pm}(t) - 1$ has exactly $2n$ zeros in $[0, T[$ for every integer n sufficiently large.

Subharmonic Solutions

With regard to Eq. (5.11), it was proved in [15, Sect. 3] that a singularity for $g(x)$ as in (g_0) produces a dynamics which is similar to that of a superlinear Duffing equation. In particular, by applying the Poincaré-Birkhoff theorem (see Chap. 2, Sect. 2.11) to the Poincaré map associated with system (5.13) the following result was obtained (cf. [15, Theorem 3.1]):

Theorem 5.2 *Let $g : \mathbb{R}_0^+ \to \mathbb{R}$ be a locally Lipschitz continuous function satisfying (g_0) and (g_∞) and let $p : \mathbb{R} \to \mathbb{R}$ be a nonconstant T-periodic function with $p \in L^1([0, T])$ and minimal period $T > 0$. Let us fix a constant $c > 0$. Then, for every integer $m \geq 1$, Eq.* (5.11) *has infinitely many periodic solutions of minimal period mT. More precisely, for each $m \geq 1$, there exists an integer $\nu_m^* \geq 1$, such that for every $\ell \in \mathbb{N}$ relatively prime to m and with $\ell \geq \nu_m^*$, Eq.* (5.11) *has at least two periodic solutions $u^+ = u_{m,\ell}^+$ and $u^- = u_{m,\ell}^-$ of minimal period mT, which do not belong to the same periodicity class, and such that $u^{\pm}(t) - c$ has exactly 2ℓ simple zeros in the interval* $[0, mT[$. *Moreover,* $\min u_{m,\ell}^{\pm}(t) \to 0^+$, $\max u_{m,\ell}^{\pm}(t) \to +\infty$ *and* $\max \dot{u}_{m,\ell}^{\pm}(t) \to -\infty$, $\min \dot{u}_{m,\ell}^{\pm}(t) \to +\infty$ *as $\ell \to \infty$.*

The above statement corresponds to [15, Theorem 3.1] modified to suit our case. Notice that in [15] the local Lipschitz condition for the map g was not assumed and the existence of at least one solution $u_{m,\ell}(t)$ was proved. On the other hand, the same proof as in [15], via the Poincaré-Birkhoff theorem, guarantees the existence of pairs of solutions $u_{m,\ell}^{\pm}(t)$ as stated in Theorem 5.2. In [15] the analysis is performed by the study of system (3.15) for $P(t) = \int_0^t \tilde{p}(\xi)\, d\xi$. Our choice of $P(t)$ as the primitive of $\tilde{p}(t)$ with mean value zero does not effect the proof. Notice also that, given any mT-periodic solution $u(\cdot)$, also $u(\cdot + jT)$ is mT-periodic, for each $j = 1, \dots, m-1$. Accordingly, any time we find a periodic solution of minimal period mT, we find indeed m of these solutions, which, however, are considered as equivalent. In this case, we say that they belong to the same *periodicity class*. Finally, we recall the interesting work of Fonda and Ureña [20] where a higher dimensional version of the Poincaré-Birkhoff was obtained. This generalized version of the theorem allows to extend results like Theorem 5.2, from the single second-order Eq. (5.11) to systems of second-order equations (at least in the weakly coupled situation). For relevant contributions in this area we refer to recent articles by A.Fonda and collaborators, see [16–19]. Moreover, using the result in [20], one can avoid assuming the local Lipschitz condition on nonlinearity. A result analogous to Theorem 5.2 holds for the forced superlinear Duffing equation $\ddot{x} + g(x) = p(t)$ where $g : \mathbb{R} \to \mathbb{R}$ is a locally Lipschitz continuous function satisfying the superlinear growth condition at infinity

$$\lim_{s \to \pm\infty} \frac{g(s)}{s} = +\infty.$$

and $p(t)$ is a T-periodic perturbation with $p \in L^1([0, T])$ (see [14] and the references therein).

Chaotic Dynamics

In Sect. 5.2 (see also [5]) we proved the presence of chaos, for $\ddot{x} + f(x) = e(t)$ under some growth assumptions of superlinear type for $f(x)$. The analysis was carried out for $e(t)$ a stepwise function or a function which is close to a stepwise one in the L^1-norm on $[0, T]$, cf. equation (H). Also when dealing with Eq. (5.11) we adopt a similar approach to our current problem and consider T-periodic forcing terms of the form

$$p_{k_1,k_2}(t) := \begin{cases} k_1\,, & 0 \le t < T_1 \\ k_2\,, & T_1 \le t < T_1 + T_2 = T \end{cases} \tag{5.16}$$

with $k_1 \neq k_2$.

The existence of chaotic dynamics for equations with singularities has been proved only recently by Margheri and Torres for the Gyldén problem in Celestial Mechanics [29] and by the authors in [6] for the Rayleigh-Plesset equation from fluid dynamics. Both the examples considered in [29] and [6] deal with equations involving singularities at the origin and a sublinear growth at infinity. The study of (5.11) with $g(x)$ satisfying (g_0) and (g_∞) corresponds to a new case from the point of view of chaotic dynamics which we follow from our recent work in [7]. Accordingly, the following result holds,

Theorem 5.3 *Let $g : \mathbb{R}_0^+ \to \mathbb{R}$ be a locally Lipschitz continuous function satisfying (g_0) and (g_∞) and let $p : \mathbb{R} \to \mathbb{R}$ be a nonconstant stepwise T-periodic function with minimal period $T > 0$. Then, for each $m \geq 2$, there are positive constants T_1^* and T_2^* such that for every $T_1 > T_1^*$ and $T_2 > T_2^*$, the Poincaré map associated with system* (5.11) *exhibits chaotic dynamics on m symbols. The result is robust with respect to small perturbations.*

As is usual, in our applications, the term chaotic dynamics is taken in the same sense as described in Chap. 4. We observe that if $g : \mathbb{R}_0^+ \to \mathbb{R}$ is locally Lipschitz continuous and satisfies (g_0) and (g_∞), then for any T-periodic function $p(\cdot) : \mathbb{R} \to \mathbb{R}$ with $p(\cdot) \in L^1(0, T)$ and any $c \in \mathbb{R}$, there is a unique solution of the differential system

$$\begin{cases} \dot{x} = y \\ \dot{y} = -cy - g(x) + p(t) \end{cases} \tag{5.17}$$

satisfying the initial condition $(x(t_0), y(t_0)) = z_0 \in \mathbb{R}_0^+ \times \mathbb{R}$ (here $t_0 \in [0, T]$ and z_0 are arbitrarily chosen initial points). We denote such a solution as $\zeta(t; t_0, z_0)$ and observe that it is continuable to the whole real line.[2] Since $p(t)$ is not necessarily continuous, solutions of (5.17) are meant in the generalized (Carathéodory) sense. Clearly, (5.17) is precisely equation

$$\ddot{x} + c\dot{x} + g(x) = p(t) \tag{5.18}$$

in the phase-plane. Obviously, system (5.17) includes also (5.12). Another way to write (5.18) as a system in the phase-plane, would be

$$\begin{cases} \dot{x} = y + P(t) \\ \dot{y} = -cy - g(x) + \bar{p} \end{cases} \tag{5.19}$$

which generalizes (5.13). The presence of chaos according to Definition 4.3 is detected in some *topological rectangles* constructed in the framework of the *Linked Twist Maps* [13, 35, 37] and *chaotic advection under a blinking protocol* [1, 2, 36]. (See Chap. 4, Sect. 4.4, where this concept is explained in detail).

5.5 Proof of Theorem 5.3

Following [7], we consider system (5.12) for $p(t) = p_{k_1,k_2}(t)$ defined as in (5.16). In this case, (5.12) is a switched system consisting of two autonomous systems of the form

$$\dot{x} = y, \quad \dot{y} = -g(x) + k_i\,, \quad i = 1, 2.$$

For convenience, we analyse the system

$$\begin{cases} \dot{x} = y \\ \dot{y} = -g(x) + \mu \end{cases} \tag{5.20}$$

where μ is any real parameter. The phase portrait of (5.20) is that of a centre outside a compact set in the open right half-plane. More precisely, we can find a closed trajectory Γ_0 in $\mathbb{R}_0^+ \times \mathbb{R}$ including in its interior region all the equilibrium points

[2] This fact follows from an identical argument as in [15, Proposition 3.1].

and such that any trajectory outside Γ_0 is also a closed orbit. These closed orbits are actually energy level lines of the form

$$\frac{y^2}{2} + G(x) - \mu x = E$$

for E fixed and sufficiently large, say $E \geq E_0$ where E_0 is the energy corresponding to Γ_0.

Remark 5.1 In the special case in which g is a strictly monotone function, the phase portrait is that of a global centre at the equilibrium point $(x_\mu, 0)$, where we have set

$$g(x_\mu) = \mu, \quad \text{with } x_\mu > 0.$$

The potential function $G(x) - \mu x$ has a strict minimum at $g^{-1}(\mu) = x_\mu$.

Furthermore, if we denote by $\tau(E)$ the period of these closed orbits, we know that $\tau(\cdot)$ is a continuous function such that $\tau(E) \to 0$ as $E \to +\infty$ (see, for instance, [15]). The asymptotic behaviour of the period function is a consequence of the superlinear condition at infinity and the assumptions on the singularity at zero. When g is strictly monotone and thus we have a global centre, we can look at conditions implying that the period map is strictly decreasing, using results from [10] or [9]. In this case, it is useful to shift the equilibrium point to the origin and take as a potential a function having a minimum at the origin as well. Accordingly, we set $\tilde{g}(x) := g(x + x_\mu) - g(x_\mu) = g(x + x_\mu) - \mu$ and $\tilde{G}(x) := \int_0^x \tilde{g}(s)\,ds = G(x + x_\mu) - G(x_\mu) - \mu x$. For instance, if $g'(x_\mu) > 0$ and $g''(x) > 0$ for $0 < x < x_\mu$, whilst $g''(x) < 0$ for $x > x_\mu$, we conclude that $\tau(E)$ is strictly decreasing, following [10, Corollary 2.3]. A simple example in which this condition is applicable is given by

$$g(x) = -\frac{1}{x^3} + 6x^2, \quad \mu = 5, \; x_\mu = 1.$$

For such a choice of $g(x)$ and μ the phase portrait of (5.20) in the right half-plane is that of a global centre at $(1, 0)$ with strictly decreasing time mapping. ◁

For any given μ we can construct annular regions in the right-half phase-plane between two closed orbits Γ' and Γ'' corresponding to energy levels E' and E'' with $E_0 \leq E' < E''$ in such a way that $\tau(E'') < \tau(E')$. We make this construction twice, for $\mu = k_1$ and for $\mu = k_2$ and denote by E_i' and E_i'' the energies corresponding to k_i for $i = 1, 2$. Choosing appropriately the energy levels, we can manage to have points of Γ_1', Γ_1'' internal to Γ_2' and viceversa. In this manner we obtain two interlinked annuli, as before. Let us call $\mathcal{A}_i$ the closed annular region having Γ_i' and Γ_i'' as inner and outer boundaries, respectively. The intersection of the two annular regions $\mathcal{A}_1$ and $\mathcal{A}_2$ consists into two topological rectangles, namely, two compact planar sets which are homeomorphic to the unit square $[0, 1]^2$. These rectangular regions are

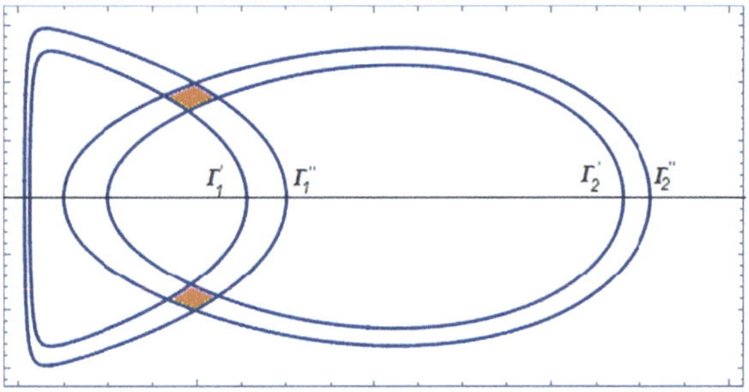

Fig. 5.4 Example of interlinked annuli for Eq. (5.20), putting in evidence the resulting topological rectangles

put in evidence in Fig. 5.4. The example has been obtained for $g(x) = -x^{-3} + x^2$ and for $\mu = \pm 18$.

Having assumed

$$\tau(E_1'') < \tau(E_1') \quad \text{and} \quad \tau(E_2'') < \tau(E_2'),$$

we find that $\mathcal{A}_1$ and $\mathcal{A}_2$ are *linked twist annuli* with respect to system (5.16), namely, a twist condition at the boundary of each annulus can be imposed provided that T_1 and T_2 are large enough. Thus we enter in the theory of linked twist maps for annular regions, as described in detail in Chap. 4 as well as in [13, 28, 37] and obtain the existence of chaotic dynamics according to Definition 4.3, via [28, Theorem 4.1]. For more numerical simulations exhibiting chaos in the Duffing equation with singularities see Chap. 3, Sect. 3.3.3.

References

1. H. Aref, Stirring by chaotic advection. J. Fluid Mech. **143**, 1–21 (1984)
2. H. Aref, Point vortex dynamics: a classical mathematics playground. J. Math. Phys. **48**(6), 065401 (2007)
3. A. Bacciotti, Stability of switched systems: an introduction. In *International Conference on Large-Scale Scientific Computing* (Springer, 2013), pp. 74–80
4. L. Burra, F. Zanolin, Chaotic dynamics in a vertically driven planar pendulum. Nonlinear Anal. **72**(3–4), 1462–1476 (2010)
5. L. Burra, F. Zanolin, Complex dynamics in a planar hamiltonian system for an equation of the duffing type. EPAM J. **2**(1), 3–24 (2016)
6. L. Burra, F. Zanolin, Non-singular solutions of a Rayleigh-Plesset equation under a periodic pressure field. J. Math. Anal. Appl. **435**(2), 1364–1381 (2016)
7. L. Burra, F. Zanolin, Periodically perturbed superlinear Duffing type equations with singularities: subharmonic solutions and complex dynamics. Dyn. Contin. Discrete Impuls. Syst. Ser. A Math. Anal. **26**(5), 349–364 (2019)
8. E.I. Butikov, Square-wave excitation of a linear oscillator. American J. Phys. **72**(4), 469–476 (2004)

9. C. Chicone, The monotonicity of the period function for planar Hamiltonian vector fields. J. Differ. Equ. **69**(3), 310–321 (1987)
10. S. Chow, D. Wang, On the monotonicity of the period function of some second order equations. Časopis pro pěstování matematiky **111**(1), 14–25 (1986)
11. L. Cveticanin, M. Zukovic, Melnikov's criteria and chaos in systems with fractional order deflection. J. Sound Vibr. **326**(3–5), 768–779 (2009)
12. M.A. Del Pino, R.F. Manásevich, Infinitely many T-periodic solutions for a problem arising in nonlinear elasticity. J. Differ. Equ. **103**(2), 260–277 (1993)
13. R.L. Devaney, Subshifts of finite type in linked twist mappings. Proceed. American Math. Soc. **71**(2), 334–338 (1978)
14. T. Ding, F. Zanolin, Periodic solutions of Duffing's equations with superquadratic potential. J. Differ. Equ. **97**(2), 328–378 (1992)
15. A. Fonda, R. Manásevich, F. Zanolin, Subharmonic solutions for some second-order differential equations with singularities. SIAM J. Math. Anal. **24**(5), 1294–1311 (1993)
16. A. Fonda, A. Sfecci, Periodic solutions of weakly coupled superlinear systems. J. Differ. Equ. **260**(3), 2150–2162 (2016)
17. A. Fonda, A. Sfecci, Multiple periodic solutions of Hamiltonian systems confined in a box. Discr. Contin. Dyn. Syst. **37**(3), 1425–1436 (2017)
18. A. Fonda, A. Sfecci, On a singular periodic Ambrosetti-Prodi problem. Nonlinear Anal. **149**, 146–155 (2017)
19. A. Fonda, R. Toader, Subharmonic solutions of Hamiltonian systems displaying some kind of sublinear growth. Adv. Nonlinear Anal. **8**(1), 583–602 (2019)
20. A. Fonda, A.J. Ureña, A higher-dimensional Poincaré-Birkhoff theorem without monotone twist. Comptes Rendus Mathematique **354**(5), 475–479 (2016)
21. W.B. Gordon, Conservative dynamical systems involving strong forces. Trans. American Math. Soc. **204**, 113–135 (1975)
22. H. Gottlieb, J.C. Sprott, Simplest driven conservative chaotic oscillator. Phys. Lett. A **291**(6), 385–388 (2001)
23. J. Hale, *Ordinary Differential Equations*, 2nd edn. (R.E. Krieger Publishing Co., Inc, Huntington, N.Y., 1980)
24. A.C. Lazer, S. Solimini, On periodic solutions of nonlinear differential equations with singularities. Proceed. American Math. Soc. **99**(1), 109–114 (1987)
25. M. Levi, On Littlewood's counterexample of unbounded motions in superquadratic potentials. In *Dynamics Reported* (Springer, 1992), pp. 113–124
26. M. Levi, J. You, Oscillatory escape in a Duffing equation with a polynomial potential. J. Differ. Equ. **140**(2), 415–426 (1997)
27. H. Li, X. Liao, S. Ullah, L. Xiao, Analytical proof on the existence of chaos in a generalized Duffing-type oscillator with fractional-order deflection. Nonlinear Anal. Real World Appl. **13**(6), 2724–2733 (2012)
28. A. Margheri, C. Rebelo, F. Zanolin, Chaos in periodically perturbed planar Hamiltonian systems using linked twist maps. J. Differ. Equ. **249**(12), 3233–3257 (2010)
29. A. Margheri, P.J. Torres, Chaotic dynamics of the Kepler problem with oscillating singularity. Adv. Nonlinear Stud. **16**(3), 551–567 (2016)
30. K.R. Meyer, G.R. Sell, Melnikov transforms, Bernoulli bundles, and almost periodic perturbations. Trans. American Math. Soc. **314**(1), 63–105 (1989)
31. C. Obi, Analytical theory of nonlinear oscillations. vii. the periods of the periodic solutions of the equation $\ddot{x} + g(x) = 0$. J. Math. Anal. Appl. **55**(2), 295–301 (1976)
32. A.V. Osipov, V.A. Pliss, Relaxation oscillations in a system of Duffing type. Differentsial'nye Uravneniya **25**(3), 435–446 (1989)
33. M. Pireddu, F. Zanolin, Chaotic dynamics in the Volterra predator-prey model via linked twist maps. Opuscula Mathematica **28**(4), 567–592 (2008)
34. M. Pireddu, F. Zanolin, Fixed points, periodic points and chaotic dynamics for continuous maps with applications to population dynamics. Lotka-Volterra Related Syst. Recent Develop. Popul. Dyn. **2**, 123 (2013)

35. R. Sturman, J.M. Ottino, S. Wiggins, *The Mathematical Foundations of Mixing: The Linked Twist Map as a Paradigm in Applications: Micro to Macro, Fluids to Solids*, vol. 22. (Cambridge University Press, 2006)
36. P.J. Torres, *Mathematical Models with Singularities: A Zoo of Singular Creatures, Atlantis Briefs in Differential Equations*, vol. 1 (Atlantis Press, Paris, 2015)
37. S. Wiggins, J.M. Ottino, Foundations of chaotic mixing. Phil. Trans. Royal Soc. London Ser. A: Math. Phys. Eng. Sci. **362**(1818), 937–970 (2004)

Chapter 6
Laser

The laser is one of the most important elements in fibre optic links since it generates the coherent optical wave that carries the signal. Applications of direct analogue laser modulation include cable TV, base station links for mobile communication, and antenna remoting [10]. A numerical study of bifurcations of periodically forced lasers have been studied in the past [12]. It was observed that chaotic dynamics would emerge as the amplitude of modulations became sufficiently large. However, the equations we consider in this chapter for the presence of chaotic dynamics are those with relatively low values of forcing [13]. Specifically, we consider a nonlinear problem which is obtained as an approximation for a large number of lasers.

6.1 A Nonlinear Equation Modelling Laser Dynamics

In this section we consider a mathematical model for laser dynamics involving a planar system of Duffing type. We give now the hypotheses on the nonlinear terms and the other parameters which will be assumed throughout the chapter.

Let $f : \mathbb{R}^2 \to \mathbb{R}$ be a locally Lipschitz continuous function and let $p : \mathbb{R} \to \mathbb{R}$ be a (nonconstant) periodic function of period $T > 0$. The first-order differential system

$$\begin{cases} \dot{x} = -y + \varepsilon f(x, y) \\ \dot{y} = (1 + y)(x - p(t)), \end{cases} \tag{6.1}$$

has been considered by several authors in connection with the study of CO_2 lasers of class B (see [7]). In most models, the role of the function f is that of a dissipative perturbation and it has a linear form. The forcing term $p(t)$ takes into account the effects of periodic modulations. The parameter $\varepsilon << 1$ is usually considered as negligible (see [2, 3, 6, 7, 13]), reducing the analysis of (6.1) to the study of the simplified associated system

L. Burra and F. Zanolin, *The Duffing Equation*, Infosys Science Foundation Series in Mathematical Sciences, https://doi.org/10.1007/978-981-97-8301-4_6

$$\begin{cases} \dot{x} = -y \\ \dot{y} = (1+y)(x - p(t)), \end{cases} \tag{6.2}$$

subject to a small perturbation. In the above-quoted articles, the authors assume

$$p(t) = \delta \cos(\omega t).$$

Some bifurcation diagrams [6, 13] and the possibility of period doubling bifurcations leading to chaos [3] for progressively increasing values of δ and changes in the parameter ω, have been investigated in previous works. Some authors use approximate maps from asymptotic expansions, and subharmonic Melnikov theory [2]. Others, (see for example, [15]) have used numerical approaches. In [14], the procedure starts with a periodically driven system having a drive amplitude as an adjustable parameter.

The aim of this example is to propose a different point of view to prove the existence of chaotic-like dynamics for the study of Eq. (6.1). Our approach is topological in nature and "elementary" in the sense that the proofs are mainly based on classical phase-plane analysis and the observation of some stretching along the path properties for the Poincaré map, as detailed in Chap. 4. As was already stressed with respect to different applications, a special property of our method is that it is robust, in the sense that the results are valid for sufficiently small perturbations. This justifies the study of system (6.2), and then extrapolate rigorously the results regarding chaotic dynamics to system (6.1), provided ε is small enough. Actually, we shall study the simplified system

$$\begin{cases} \dot{x} = -y \\ \dot{y} = (1+y)(x - d(t)), \end{cases} \tag{6.3}$$

where $d : \mathbb{R} \to \mathbb{R}$ is a periodic stepwise function. The motivation for a periodic stepwise function comes from the work of Erneux and Glorieux [7]. The authors study a CO_2 laser, where damped relaxation oscillations were observed at the output of this laser. In their study, they note that the laser undergoes step periodic changes in its losses. They observe that as seen from the long-time behaviour, these changes were sufficiently small and did not modify the average intensity. As a conclusion we will prove that the complex behaviour for the solutions of system (6.3) will be valid also for the solutions of Eq. (6.1) provided that

$$\int_0^T |p(t) - d(t)|\, dt < \varepsilon$$

and ε is sufficiently small.

6.2 Topological Approach and Abstract Results

For a detailed description of the topological approach used to prove the chaotic-like dynamics for (6.3) please see Chap. 4.

6.3 Chaotic Dynamics for the Poincaré Map Associated to Eq. (6.3)

Within this framework, we consider the Poincaré map associated with the system (6.1). From the fundamental theory of ODEs we have for every $t_0 \in \mathbb{R}$ and $z_0 = (x_0, y_0) \in \mathbb{R}^2$, that there exists a unique solution $\zeta(\cdot) = \zeta(\cdot; t_0, z_0)$ of (6.1), satisfying the initial condition $x(t_0) = x_0$, $y(t_0) = y_0$, which is defined on a right maximal interval of existence $[t_0, \omega[\subseteq [t_0, +\infty[$. Recall that the Poincaré map for the T-periodic problem is defined as

$$\Phi : \mathbb{R}^2 \supseteq \mathrm{dom}(\Phi) \to \mathbb{R}^2, \quad \Phi : z_0 \mapsto \zeta(t_0 + T; t_0, z_0),$$

that is the function which maps any initial point z_0 (at the time $t = t_0$) to the point after a period T, on the trajectory of (6.1) departing from z_0. Usually, we take $t_0 = 0$ (this will be always assumed, unless otherwise mentioned). The domain $\mathrm{dom}(\Phi)$ of the Poincaré map is an open subset of the plane and Φ is a homeomorphism onto its image. If we choose $p(t) = \delta \cos(\omega t)$ and $\varepsilon = 0$, the system becomes (6.2). To get a geometric feel of Eq. (6.2), we have a simulation of the phase portrait of this system with $p(t) = \delta \cos(\omega t)$, as well as the graphs of the solutions of x *versus* t and y *versus* t. in Figs. 6.1 and 6.2.

Now we come back to system (6.3) which is precisely (6.2) where $p(t) = d(t)$ is a periodic stepwise function. Introducing change of variable $y = z - 1$, changes the system (6.3) to

$$\begin{cases} \dot{x} = 1 - z \\ \dot{z} = z(x - d(t)). \end{cases} \tag{6.4}$$

We are interested in solutions with $y > -1$, that is $z > 0$. Setting $w = \log z, \quad z = e^w$, system (6.4) becomes

$$\begin{cases} \dot{x} = 1 - e^w \\ \dot{w} = x - d(t). \end{cases} \tag{6.5}$$

Having assumed that $d(t)$ is piecewise constant, we choose a time interval where $d(t)$ is constant and set the parameter $\mu = d(t)$. The system for $\mu = \mu_0 > 0$ is

$$\begin{cases} \dot{x} = 1 - e^w \\ \dot{w} = x - \mu_0 \end{cases} \tag{E_0}$$

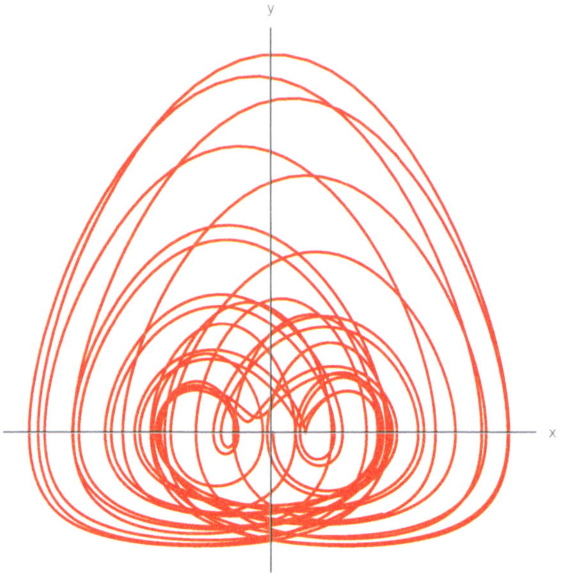

Fig. 6.1 A Phase portrait for system (6.2) for $\delta = .58$ and $\omega = .5$ for initial values $x(0) = .02,\ y(0) = 0$

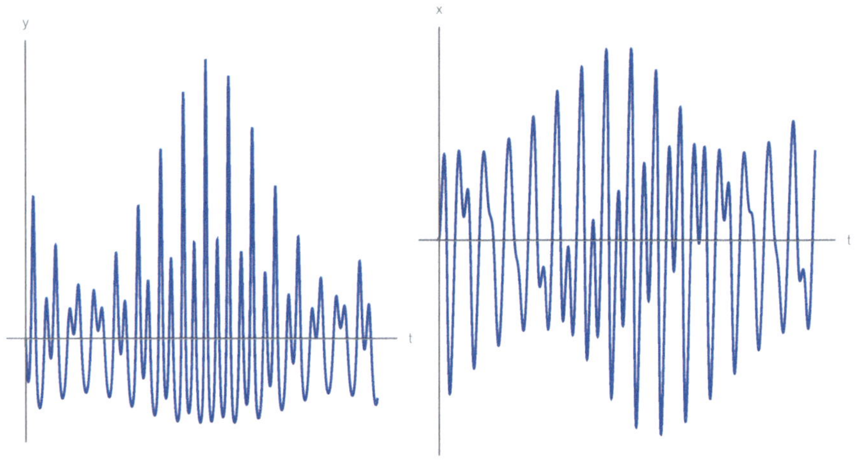

Fig. 6.2 The solution curves for y versus t and x versus t for the system (6.2) for $d = .58$ and $\omega = .5$ for initial values $x(0) = .02,\ y(0) = 0$

which is a planar Hamiltonian system. The level lines of (E_0) for some particular value of μ_0 are shown in Fig. 6.3.

The system for $\mu = -\mu_c$ where $\mu_c > 0$ is

$$\begin{cases} \dot{x} = 1 - e^w \\ \dot{w} = x + \mu_c. \end{cases} \tag{E_c}$$

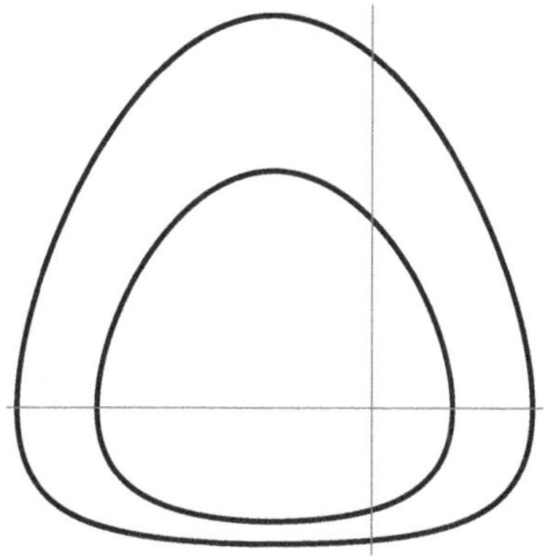

Fig. 6.3 The level lines for system (E_0) for $\mu_0 = .6$ for initial values $x_0 = .5,\ w_0 = 0$ and $x_0 = 1,\ w_0 = 0$

We consider the *switched system* of equations given by (E_0) and (E_c) and look for some T-periodic solutions. As before we split the time period T as

$$T = r_0 + r_c,$$

that is, to be explicit we consider the system (E_0) for $t \in [0, r_0[$. Next we switch to system (E_c) for $t \in [r_0, T[$. At time T we switch back to system (E_0) and repeat the cycle with T-periodicity. In other words, $d(t)$ is extended to the entire real line by T-periodicity.

Such two-state alternating behaviour can be concisely described by assuming

$$d(t) := \begin{cases} \mu_0 & \text{for } 0 \le t < r_0 \\ -\mu_c & \text{for } r_0 \le t < T, \end{cases}$$

μ_0 and μ_c are parameters with $\mu_0,\ \mu_c\ >\ 0.$

6.4 Linked Twist Maps

It is our aim now to prove that system (6.3) has very rich dynamics. In fact, as detailed in Chap. 4 we show the presence of a *topological horseshoe* for the Poincaré map $\Phi : \mathbb{R}^2 \to \mathbb{R}^2$. As a consequence, we will show that all the complexity which is associated with a horseshoe geometry, namely, the semi-conjugation to the Bernoulli shift, sensitivity to initial conditions, and positive topological entropy, can be implied. As in the previous applications, we enter in the framework of linked twist maps by constructing suitably overlapping annular regions. With respect to the example considered, the two annular regions $\mathcal{A}$ and $\mathcal{B}$ cross in two disjoint topological rectangles $\mathcal{R}_1$ and $\mathcal{R}_2$. Each annulus is turned onto itself by a homeomorphism which leaves the boundaries of the annulus invariant.

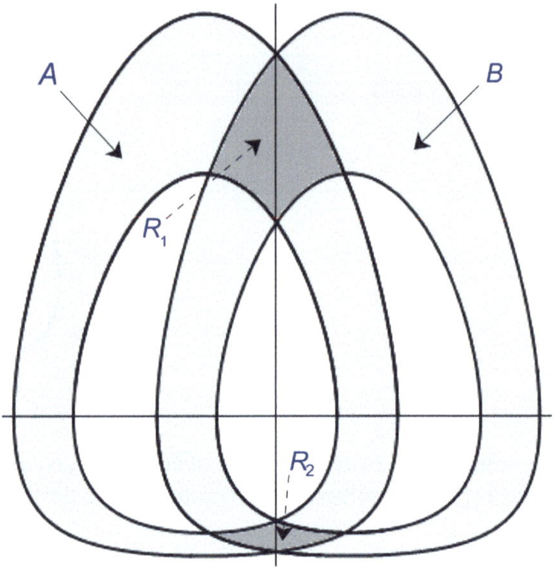

Fig. 6.4 Two annuli formed by linking level lines of (E_0) and (E_c)

With the aid of Fig. 6.4 we try now to explain the horseshoe-type geometry which we alluded to earlier.

To proceed in this direction, we take two closed overlapping annuli which we label as $\mathcal{A}$ and $\mathcal{B}$ made from the level lines of the first integrals associated to the system (E_0) and (E_c), respectively. The inner and outer boundaries of each annulus are closed trajectories surrounding the equilibrium points of both systems. Such annuli intersect in two compact disjoint rectangular sets $\mathcal{R}_1$ and $\mathcal{R}_2$ as illustrated in Fig. 6.4. In a scenario as depicted in Fig. 6.4, we say that the annuli are *linked together*. As a second step, we give an "orientation" to $\mathcal{R}_i$ (for $i = 1, 2$) by selecting, two disjoint compact arcs in the boundary, denoted by $\mathcal{R}^-_{i,left}$ and $\mathcal{R}^-_{i,right}$. We also label the *left* and the *right* sides of $\partial\mathcal{R}_i$. Then we set

$$\mathcal{R}^-_i := \mathcal{R}^-_{i,left} \cup \mathcal{R}^-_{i,right}.$$

The closure of $\partial\mathcal{R}_i \setminus \mathcal{R}^-_i$ is denoted by $\mathcal{R}^+_i$.

This consists of two disjoint components which are compact arcs that we name the **up** and **down** parts of $\partial\mathcal{R}_i$.

In the specific example of Fig. 6.4, we orientate $\mathcal{R}_1$ and $\mathcal{R}_2$ as follows: We take as $\mathcal{R}^-_1$ the intersection of $\mathcal{R}_1$ with the inner and outer boundaries of $\mathcal{A}$ and as $\mathcal{R}^-_2$ the intersection of $\mathcal{R}_2$ with the inner and outer boundaries of $\mathcal{B}$. The way in which we choose to name (as left/right) the two components of $\mathcal{R}^-_i$ is not relevant to the rest of the analysis. As a third step, we observe that the Poincaré map associated to system (6.5) can be decomposed as

$$\Phi = \phi_0 \circ \phi_c,$$

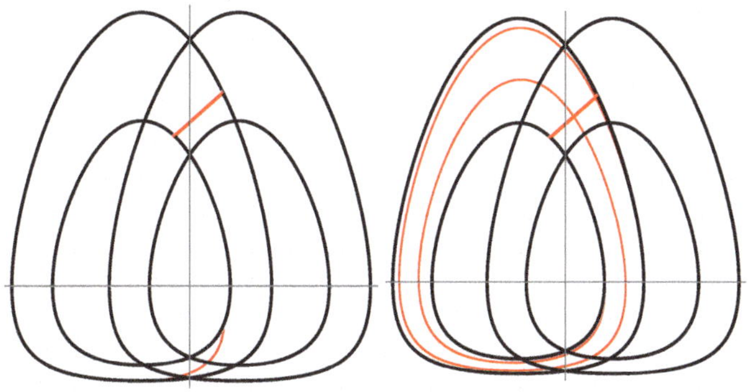

Fig. 6.5 The image of the path "gamma" after $t = 5$, and $t = 250$

where ϕ_0 is the Poincaré map of system (E_0) on the time interval $[0, r_0]$ and ϕ_c is the Poincaré map for (E_c) on the time interval $[0, r_c] = [0, T - r_0]$.

Consider also a path $\gamma : [0, 1] \to \mathcal{R}_1$ with $\gamma(0) \in \mathcal{R}^-_{1,left}$ and $\gamma(1) \in \mathcal{R}^-_{1,right}$.

As can be illustrated in Fig. 6.5 that follows, the points of $\mathcal{R}^-_{1,left}$ move faster than those belonging to $\mathcal{R}^-_{1,right}$ under the action of system (E_0). Hence, for a choice of the first switching time r_0 large enough, it is possible to make the path $[0, 1] \ni \theta \mapsto \phi_0(\gamma(\theta))$ turn in a spiral-like fashion inside the annulus $\mathcal{A}$ and cross at least twice the rectangular region $\mathcal{R}_2$ from $\mathcal{R}^-_{2,left}$ to $\mathcal{R}^-_{2,right}$.

Thus we can select two sub-intervals of $[0, 1]$ such that $\phi_0 \circ \gamma$ restricted to each of these intervals is a path contained in $\mathcal{R}_2$ which connects the two components of $\mathcal{R}^-_2$.

Now, we observe that the points of $\mathcal{R}^-_{2,left}$ move faster than those belonging to $\mathcal{R}^-_{2,right}$ under the action of system (E_0). Therefore, we can repeat the same argument as above and conclude that for a suitable choice of $r_c = T - r_0$ large enough we can transform, via ϕ_c, any path in $\mathcal{R}_2$ joining the two components of $\mathcal{R}^-_2$ onto a path which crosses at least once $\mathcal{R}_1$ from $\mathcal{R}^-_{1,left}$ to $\mathcal{R}^-_{1,right}$.
In conclusion, our main result can be stated as follows:

Theorem 6.1 *For any choice of* μ *with* $\mu > 0$ *and for every pair* $(\mathcal{A}, \mathcal{B})$ *of annuli linked together, the following conclusion holds:*

For every integer $m \geq 2$ *there exist two positive constants* α *and* β, *such that for each*

$$r_0 > \alpha \quad and \quad r_c > \beta,$$

the Poincaré map associated to system (6.5) *induces chaotic dynamics on* m *symbols in* $\mathcal{R}_1$ *and in* $\mathcal{R}_2$.

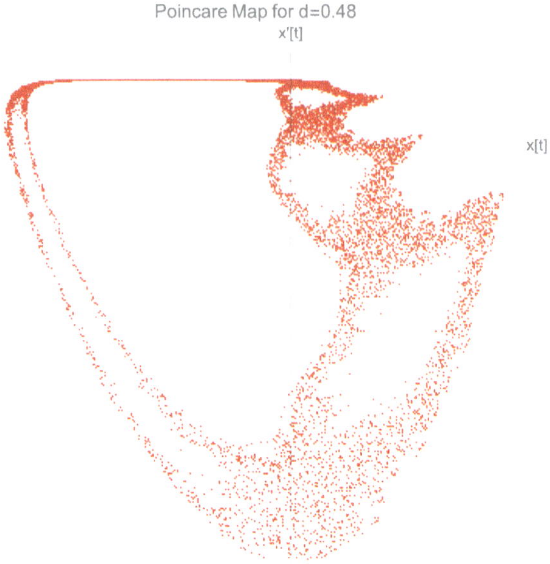

Fig. 6.6 The stroboscopic mapping associated with $\ddot{x} + (x - \delta\ \cos\ \omega t) + \dot{x}(\delta\ \cos\ \omega t - x) = 0$ for $\delta = 0.48,\ \omega = 0.9,\ x(0) = 0.1,\ \dot{x}(0) = 0$

The period map related to the energy level lines associated with equations of type (E_0) or (E_c) has been shown to be monotonic (see for example [4, 5, 8, 9]). Indeed, if we look at Sect. 6.3, we find that the system (E_0) is similar to Eq. (1.31) in Chap. 1, Sect. 1.4.2. This was shown to satisfy the Chicone condition of Theorem 1.3, for monotonic increasing period of the period function.

Linked twist maps (on long periods) arise as a consequence of the monotonicity of the period map. In view of our result in Theorem 6.1 one could conclude that Eq. (6.1) can, under certain assumptions, be shown to possess extremely complex dynamics. In fact it can be shown to exhibit chaotic dynamics.

We end this introductory section with a few observations about our main result. In our approach, which is purely topological, we just have to check a twist hypothesis on the boundary, without the necessity of verifying any hyperbolicity condition (which is required by the "Smale" horseshoe). Hence, our technique allows one to detect the presence of chaotic dynamics by means of elementary tools. We also observe that our result is stable with respect to small perturbations of the coefficients. We include here a Poincaré return map of (6.2) with some particular values of the parameters. Figures 6.6 and 6.7 correspond to system (6.2) with $p(t) = \delta \cos \omega t$, with a slight variation of the initial values.

In Fig. 6.7 two Poincaré return maps are shown with negative values for $x(0)$, and it can be seen that with a very small change in the initial condition, the Poincaré return map exhibits a difference.

Fig. 6.7 The stroboscopic mapping associated with the scalar equation $\ddot{x} + (x - \delta \cos \omega t) + \dot{x}(\delta \cos \omega t - x) = 0$ for $\delta = 0.48$, $\omega = 0.9$, with different initial values $x(0) = -0.1$, $\dot{x}(0) = 0$, (left) and $x(0) = -0.19$, $\dot{x}(0) = 0$ (right)

6.5 Technical Estimates and Proof of the Main Result

Let us consider system (E_0). The first integral of this system is

$$\mathcal{E}_0(x(t), w(t)) := e^{w(t)} - w(t) + \frac{1}{2}x^2(t) - \mu_0 x(t) = C_0^*.$$

Similarly the first integral of the system (E_c)

$$\mathcal{E}_c(x(t), w(t)) := e^{w(t)} - w(t) + \frac{1}{2}x^2(t) + \mu_c x(t) = C_c^*.$$

The level lines Γ_0 and Γ_c of the two systems (E_0) and (E_c), respectively, are closed orbits which run in a counter-clockwise sense, and complete one turn for their fundamental periods which we denote by $\tau_0(l)$ and $\tau_c(l)$, respectively, l denoting the energy level. To be a little more precise we denote

$$\Gamma_0(l) := \{(x, y) \in (\mathbb{R}^2) : \mathcal{E}_0(x(t), w(t)) = l\}.$$

$\Gamma_c(l)$ is defined in a similar manner. The fundamental period of these level lines, for different values of the parameter $\mu = \mu_0$ or $\mu = -\mu_c$ can be easily calculated to be

$$2\int_{w_-}^{w_+} \frac{dw}{\sqrt{\mu^2 + k - 2(e^w - w - 1)}}$$

where k is some constant related to l.

Before giving the details for the proof of our main result, we give the conditions on the energy level lines of two annuli $\mathcal{A}$ and $\mathcal{B}$, to ensure that they are linked together. Assume now we have two closed orbits $\Gamma_0(\ell_1)$ and $\Gamma_0(\ell_2)$ for system

(E_0), with $\ell_1 < \ell_2$. Let the level line $\Gamma_0(\ell_1)$ intersect the $x-$ axis at P_{l_1}, P'_{l_1}, such that $P_{l_1} < P'_{l_1}$, and the level line $\Gamma_0(\ell_2)$ intersect the same line at the points P_{l_2}, P'_{l_2}, $P_{l_2} < P'_{l_2}$, respectively, such that $P_{l_2} < P_{l_1} < P'_{l_1} < P'_{l_2}$. The formula holds because the level lines of energy ℓ_1 are inside the level line of energy ℓ_2.

Similarly let $\Gamma_c(h_1)$ and $\Gamma_c(h_2)$ for system (E_c), with $h_1 < h_2$. Let the level line $\Gamma_c(h_1)$ intersect the $x-$ axis at Q_{h_1}, Q'_{h_1}, such that $Q_{h_1} < Q'_{h_1}$, and the level line $\Gamma_c(h_2)$ intersect the same line at the points Q_{h_2}, Q'_{h_2} such that, $Q_{h_2} < Q'_{h_2}$, respectively, and such that $Q_{h_2} < Q_{h_1} < Q'_{h_1} < Q'_{h_2}$.

Then the two annuli $\mathcal{A}$ and $\mathcal{B}$ will be linked together if

$$P_{l_2} < P_{l_1} < Q_{h_2} < Q_{h_1} < P'_{l_1} < P'_{l_2} < Q'_{h_1} < Q'_{h_2}.$$

Consistent with the notation introduced above, we can denote the level lines in $\mathcal{A}$ and $\mathcal{B}$ by

$$\bigcup_{\ell_1 \le \ell \le \ell_2} \Gamma_0(\ell) \text{ and } \bigcup_{h_1 \le h \le h_2} \Gamma_c(h),$$

respectively. To simplify the explanation of the concept of "stretching" we introduce generalized polar coordinates, so that we can express the solution $\zeta(\cdot, z) = (x(\cdot, z), w(\cdot, z))$ of system (E_0) with initial point $z = (x_0, w_0)$, the angular coordinate $\theta(t, z)$ with the origin shifted to a critical point say P of the system (E_0), for any time t. Also, it can be seen that $\theta(t, z) - \theta(0, z) = 2m\pi$ (with m a non-negative integer) if and only if $t = m\tau_0(\ell)$. Let $m_1 \ge 2$ and $m_2 \ge 1$ be two fixed integers (or equivalently $m_1 = 1$ and $m_2 \ge 2$ can be treated in a similar manner.)

With polar coordinates we can express the solution $\zeta(\cdot, z) = (x(\cdot, z), w(\cdot, z))$ of system (E_0) with initial point in $z = (x_0, w_0) \in \mathcal{A}$ through the radial coordinate $\rho(t, z)$ and the angular coordinate $\theta(t, z)$. Therefore we can assume that $\theta(0, z) \in [\frac{\pi}{4}, \frac{3\pi}{4}]$. For any $t \in [0, r_0]$ and $z \in \mathcal{A}$, we introduce again the *rotation number*, for convenience, the quantity

$$\mathrm{rot}_0(t, z) := \frac{\theta(t, z) - \theta(0, z)}{2\pi},$$

that indicates the normalized angular displacement along the orbit of system (E_0) starting at z, during the time interval $[0, t]$.

From the fact that the time-map τ_0 is strictly increasing, we know that $\tau_0(\ell_1) < \tau_0(\ell_2)$. We shall use this condition to show that a twist property for the rotation number holds for sufficiently large time intervals.

In order for there to be a stretching we claim that the angular separation between initial points on $\Gamma_0(\ell_1)$ and $\Gamma_0(\ell_2)$ after a sufficient time has elapsed is $2\pi + \pi$. That is,

$$\left(\frac{2\pi}{\tau_0(\ell_1)} - \frac{2\pi}{\tau_0(\ell_2)}\right) t \ge 2\pi + \pi$$

Indeed, we claim that if we choose a switching time $r_0 \geq \alpha$, where

$$\alpha := \frac{(m_1 + 3 + \frac{1}{2})\, \tau_0(\ell_1)\, \tau_0(\ell_2)}{\tau_0(\ell_2) - \tau_0(\ell_1)}, \tag{6.6}$$

then, for any path $\gamma : [0, 1] \to \mathcal{R}_1$, with $\gamma(0) \in \Gamma_0(\ell_1)$ and $\gamma(1) \in \Gamma_0(\ell_2)$, the following interval inclusion is fulfilled:

$$[\theta(r_0, \gamma(1)), \theta(r_0, \gamma(0))] \supseteq [2\pi(n + m_1) - \frac{5}{4}\pi,\ (2n + \frac{1}{4})\pi] \quad \text{for some } n = n(r_0) \in \mathbb{N}. \tag{6.7}$$

To check our claim, at first, we note that for a path $\gamma(s)$ as above, it holds that $\mathrm{rot}_0(t, \gamma(0)) \geq \lfloor t/\tau_0(\ell_1) \rfloor$ and $\mathrm{rot}_0(t, \gamma(1)) \leq \lceil t/\tau_0(\ell_2) \rceil$, for every $t > 0$ and so

$$\mathrm{rot}_0(t, \gamma(0)) - \mathrm{rot}_0(t, \gamma(1)) > t\, \frac{\tau_0(\ell_2) - \tau_0(\ell_1)}{\tau_0(\ell_1)\, \tau_0(\ell_2)} - 2, \quad \forall t > 0.$$

Hence, for $t \geq \alpha$, with α defined as in (6.6), we obtain

$$\mathrm{rot}_0(t, \gamma(0)) > m_1 + 1 + \tfrac{1}{2} + \mathrm{rot}_0(t, \gamma(1)),$$

which, in turns, implies

$$\theta(t, \gamma(0)) - \theta(t, \gamma(1)) > 2\pi(m_1 + 1), \quad \forall t \geq \alpha.$$

Therefore, recalling the bound $2\pi(\lceil t/\tau_0(\ell_2) \rceil - \frac{3}{2}) < \theta(t, \gamma(1)) \leq 2\pi \lceil t/\tau_0(\ell_2) \rceil$, the interval inclusion (6.7) is achieved, for

$$n = n(r_0) := \left\lceil \frac{r_0}{\tau_0(\ell_2)} \right\rceil. \tag{6.8}$$

This proves our claim. By the continuity of the composite mapping $[0, 1] \ni s \mapsto \mathrm{rot}_0(r_0, \gamma(s))$, it follows now that

$$\{\theta(r_0, \gamma(s)),\ s \in [0, 1]\} \supseteq [(2n + \frac{1}{4})\pi,\ (n + m_1 - 1)2\pi - 5\frac{\pi}{4}].$$

As a consequence, by the Bolzano theorem, there exist m_1 pairwise disjoint maximal intervals $[t_i{}', t_i{}''] \subseteq [0, 1]$, for $i = 0, \ldots, m_1 - 1$, such that

$$\theta(r_0, \gamma(s)) \in [(2n + 1)\frac{\pi}{4} + 2\pi i,\ 2\pi n - \frac{5\pi}{4} + 2\pi i], \quad \forall s \in [t_i{}', t_i{}''],\ i = 0, \ldots, m_1 - 1,$$

and $\theta(r_0, \gamma(t_i')) = (2n+1)\frac{\pi}{4} + 2\pi i$, as well as $\theta(r_0, \gamma(t_i'')) = 2\pi n - \frac{5\pi}{4} + 2\pi i$.

Having oriented the rectangular regions by choosing

$$\mathcal{R}^-_{1,left} := \mathcal{R}_1 \cap \Gamma_0(\ell_1) \quad \text{and} \quad \mathcal{R}^-_{1,right} := \mathcal{R}_1 \cap \Gamma_0(\ell_2),$$

as well as

$$\mathcal{R}^-_{2,left} := \mathcal{R}_2 \cap \Gamma_c(h_1) \quad \text{and} \quad \mathcal{R}^-_{2,right} := \mathcal{R}_2 \cap \Gamma_c(h_2),$$

we introduce the m_1 nonempty and pairwise disjoint compact sets

$$\begin{aligned}\mathcal{H}_i :=& \{z \in \mathcal{R}_1 : \theta(r_0, z) \in [(2n+1)\frac{\pi}{4} + 2\pi i,\ 2\pi n - \frac{5\pi}{4} + 2\pi i]\},\\ & i = 0, \dots, m_1 - 1.\end{aligned}$$

We are ready to prove that

$$(\mathcal{H}_i, \phi_0) : \widetilde{\mathcal{R}}_1 \Leftrightarrow\!\!\!\!\to \widetilde{\mathcal{R}}_2, \quad i = 0, \dots, m_1 - 1, \tag{6.9}$$

where we recall that ϕ_0 is the Poincaré map associated to system (E_0). Indeed, let's take a path $\gamma : [0, 1] \to \mathcal{R}_1$, with $\gamma(0) \in \mathcal{R}^-_{1,left}$ and $\gamma(1) \in \mathcal{R}^-_{1,right}$. For $r_0 \geq \alpha$ and fixing $i \in \{0, \dots, m_1 - 1\}$, we know that there exists a sub-interval $[t_i', t_i''] \subseteq [0, 1]$, such that $\gamma(t) \in \mathcal{H}_i$ and $\phi_0(\gamma(t)) \in \mathcal{A}$, $\forall t \in [t_i', t_i'']$. Noting now that $\Gamma_\mu(\phi_0(\gamma(t_i'))) \leq h_1$ and $\Gamma_\mu(\phi_0(\gamma(t_i''))) \geq h_2$, it follows that there exists a sub-interval $[t_i^*, t_i^{**}] \subseteq [t_i', t_i'']$ such that $\phi_0(\gamma(t)) \in \mathcal{R}_2$, $\forall t \in [t_i^*, t_i^{**}]$ and $\phi_0(\gamma(t_i^*)) \in \mathcal{R}^-_{2,left}$, as well as $\phi_0(\gamma(t_i^{**})) \in \mathcal{R}^-_{2,right}$. Therefore condition (6.9) is fulfilled.

We now consider the system (E_c). This time we focus our attention on the solutions of such a system starting from $\mathcal{R}_2$ and crossing $\mathcal{R}_1$ at least m_2 times. We define as in the previous case the angular coordinate $\tilde{\theta}(t, z)$ with the origin shifted to a critical point say Q of the system (E_c). The normalized rotation number is now defined as

$$\mathrm{rot}_c(t, w) := \frac{\tilde{\theta}(t, w) - \tilde{\theta}(0, w)}{2\pi}.$$

Since the time-map τ_c is strictly increasing, by choosing a switching time $r_c \geq \beta$, with

$$\beta := \frac{(m_2 + 3 + \frac{1}{2})\,\tau_c(h_1)\,\tau_c(h_2)}{\tau_c(h_2) - \tau_c(h_1)}, \tag{6.10}$$

then, for any path $\sigma : [0, 1] \to \mathcal{R}_2$, with $\sigma(0) \in \Gamma_c(h_1)$ and $\sigma(1) \in \Gamma_c(h_2)$, the following interval inclusion is satisfied:

$$\begin{aligned}[\tilde{\theta}(r_c, \sigma(1)), \tilde{\theta}(r_c, \sigma(0))] \supseteq& [\pi(2n' + \frac{5}{4}), 2\pi(2n' + m_2 + \frac{3}{4})],\\ & \text{for some } n' = n'(r_c) \in \mathbb{N}.\end{aligned} \tag{6.11}$$

The claim can be proved with arguments analogous to the previous ones. The non-negative integer n' has now to be chosen as

$$n' := \left\lceil \frac{r_c}{\tau_c(h_2)} \right\rceil. \tag{6.12}$$

For $\widetilde{\mathcal{R}}_1$ and $\widetilde{\mathcal{R}}_2$ as above and introducing the m_2 nonempty, compact and pairwise disjoint sets

$$\mathcal{K}_i := \{w \in \mathcal{R}_2 : \tilde{\theta}(r_c, w) \in [(2n + \frac{5}{4})\pi + 2\pi i,\ (2n + \frac{7}{4})\pi + 2\pi i]\},$$
$$i = 0, \ldots, m_2 - 1,$$

we claim that

$$(\mathcal{K}_i, \phi_c) : \widetilde{\mathcal{R}}_2 \Bumpeq\!\!\!\!\!\!\longrightarrow \widetilde{\mathcal{R}}_1,\ i = 0, \ldots, m_2 - 1, \tag{6.13}$$

where ϕ_c is the Poincaré map associated to system (E_c). The proof of this claim follows the same argument as in the proof of Eq. (6.9), and therefore we omit it. The stretching properties in (6.9) and (6.13) allow us to apply Theorem 4.5, Chap. 4 and the thesis follows immediately. Note that the classical Melnikov method to detect chaos does not apply to these equations, as pointed out in [13, pg.1084].

We conclude this chapter with a simulation of the Poincaré return map of the equation $\dot{x}(t) = -y(t) - cx(t)(1 + (A-1)(1 + y(t)))\dot{y}(t) = (1 + y(t))(x(t) - d(t))$, the values of the various parameters are shown in the caption of Fig. 6.8 [7].

It is interesting to point out that the same geometric configuration may be seen in the study of a SIR infection model with seasonal effects (see [1]). For further investigations in this direction see also [11].

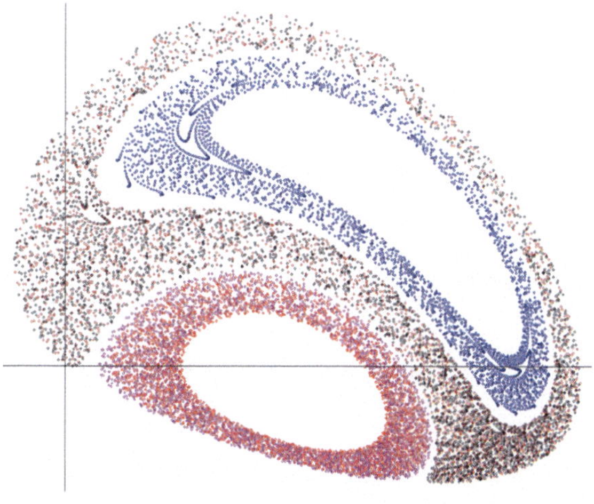

Fig. 6.8 Poincaré return map of equation $\dot{x} = -y - .00001x(t)(2 + y(t)),\ \dot{y} = (1 + y(t))(x(t) - 0.5\tanh(6\cos(.47t)))$

References

1. G. Barrientos, P.J.Á. Rodríguez, A. Ruiz-Herrera, Chaotic dynamics in the seasonally forced SIR epidemic model. J. Math. Biol. **75**(6–7), 1655–1668 (2017)
2. L. Billings, E. Bollt, D. Morgan, I.B. Schwartz, Stochastic global bifurcation in perturbed Hamiltonian systems. Discrete Contin. Dyn. Syst. **2003**(suppl.), 123–132 (2003) Dynamical systems and differential equations (Wilmington, NC, 2002)
3. T.W. Carr, L. Billings, I.B. Schwartz, I. Triandaf, Bi-instability and the global role of unstable resonant orbits in a driven laser. Physica D: Nonlinear Phenomena **147**(1–2), 59–82 (2000)
4. C. Chicone, The monotonicity of the period function for planar Hamiltonian vector fields. J. Differ. Equ. **69**(3), 310–321 (1987)
5. S. Chow, D. Wang, On the monotonicity of the period function of some second order equations. Časopis pro pěstování matematiky **111**(1), 14–25 (1986)
6. T. Erneux, S.M. Baer, P. Mandel, Subharmonic bifurcation and bistability of periodic solutions in a periodically modulated laser. Phys. Rev. A **35**(3), 1165 (1987)
7. T. Erneux, P. Glorieux. *Laser Dynamics* (Cambridge University Press, 2010)
8. E. Freire, A. Gasull, A. Guillamon, First derivative of the period function with applications. J. Differ. Equ. **204**(1), 139–162 (2004)
9. A. Gasull, A. Guillamon, J. Villadelprat, The period function for second-order quadratic odes is monotone. Qual. Theory Dyn. Syst. **4**(2), 329–352 (2004)
10. J. Piprek, J.E. Bowers, *Analog Modulation of Semiconductor Lasers*, vol. 234 (Cambridge University Press, 2002)
11. A. Ruiz Herrera, Paradoxical phenomena and chaotic dynamics in epidemic models subject to vaccination. Commun. Pure Appl. Anal. **19**(5), 2533–2548 (2020)
12. I.B. Schwartz, Infinite primary saddle-node bifurcation in periodically forced systems. Phys. Lett. A **126**(7), 411–418 (1988)
13. I.B. Schwartz, T. Erneux, Subharmonic hysteresis and period doubling bifurcations for a periodically driven laser. SIAM J. Appl. Math. **54**(4), 1083–1100 (1994)
14. I.B. Schwartz, I. Triandaf, R. Meucci, T.W. Carr, Open-loop sustained chaos and control: A manifold approach. Phys. Rev. E **66**(2), 026213 (2002)
15. J.R. Tredicce, F.T. Arecchi, G. Lippi, G. Puccioni, Instabilities in lasers with an injected signal. JOSA B **2**(1), 173–183 (1985)

Chapter 7
The Forced Pendulum

We study the existence of subharmonic solutions and solutions with complicated dynamics in a pendulum equation subjected to a periodic forcing term which is allowed to have a nonzero mean value. Our choice of the periodic perturbation is that of a step function which can be interpreted as two alternating torques. The proofs are based on the theory of topological horseshoes in connection with some geometric features associated to linked twist maps along the lines of Chap. 4.

7.1 The Periodically Forced Pendulum Equation

The title of a famous article by Professor Mawhin *"The forced pendulum: a paradigm for nonlinear analysis and dynamical systems"* [20] expresses very well the importance of the forced pendulum equation

$$\ddot{x} + A \sin x = h(t), \quad A > 0 \tag{7.1}$$

as well as its damped counterpart

$$\ddot{x} + c\dot{x} + A \sin x = h(t), \quad A,\, c > 0, \tag{7.2}$$

as an inexhaustible source of challenging problems which have been tackled with diverse approaches. In some of his fundamental works on this subject, like [18–24], Jean Mawhin has thoroughly illustrated a wealth of results which emerge from these deceptively simple looking equations.

We assume that $h : \mathbb{R} \to \mathbb{R}$ is a T-periodic function which is locally integrable. We also employ a standard notation splitting $h(t)$ as

$$h(t) = \bar{h} + \tilde{h}(t), \quad \text{with } \bar{h} := \frac{1}{T} \int_0^T h(t)\, dt.$$

L. Burra and F. Zanolin, *The Duffing Equation*, Infosys Science Foundation
Series in Mathematical Sciences, https://doi.org/10.1007/978-981-97-8301-4_7

In 1984 Mawhin and Willem [25], using critical point theory for functionals defined on Hilbert spaces, proved a very general result on the existence of multiple periodic solutions for Eq. (7.1). Namely, for an arbitrary T-periodic forcing term $h(\cdot)$ with $h \in L^1([0, T])$ and mean value zero, they proved that *at least two geometrically distinct T-periodic solutions* always exist. For more recent work in this area, including the case of subharmonic solutions, see [9, 15], as well as [4] and the references therein. These latter articles apply different versions of the Poincaré-Birkhoff theorem and the study of rotation numbers as detailed in Chap. 2, Sect. 2.11.

The case in which the mean value of $h(t)$ is nonzero, has been widely investigated, starting with [18], for both Eqs. (7.1) and (7.2). Although in this area of research important progress has been made, there are still some open problems to be investigated, especially, with respect to the multiplicity of periodic solutions (see [11, 12, 16, 26, 27, 29, 32, 34] as well as [23] for contributions in this direction).

A particular case of the above equations, where the mean value of $h(t)$ is not equal to zero, is given by the pendulum equation with a constant (or nearly constant) torque. This latter kind of equations appear as a relevant mathematical model in different contexts. In its simplest formulation, a pendulum with constant torque $r \neq 0$ can be described by the equation

$$\ddot{x} + A \sin x = r \tag{7.3}$$

or by its dissipative counterpart

$$\ddot{x} + c\dot{x} + A \sin x = r, \qquad c > 0. \tag{7.4}$$

In 1931-1933, Tricomi (see [22, 33]), motivated by the researches of Vallauri on synchronous electrical motors, performed an extensive study of Eq. (7.4). The results of Tricomi were well recognized as significant contributions in the theory of nonlinear oscillations and received wide mention in classical books like those of Andronov-Vitt-Khaĭkin [1], Sansone-Conti [28] and Stoker [31]. More recently, generalizations of equation (7.3) and (7.4) have been considered in connection with Josephson's junction equation arising in the studies of superconductivity. For an analysis of a variant of (7.4) in the cylindrical phase space, see [17]. We also recommend [22, 23] and the reference therein for a comprehensive research and survey of these equations.

In the past two decades several authors, using different approaches, have investigated the presence/absence of chaos for the periodically perturbed pendulum Eqs. (7.1) or (7.2). Typical results, as well as numerical simulations, suggest the presence of chaos for small values of the damping and its absence for $c > 0$ large (see, for instance, [14]). In the frictionless case ($c = 0$) the presence of complex dynamics has been proved in [5–7] by exploiting the variational structure of (7.1). Although these results apply to a broad class of equations $\ddot{x} + g(t, x) = h(t)$ with a periodic nonlinearity, including the case $g(t, x) = A \sin(x)$, the zero mean value assumption $\bar{h} = 0$ is required. On the other hand, a Melnikov-type approach was employed in [3] for a second-order equation $\ddot{x} + a^2 g(x) = h(t)$ generalizing (7.1). In [3] the main assumptions require that the autonomous equation has a homoclinic (respectively, heteroclinic) orbit and that the forcing term is sufficiently smooth and such that

$h'(t)$ (respectively, $h(t)$) has a simple zero. The existence of a transversal homoclinic (respectively, heteroclinic) point for the Poincaré map is proved for $A = a^2$ sufficiently large. The forced damped pendulum Eq. (7.2) for $c = 0.1$, $A = 1$ and $h(t) = \cos(t)$ is studied in [13] and the evidence of chaotic dynamics is indicated. A computer assisted proof of the presence of chaos is then provided in [2, 8], with rigorous numerics.

Another important issue about (7.1) and (7.2) is the search of *subharmonic solutions*. Results in this direction have been obtained for the Hamiltonian case ($c = 0$) in [10, 30], again for a zero mean value forcing term.

Statement of the Main Result

The purpose of this chapter is to provide as we stressed in earlier chapters, a simple topological mechanism, different from the approaches adopted in the papers quoted before, which ensures the presence of complex dynamics as well as the existence of multiple harmonic and subharmonic solutions when in (7.3) the torque is given by a periodic function which is approximated by a step function. Again, as mentioned in other applications, our results are stable with respect to small perturbations and can therefore be applied to the dissipative counterpart of the equation, provided that the damping coefficient c is sufficiently small. When applied to Eq. (7.1) (respectively, to (7.2) for c small) our results take into account both the cases $\bar{h} = 0$ and $\bar{h} \neq 0$.

Let $A > 0$ and p, q be fixed constants with

$$-A < p, q < A, \qquad p \neq q. \tag{7.5}$$

We consider the periodically forced pendulum equation

$$\ddot{x} + A \sin x = h_{q,p}(t), \tag{7.6}$$

where $h_{q,p}(t)$ is a T-periodic step function depending on the pair (p, q) and defined as follows:

$$h_{q,p}(t) = \begin{cases} q & \text{for } 0 \leq t < T_1 \\ p & \text{for } T_1 \leq t < T_1 + T_2 = T \,. \end{cases} \tag{7.7}$$

Due to the special nature of the forcing term, we can interpret Eq. (7.6) as a pendulum which is first subjected to a constant torque q during the time interval $[0, T_1[$. At the time T_1 it is subjected to another value of the torque, called p, which lasts for a duration of time T_2. Such a procedure is repeated with T-periodicity, with

$$T = T_1 + T_2 \,.$$

The mean value of the forcing term $h(t) = h_{q,p}(t)$ is given by

$$\bar{h} := \frac{1}{T}\int_0^T h(t)\,dt = \frac{T_1}{T}q + \frac{T_2}{T}p,$$

hence, it is not necessarily equal to zero and can take any value between p and q by suitably tuning T_1 and T_2 .

We shall prove the presence of chaotic-like dynamics for Eq. (7.6), as well as for (7.1) with $h \sim h_{q,p}$ and for (7.2) with $h \sim h_{q,p}$ and c small, under a wide range of values for p and q. For instance, we can take p and q with the same sign or with opposite sign. The case in which $p = 0$ (or $q = 0$) can be handled as well. The only assumption we require on the parameters p and q is a geometric condition of linked annuli type which can be easily determined from the triplet (A, p, q). We shall label as *admissible triplets* in the setting of Eq. (7.6) all those triplets (A, p, q) for which a "geometric linking condition" is satisfied (See Sect. 7.2 for the precise explanation). As an example of admissible triplets, we notice that our geometric condition is always satisfied where p and q have the same sign, whilst for $q < 0 < p$, it holds for $q = -p$ as well as for a wide range of parameters not too far from the symmetry. For any configuration of linked annuli for which our theory applies, we prove the presence of complicated dynamics, including the existence of infinitely many subharmonics, when T_1 and T_2 are sufficiently large. The precise meaning of chaotic dynamics that we use in this chapter is elaborated in Chap. 4, and, in particular, we recall it implies the existence of compact invariant sets where the Poincaré map is semi-conjugate to a Bernoulli shift on m symbols.

All the Eqs. (7.1), (7.2) and (7.6) are of the form

$$\ddot{x} + g(\dot{x}, x) = e(t) \tag{7.8}$$

with $g : \mathbb{R} \times \mathbb{R} \to \mathbb{R}$ a globally Lipschitz continuous functions and $e : \mathbb{R} \to \mathbb{R}$ a T-periodic function with $e \in L^1([0, T])$. Equation (7.8) can be written as the equivalent first-order system

$$\dot{x} = y, \quad \dot{y} = -g(y, x) + e(t). \tag{7.9}$$

From the fundamental theory of ODEs we have, for every $t_0 \in \mathbb{R}$ and $z_0 = (x_0, y_0) \in \mathbb{R}^2$, that there exists a unique solution $\zeta(\cdot) = \zeta(\cdot; t_0, z_0)$ of (7.9) satisfying the initial condition $x(t_0) = x_0$, $y(t_0) = y_0$, which is globally defined on $\mathbb{R}$. The Poincaré map for the T-periodic problem is defined as before as

$$\Phi : \mathbb{R}^2 \to \mathbb{R}^2, \quad \Phi : z_0 \mapsto \zeta(t_0 + T; t_0, z_0), \tag{7.10}$$

Usually, we take $t_0 = 0$ (this will be always assumed unless otherwise mentioned). In our case, Φ is a global homeomorphism of $\mathbb{R}^2$. When we speak of the Poincaré map with reference to a scalar second-order equation with a T-periodic forcing term like (7.8), we implicitly mean the Poincaré map for the associated first-order planar system.

We now state formally our main result.

Theorem 7.1 *Let (A, p, q) be a given admissible triplet. Then, for every integer $m \geq 2$ there exist two positive constants T_1^* and T_2^* such that for every T-periodic forcing term $h_{q,p}(t)$ defined with $T_1 > T_1^*$ and $T_2 > T_2^*$ and $T = T_1 + T_2$, the Poincaré map associated to* (7.6) *induces a full shift on m symbols.*

As mentioned earlier, our result is robust (i.e. persistent under small perturbations of the coefficients). Accordingly, the same theorem applies to (7.2), namely, we have

Theorem 7.2 *Under the assumptions of Theorem 7.1 and for any fixed pair (T_1, T_2) with $T_1 > T_1^*$ and $T_2 > T_2^*$, there exists $\varepsilon > 0$ such that for any $c \in [-\varepsilon, \varepsilon]$ and any T-periodic forcing term $h(t)$ with $h \in L^1([0, T])$, with $||h - h_{q,p}||_{L^1([0,T])} \leq \varepsilon$, the Poincaré map associated to $\ddot{x} + c\dot{x} + A \sin x = h(t)$ induces a full shift on m symbols.*

The conditions on the length of the period means that with our method we can detect chaotic dynamics for a slowly oscillating forcing term. On the other hand, our assumptions could be equivalently rephrased as hypotheses of the largeness of the coefficient of $\sin x$ for fixed T_1 and T_2. Hence, the following dual version of Theorem 7.1 can be obtained:

Theorem 7.3 *Let (A, p, q) be a given admissible triplet and let $h_{q,p}(t)$ be a fixed T-periodic forcing term. for $T = T_1 + T_2$. Then, for every integer $m \geq 2$ there exists Λ_m^* such that for every $\lambda > \Lambda_m^*$ the Poincaré map associated to*

$$\ddot{x} + \lambda A \sin x = \lambda h_{q,p}(t)$$

induces a full shift on m symbols.

A dual version of Theorem 7.2 taking into account damping can be similarly stated.

It may be interesting to compare the results in the present chapter with those in [11]. In [11] the more general periodically perturbed equation

$$\ddot{x} + Kf(x) = h(t) \tag{7.11}$$

was considered, with $f : \mathbb{R} \to \mathbb{R}$ a nonlinear continuous function satisfying suitable properties compatible with the choice $f(x) = \sin x$. Therein multiplicity results for periodic solutions were obtained, with a growing number of subharmonics, as $K \to +\infty$. With a change of variable, such a result corresponds to the multiplicity of subharmonics for the equation

$$\ddot{x} + f(x) = K^{-1} h(t/\sqrt{K}). \tag{7.12}$$

Hence, a large $K > 0$ produces not only a larger period for the forcing term (this will be consistent with our assumptions in Theorems 7.1 and 7.2), but it also has the effect of making the original forcing term smaller in amplitude, which is not our case. With this respect, we also recall the work of Battelli and Palmer [3] described before and point out that, for the same reason, our results differ from their work.

The plan of the chapter is as follows. In the first part of Sect. 7.2 we introduce the topological concepts and definitions used in the article; then we describe the geometric conditions of the linked twist maps type which play a crucial role in our theorems. Section 7.3 is devoted to developing the precise technical estimates and the proof of the main results.

7.2 Topological Lemmas

For the topological tools and the concept of chaos which we use here, we refer to Chap. 4. In particular, let us recall the concept of *generalized rectangles* and their orientation.

Now, from Chap. 4, we have

Theorem 7.4 *Let* $(\mathcal{H}, \chi) : \widetilde{\mathcal{R}} \Leftrightarrow\!\!\!\to \widetilde{\mathcal{R}}$*, where* $\widetilde{\mathcal{R}}$ *is an oriented rectangle of a metric space* X *and* $\mathcal{H} \subseteq \mathcal{R} \cap D_\chi$ *is a compact set. Then* χ *has a fixed point in* $\mathcal{H}$.

As was already explained, this Theorem 7.4 is the basic result to prove the multiplicity of fixed points for a given map, provided that the stretching condition is verified with respect to different (and disjoint) compact subsets (and not just one) of $\mathcal{R} \cap D_\chi$. In order to prove the existence of periodic points, one has to apply Theorem 7.4 to the iterates of χ. The k-th order periodic points which are found with this approach are non-trivial, i.e. they are not wrongly understood points of the lower period. This clearly follows from the Definition 4.4, Chap. 4.

To prove the main results, we use Theorem 4.5 explained in Chap. 4. In the applications of Theorem 4.5 to the ODE models considered in the present chapter, we take $X \equiv \mathbb{R}^2$ and the maps ψ_1, ψ_2 to be the Poincaré maps associated with some planar systems for different time intervals. The *oriented rectangles* will be constructed by a suitable choice of pairs of level lines. Our construction fits within the framework of the theory of the *linked twist maps*.

Homoclinic Orbits and Linking Rectangles

In order to introduce our main result, we need first to recall briefly the phase portraits associated with the system

$$\dot{x} = y, \qquad \dot{y} = -A \sin x + r \tag{7.13}$$

for different values of the constant $r \in]-A, A[$. System (7.13), which is actually Eq. (7.3) written in the phase-plane, is a conservative system and its orbits are contained in the level lines of the energy

$$E_r(x, y) := \frac{1}{2}y^2 + A(1 - \cos x) - r\,x.$$

By the periodicity of the vector field in the x-variable, we can confine ourselves to the strip

$$\mathcal{S} := [-\pi, \pi] \times \mathbb{R}.$$

Note, however, that $E_r(x, y)$ is not 2π-periodic in the x-variable (except when $r = 0$).

• For $r = 0$ we have the well-known phase portrait of the pendulum, with $(0, 0)$ a centre surrounded by periodic orbits filling the open region

$$\mathcal{W}_0 := \{(x, y) \in \mathcal{S} : E_0(x, y) < 2A\}.$$

(See Chap. 1, Fig. 1.3). The boundary of $\mathcal{W}_0$ is made by the saddle points $(\pm\pi, 0)$ connected by two heteroclinic orbits

$$\mathcal{O}^+ := \{(x, y) \in]-\pi, \pi[\, \times \mathbb{R}_0^+ \,:\, E_0(x, y) = 2A\}$$

and

$$\mathcal{O}^- := \{(x, y) \in]-\pi, \pi[\, \times \mathbb{R}_0^- \,:\, E_0(x, y) = 2A\},$$

contained, respectively, in the upper half-plane and in the lower one.

• For $0 < r < A$, we have two equilibrium points $P_r^0 := (x_r^0, 0)$ and $P_r^+ := (x_r^+, 0)$ in the strip $\mathcal{S}$, for

$$x_r^0 := \arcsin(r/A), \quad \text{and } x_r^+ := \pi - x_r^0, \quad \text{with } 0 < x_r^0 < \pi/2 < x_r^+ < \pi.$$

The point P_r^0 is a local centre surrounded by periodic orbits filling the open region

$$\mathcal{W}_r := \{(x, y) \in \mathcal{S} : E_r(x, y) < e_{r,\text{crit}}\} \text{ with } e_{r,\text{crit}} := A + \sqrt{A^2 - r^2} - r x_r^+.$$

The boundary of $\mathcal{W}_r$ is made by the saddle point P_r^+ and a homoclinic separatrix

$$\mathcal{O}_r := \{(x, y) \in \mathcal{S} \setminus \{P_r^+\} \,:\, E_r(x, y) = e_{r,\text{crit}}\}.$$

• For $-A < r < 0$, the situation is completely symmetric to the previous case, leading to a phase portrait which is a reflection of the previous case with respect to the y-axis, as can be seen in Fig. 7.1. Now we have an equilibrium point $P_r^0 := (x_r^0, 0)$ with $x_r^0 := \arcsin(r/A)$ a saddle point and $P_r^- := (x_r^-, 0)$ in the strip $\mathcal{S}$, with $x_r^- := -\pi - x_r^0$. In this case, we have $-\pi < x_r^- < -\pi/2 < x_r^0 < 0$.

In Fig. 7.1 we have drawn the different phase portraits for the corresponding two cases $r > 0$ and $r < 0$.

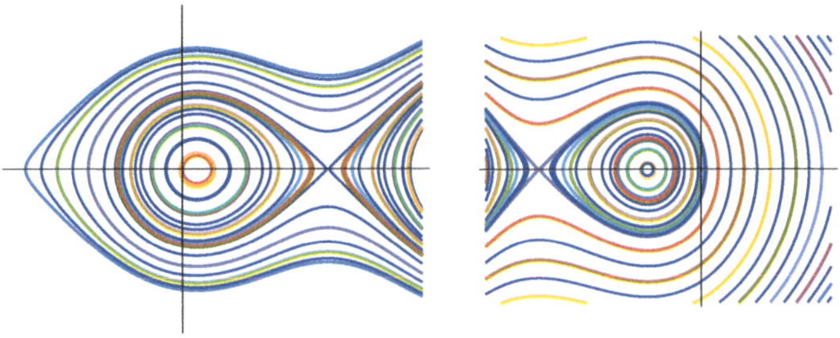

Fig. 7.1 The energy level lines in the phase-plane for Eq. (7.3) with $A = 1$, for $0 < r = 0.3$ (at the left) and $0 > r = -0.7$ (to the right)

Let us consider now Eq. (7.6) for $h_{q,p}(t)$ a T-periodic function defined as in (7.7). For the associated planar system

$$\begin{cases} \dot{x} = y \\ \dot{y} = h_{q,p}(t) - A \sin x \end{cases} \tag{7.14}$$

we can describe the dynamics as a superposition of two autonomous systems of the form (7.13), where, respectively, $r = q$ for $t \in [0, T_1]$ and $r = p$ for $t \in [0, T_2]$ (observe that, due to the autonomous nature of the systems, considering the interval $[0, T_2]$ is equivalent to considering the interval $[T_1, T_1 + T_2]$). Now, the Poincaré map for the T-periodic problem, which is defined as

$$\Phi : \mathbb{R}^2 \supseteq \mathrm{dom}(\Phi) \to \mathbb{R}^2, \quad \Phi : z_0 \mapsto \zeta(T; 0, z_0),$$

for $\zeta(t; 0, z_0)$ the solution of (7.14) with $\zeta(0; 0, z_0) = z_0$, can be expressed as the composition

$$\Phi = \Phi_p^{T_2} \circ \Phi_q^{T_1}, \tag{7.15}$$

where Φ_r^τ denotes the Poincaré map associated to (7.13) for the time interval $[0, \tau]$.

As a next step, we construct two rectangular regions $\mathcal{R}_1$ and $\mathcal{R}_2$ which will be subsequently oriented in order to enter the setting of the abstract topological results referred to in the first part of this section. Such rectangular regions are obtained as intersections of suitably chosen strips surrounding or emanating from the homoclinic loops. More precisely, when p and q have different sign, $\mathcal{R}_1$ and $\mathcal{R}_2$ come from the intersection of two annular regions $\mathcal{A}$ and $\mathcal{B}$ inside the homoclinics. When p and q have the same sign, $\mathcal{R}_1$ and $\mathcal{R}_2$ arise from the intersection of one annular region bounded by one homoclinic orbit and two "tails" emanating from the other homoclinic point. Due to the Hamiltonian structure of system (7.13) we can describe the pertinent regions for the different cases, by means of energy level lines. Accordingly, we examine different possibilities for linking such pairs of level lines.

• In the first instance we choose two values of p and q with $0 < p < q < 1$. One such a pair of level lines is shown in Fig. 7.2. The case $-1 < q < p < 0$ is completely symmetric and its treatment is hence omitted.

When these are superimposed on each other, they form the rectangles as is seen in the intersecting portions of Fig. 7.3.

• In the next instance we consider two values for p and q, with $q < 0 < p$, construct two annuli near the homoclinics and link them as before to form the two rectangles.

We explain in detail how the precise parameters required to form the topological rectangles for this second case: Let the homoclinic points of the two systems for $r = p$ and $r = q$, using the notation introduced earlier be x_p^+, and x_q^-. Let the other points of intersection of the homoclinic orbit with the x-axis be $x_{p_h}^+$, and $x_{q_h}^-$. Let the periodic orbits of the two systems intersect the x-axis at the points $x_\ell^{p_{per}}$, $x_\rho^{p_{per}}$, and, respectively, at the points $x_\ell^{q_{per}}$, $x_\rho^{q_{per}}$, with ℓ for left and ρ for right. The phase portrait of the system (7.14), has periodic orbits lying inside the homoclinic orbits,

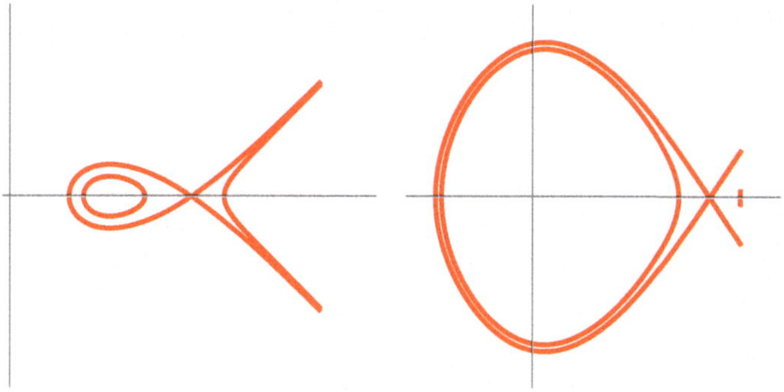

Fig. 7.2 The tails for $q = 0.9$ (left) and the annulus for $p = 0.2$ (right)

Fig. 7.3 The two regions for $q = 0.9$ (left) and $p = 0.2$ (right) superimposed to form the two rectangles

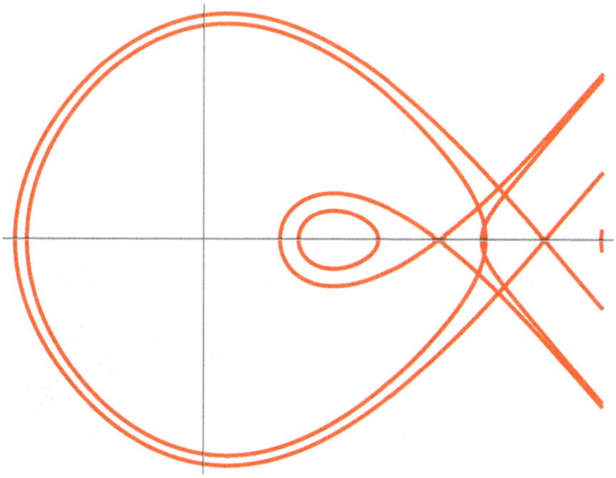

this can be seen in Fig. 7.4. We superimpose these level lines, as shown in Fig. 7.5, with the condition

$$x_{\ell}^{q_{per}} < x_{p_h}^{+} < x_{\ell}^{p_{per}} < x_{\rho}^{q_{per}}$$

to form the rectangles. In the example chosen we recall that $p > 0$ and $q < 0$.

Next we orient $\mathcal{R}_1$ and $\mathcal{R}_2$ as follows: We take as $\mathcal{R}_1^-$ the intersection of $\mathcal{R}_1$ with the inner and outer boundaries of $\mathcal{A}$ and as $\mathcal{R}_2^-$ the intersection of $\mathcal{R}_2$ with the inner and outer boundaries of $\mathcal{B}$. (See Fig. 7.6).

As always, the way in which we choose to name (as left/right) the two components of $\mathcal{R}_i^-$ is not relevant to the rest of the analysis. In the next step, we recall the splitting (7.15) of the Poincaré map associated to (7.14) and look for a deformation of a path under its influence. Consider a path $\gamma : [0, 1] \to \mathcal{R}_1$ with $\gamma(0) \in \mathcal{R}_{1,left}^-$ and $\gamma(1) \in \mathcal{R}_{1,right}^-$.

As can be illustrated in the figures that follow, the points on the periodic level line move faster than those on the homoclinic level line. In fact the points on the homoclinic level line approach the saddle point as $t \to \infty$ and therefore they can never complete a full turn (indeed, in some cases even a half turn may not be reached). Hence for the choice of the first switching time T_1 large enough, it is possible to make

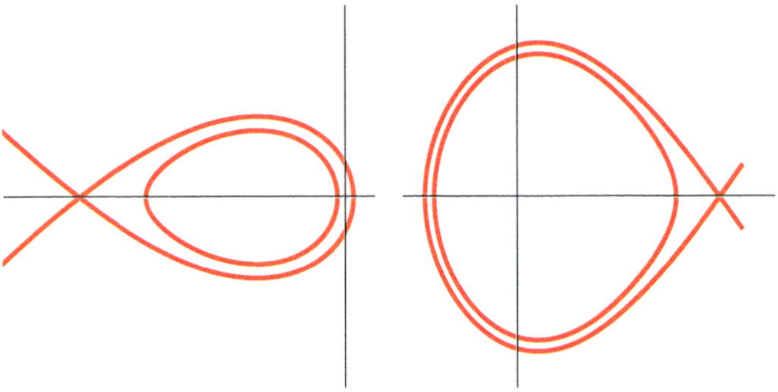

Fig. 7.4 The two annuli before linking, for $p = 0.3$ (right) and $q = -0.7$ (left), showing the annular regions $\mathcal{A}$ on the left and $\mathcal{B}$ on the right

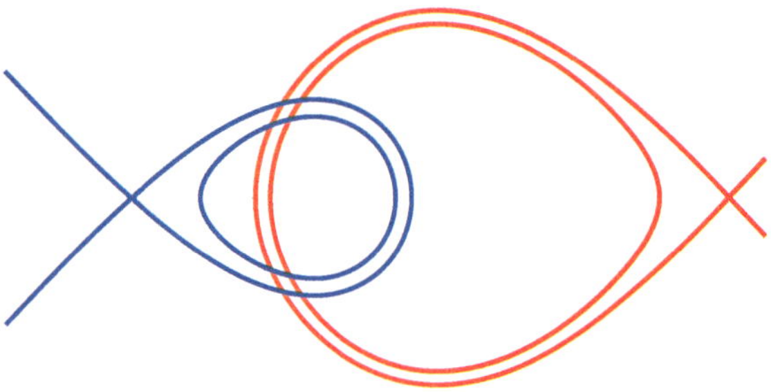

Fig. 7.5 The oriented rectangles for $p = 0.28$ (right) and $q = -0.7$ (left)

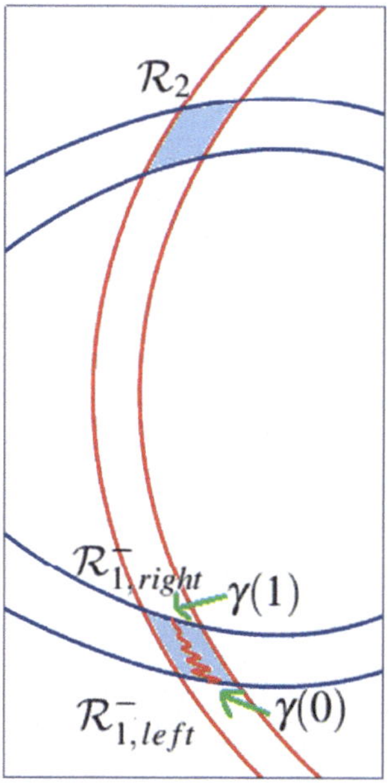

Fig. 7.6 The oriented rectangles $\mathcal{R}_1$ and $\mathcal{R}_2$ for $p = 0.28$ and $q = -0.7$ showing the boundaries of $\mathcal{R}_1$: $\mathcal{R}^-_{1,\mathrm{left}}$ and $\mathcal{R}^-_{1,\mathrm{right}}$, and the path γ

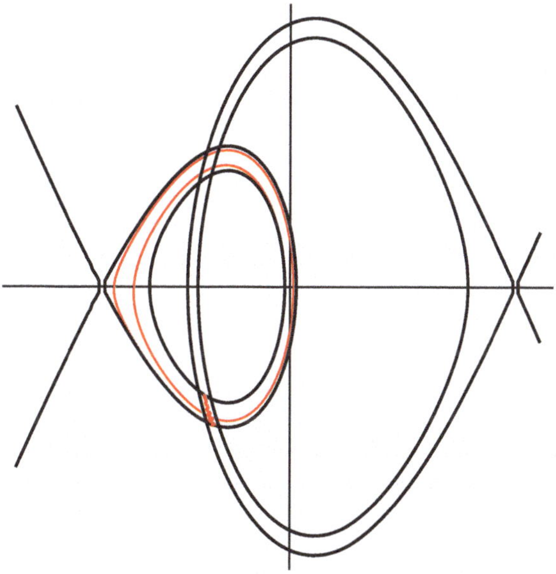

Fig. 7.7 The image of the path γ after $t = 50$

the path $[0, 1] \ni \theta \mapsto \Phi_q^{T_1}(\gamma(\theta))$ turn in a spiral-like fashion inside the annulus $\mathcal{A}$ and cross at least twice the rectangular region $\mathcal{R}_2$, as can be seen clearly in Fig. 7.7.

Thus we can select two sub-intervals of $[0, 1]$ such that $\Phi_q^{T_1} \circ \gamma$ restricted to each of these intervals is a path contained in $\mathcal{R}_2$ which connects the two components of $\mathcal{R}_2^-$.

Now, we observe that the points of $\mathcal{R}^-_{2,right}$ which are on the periodic level line move faster than those belonging to $\mathcal{R}^-_{2,left}$ under the action of the second system. Therefore, we can repeat the same argument as above and conclude that for a suitable choice of T_2 large enough we can transform, via $\Phi_p^{T_2}$, any path in $\mathcal{R}_2$ joining the two components of $\mathcal{R}_2^-$ onto a path which crosses at least once $\mathcal{R}_1$ from $\mathcal{R}^-_{1,left}$ to $\mathcal{R}^-_{1,right}$. Clearly taking a larger value for T_1 we can have $m > 2$ crossings.

In conclusion, the above graphical explanation and the results in Chap. 4 (Theorem 4.5), provides a visual proof of Theorem 7.1.

In what follows, for the technical estimates, it is convenient to represent the T-periodic system (7.14) by two different autonomous systems

$$\begin{aligned} \dot{x} &= y \\ \dot{y} &= q - A\sin x \end{aligned} \tag{7.16}$$

and

$$\begin{aligned} \dot{x} &= y \\ \dot{y} &= p - A\sin x, \end{aligned} \tag{7.17}$$

the first one for a time interval of length T_1 and the second one for a time interval of length T_2 (recall that $T = T_1 + T_2$).

7.3 Technical Estimates and Proof of the Main Result

In this section, we construct the proof of topological lemmas stated in Sect. 7.2 as applied to the Poincaré map Φ associated to system (7.14) and recalling the splitting of Φ in Eq. (7.15).

The plan of the proof is to show that Theorem 4.5, Chap. 4, applies to Φ with $\psi_1 = \Phi_q^{T_1}$ and $\psi_2 = \Phi_p^{T_2}$ for a suitable choice of oriented rectangles. As described in the visual examples of the preceding section, such oriented rectangles will be obtained by the intersection of invariant regions between the level lines of the systems (7.16) and (7.17). To this end, we first study the energy level lines associated to (7.13) focussing our attention to the strip

$$\mathcal{S} := [-\pi, \pi] \times \mathbb{R}.$$

The first integral or the energy function for the system (7.13) is given by

$$E(x, y) = E_r(x, y) := \frac{1}{2}y^2 + A(1 - \cos x) - r\,x.$$

(The trajectories of (7.13) lie on energy level lines of constant energy). We denote by

$$\Gamma^e = \Gamma_r^e := \{(x, y) \in \mathcal{S} : E_r(x, y) = e\}$$

the part of the level line of energy e which is contained in $\mathcal{S}$. We also introduce the critical energy (or the energy of the homoclinic orbit)

$$e_h = e_{r,\mathrm{crit}} := A + \sqrt{A^2 - r^2} - r x_r^+.$$

Where, as introduced earlier, $(x_r^+, 0)$ is the homoclinic point of the system (7.13). We mention here that, for simplicity, we consider only the case $r > 0$, giving us the homoclinic point $(x_r^+, 0)$. For $r < 0$, as previously observed, the treatment is completely symmetric, giving the other homoclinic point $(x_r^-, 0)$. For this reason we proceed considering only x_r^+.

Now, for every r, we define the set

$$\mathcal{W} = \mathcal{W}_r := \{(x, y) \in \mathcal{S} : E_r(x, y) \le e_{r,\mathrm{crit}}\},$$

which is a compact subset of $\mathcal{S}$ contained in the rectangle $[-\pi, \pi] \times [-y_r, y_r]$, where

$$y_r := \sqrt{2(A + \sqrt{A^2 - r^2} - r x_r^+)}$$

is the maximal height of $\Gamma_r^{e_{\mathrm{crit}}}$.

The boundary of $\mathcal{W}$ can be represented as

$$\partial\mathcal{W}_r = \{(x, y) \in \mathcal{S} : E_r(x, y) = e_{r,\mathrm{crit}}\} = \Gamma_r^{e_{r,\mathrm{crit}}} = \mathcal{O}_r \cup \{(x_r^+, 0)\},$$

where $\mathcal{O}_r$ is the homoclinic orbit

$$\mathcal{O}_r := \{(x, y) \in \mathcal{S} : E_r(x, y) = e_{r,\mathrm{crit}}\}.$$

For every fixed value $e \in \mathbb{R}$, with

$$0 < e < e_{r,\mathrm{crit}},$$

the level line Γ^e is a periodic orbit of system (7.13) (see, as an example, the internal two orbits in Fig. 7.4). We denote by $\mathcal{T}_r(e)$ the fundamental period of Γ_r^e . Such minimal period can be computed by the time-mapping formula (see Chap. 1).

$$\mathcal{T}_r(e) := 2 \int_{x^-(e)}^{x^+(e)} \frac{ds}{\sqrt{2(\sqrt{A^2 - r^2} + A \cos x + r(s - x^+(e))}}. \tag{7.18}$$

For any fixed r, the period time mapping

$$]0, e_{r,\text{crit}}[\ni e \mapsto \mathcal{T}_r(e) \in]0, +\infty)$$

is continuous and such that $\mathcal{T}_r(e) \to +\infty$ as $e \to e_{r,\text{crit}}$. As before, to simplify explanation of the concept of "stretching" we introduce generalized polar coordinates, so that we can express the solution $\zeta(\cdot, z) = (x(\cdot, z), w(\cdot, z))$ of system (7.17) with initial point in $z = (x_0, w_0) \in \mathcal{A}$ through the radial coordinate $\rho(t, z)$ and the angular coordinate $\theta(t, z)$. Therefore, taking the angular coordinates in the clockwise sense (starting from the positive $x-$axis with the usual orientation), we can assume that $\theta(0, z) \in [0, \pi]$. Let $m_1 \geq 2$ and $m_2 \geq 1$ be two fixed integers or equivalently $m_1 = 1$ and $m_2 \geq 2$. We now fix a level (for $r = q$),

$$e_0 \in]0, e_{r,\text{crit}}[$$

and focus our attention on the region in the strip $\mathcal{S}$ between $\Gamma_r^{e_0}$ and $\Gamma_r^{e_{r,\text{crit}}}$, and define the set

$$\mathcal{W}_{e_0} = \mathcal{W}_{e_0,r} := \{(x, y) \in \mathcal{S} : e_0 \leq E_r(x, y) \leq e_{r,\text{crit}}\},$$

Notice that the set $\mathcal{W}_{e_0,r}$ is invariant, that is, every solution $(x(t), y(t))$ of Eq. (7.13) with initial point $(x(0), y(0)) \in \mathcal{W}_{e_0,r}$, remains in $\mathcal{W}_{e_0,r}$ for all t. We also introduce the sets:

$$\mathcal{U} := \{(x, y) \in \mathcal{S} : e_0 \leq E(x, y) \leq e_{r,\text{crit}},\ \ x \geq 0,\ y \geq 0\},$$

$$\mathcal{V} := \{(x, y) \in \mathcal{S} : e_0 \leq E(x, y) \leq e_{r,\text{crit}},\ \ x \geq 0,\ y \leq 0\}.$$

We are interested in those solutions with initial points in $\mathcal{V}$, for which we assume

$$\theta(0, z) \in [0, \pi].$$

Note that, for $e = e_{r,\text{crit}}$ the solution lies on the homoclinic orbit $\mathcal{O}$. Let Γ_p^e and Γ_q^e denote the periodic level lines of the systems (7.16) and (7.17), respectively. Let $\Gamma_p^{e_{r,\text{crit}}}$ $\Gamma_q^{e_{r,\text{crit}}}$, be the corresponding homoclinic level lines. Our claim is that for a sufficiently large time intervals, a twist condition can be proved. Let $\gamma : [0, 1]$ be any path with $\gamma(0) \in \mathcal{R}^-_{1,\text{right}}$ and $\gamma(1) \in \mathcal{R}^-_{1,\text{left}}$. For the angular coordinate taken in the clockwise sense we have:

$\theta(\tau, \gamma(0)) < \pi\ \ \forall\, \tau \geq 0, \quad \theta(\tau, \gamma(1)) > 2k\pi$ if $\tau > k\tau_q(e)$.

For a stretching to happen with m_1 crossings of the region $\mathcal{R}_2$ we claim that the angular separation between initial points on Γ_q^{crit} and Γ_q^{e} and is $2(m_1 - 1)\pi + \pi$. (After a sufficient time has elapsed). Indeed, we claim that if we choose a switching time $\geq \alpha$, where

$$\alpha := m_1\tau_q(e) \tag{7.19}$$

then, for any path $\gamma : [0, 1] \to \mathcal{R}_1$, with $\gamma(0) \in \Gamma_q^{\text{crit}}$ and $\gamma(1) \in \Gamma_q^e$, the following interval inclusion happens:

$$[\theta(\tau, \gamma(1)), \theta(\tau, \gamma(0))] \supseteq [\pi, 2m_1\pi] \quad \text{for any} \quad \tau \geq \alpha.$$

By the continuity of the composite mapping $[0, 1] \ni s \mapsto \theta(\tau, \gamma(s)$ it follows now that)

$$\{\theta(\tau, \gamma(s)),\ s \in [0, 1]\} \supseteq [\pi,\ 2m_1\pi].$$

We now invoke the Bolzano theorem, to claim that there exist m_1 pairwise disjoint maximal intervals $[t_i{}', t_i{}''] \subseteq [0, 1]$, for $i = 0, \ldots, m_1 - 1$, such that

$$\theta(\tau, \gamma(s)) \in [\pi + i2\pi, \pi + (i + 1)2\pi],\ \forall s \in [t_i{}', t_i{}''],\ i = 0, \ldots, m_1 - 1,$$

and $\theta(\tau, \gamma(t_i{}')) = \pi + i2\pi$, as well as $\theta(\tau, \gamma(t_i{}'')) = \pi + 2(i + 1)\pi$.

Introducing m_1 nonempty and pairwise disjoint compact sets

$$\mathcal{H}_i := \{z \in \mathcal{A} : \theta(\tau, z) \in [\pi + i2\pi, \pi + (i + 1)2\pi]\},\ i = 0, \ldots, m_1 - 1,$$

we are ready to prove that

$$(\mathcal{H}_i, \psi_1) : \widetilde{\mathcal{R}}_1 \Leftrightarrow\!\!\!\!\rightarrow \widetilde{\mathcal{R}}_2, \quad i = 0, \ldots, m_1 - 1, \tag{7.20}$$

where we recall that ψ_1 is the Poincaré map associated to system (7.17). Indeed, let's take a path $\gamma : [0, 1] \to \mathcal{R}_1$, with $\gamma(0) \in \mathcal{R}^-_{1,left}$ and $\gamma(1) \in \mathcal{R}^-_{1,right}$. For $\tau \geq \alpha$ and fixing $i \in \{0, \ldots, m_1 - 1\}$, we know that there exists a sub-interval $[t_i{}', t_i{}''] \subseteq [0, 1]$, such that $\gamma(t) \in \mathcal{H}_i$ and $\psi_1(\gamma(t)) \in \mathcal{A}$, $\forall t \in [t_i{}', t_i{}'']$. By continuity it follows that there exists a sub-interval $[t_i^*, t_i^{**}] \subseteq [t_i{}', t_i{}'']$ such that $\psi_1(\gamma(t)) \in \mathcal{R}_2$, $\forall t \in [t_i^*, t_i^{**}]$ and $\psi_1(\gamma(t_i^*)) \in \mathcal{R}^-_{2,left}$, as well as $\psi_1(\gamma(t_i^{**})) \in \mathcal{R}^-_{2,right}$. Therefore condition (7.20) is fulfilled.

We now consider the system (7.16). By a completely symmetric argument we find a time $\beta := m_2 T_p(e)$. In a similar manner, for a switching time $T_2 > \beta$ we can check that

$$(\mathcal{K}_i, \psi_2) : \widetilde{\mathcal{R}}_2 \Leftrightarrow\!\!\!\!\rightarrow \widetilde{\mathcal{R}}_1,\ i = 0, \ldots, m_2 - 1, \tag{7.21}$$

where ψ_2 is the Poincaré map associated to system (7.17). The stretching properties in (7.20) and (7.21) allow to apply Theorem 4.5, Chap. 4, and the thesis follows immediately.

We now include here some Poincaré return maps for the equation of the damped and forced pendulum, Eq. (7.2)

$$\ddot{x} + c\dot{x} + A \sin x = h(t),$$

with $h(t) = .2\cos(t) + .3$, $A = 1$, and $c = .00015$.

In the next simulation, for the perturbation, we choose instead of a continuous function, a truncated cosine function, which, in our opinion, may be an acceptable approximation to a simulation of the pendulum with the constant torque in the discontinuous steps of the truncated cosine function. We set

$$\cos_k(s) := \max\{-k, \min\{k, \cos(s)\}\}, \quad \text{for } 0 < k < 1.$$

Observing Figs. 7.8 and 7.9 it appears that the presence of even very small friction produces some stable periodic (harmonic or subharmonic) solutions; on the other

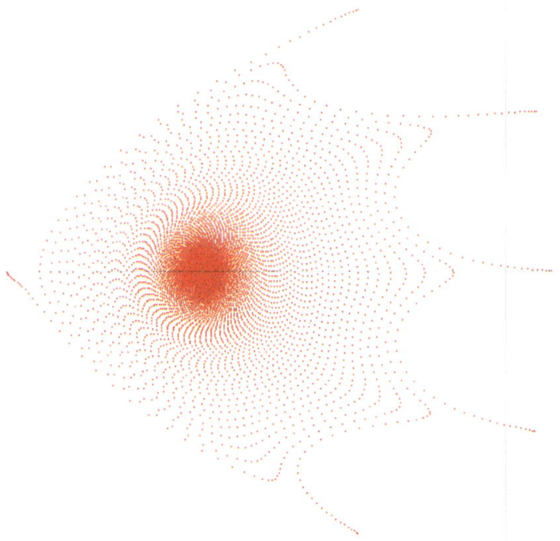

Fig. 7.8 The Poincaré return map for Eq. (7.2), $x(0) = 0.1$, $\dot{x}(0) = 0$

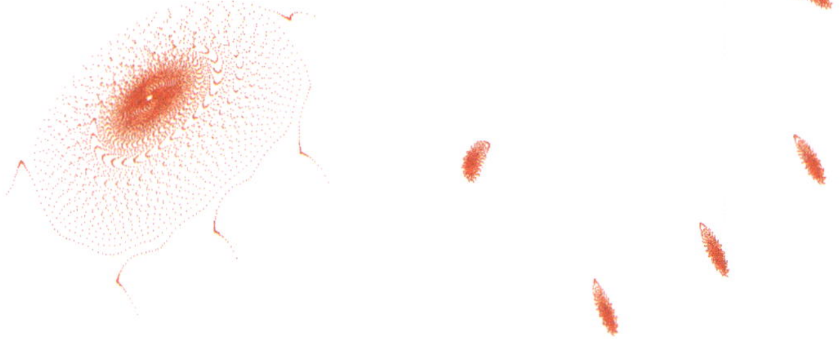

Fig. 7.9 The Poincaré return map for Eq. (7.2), $x(0) = 0.021367$, $\dot{x}(0) = 0$; left and $x(0) = 0.021366$, $\dot{x}(0) = 0$ on the right, replacing $h(t) = \cos_k(t + 1)$

hand these simulations also reveal geometrically the chaotic nature of Eq. 7.2, as it is evident that an extremely small change in the initial conditions results in two completely different Poincaré return maps.

References

1. A.A. Andronov, A.A. Vitt, S.È. Khaĭkin, *Theory of Oscillators*. (Dover Publications, Inc., New York, 1987). Translated from the Russian by F. Immirzi, Reprint of the 1966 translation
2. B. Bánhelyi, T. Csendes, B.M. Garay, L. Hatvani, A computer-assisted proof of σ_3-chaos in the forced damped pendulum equation. SIAM J. Appl. Dyn. Syst. **7**(3), 843–867 (2008)
3. F. Battelli, K.J. Palmer, Chaos in the Duffing equation. J. Differ. Equ. **101**(2), 276–301 (1993)
4. A. Boscaggin, R. Ortega, F. Zanolin, Subharmonic solutions of the forced pendulum equation: a symplectic approach. Arch. Math. (Basel) **102**(5), 459–468 (2014)
5. E. Bosetto, E. Serra, A variational approach to chaotic dynamics in periodically forced nonlinear oscillators. Ann. Inst. H. Poincaré Anal. Non Linéaire **17**(6), 673–709 (2000)
6. E. Bosetto, E. Serra, S. Terracini, Density of chaotic dynamics in periodically forced pendulum-type equations. Atti Accad. Naz. Lincei Cl. Sci. Fis. Mat. Natur. Rend. Lincei (9) Mat. Appl. **12**, 107–113 (2002)
7. E. Bosetto, E. Serra, S. Terracini, Generic-type results for chaotic dynamics in equations with periodic forcing terms. J. Differ. Equ. **180**(1), 99–124 (2002)
8. T. Csendes, B. Bánhelyi, L. Hatvani, Towards a computer-assisted proof for chaos in a forced damped pendulum equation. J. Comput. Appl. Math. **199**(2), 378–383 (2007)
9. A. Fonda, R. Toader, Periodic solutions of pendulum-like Hamiltonian systems in the plane. Adv. Nonlinear Stud. **12**(2), 395–408 (2012)
10. A. Fonda, M. Willem, Subharmonic oscillations of forced pendulum-type equations. J. Differ. Equ. **81**(2), 215–220 (1989)
11. A. Fonda, F. Zanolin, Periodic oscillations of forced pendulums with very small length. Proc. Roy. Soc. Edinburgh Sect. A **127**(1), 67–76 (1997)
12. G. Fournier, J. Mawhin, On periodic solutions of forced pendulum-like equations. J. Differ. Equ. **60**(3), 381–395 (1985)
13. J. Hubbard, The forced damped pendulum: chaos, complication and control. American Math. Mon. **106**(8), 741–758 (1999)
14. M. Levi, Nonchaotic behavior in the Josephson junction. Phys. Rev. A (3), **37**(3), 927–931 (1988)
15. S. Marò, Periodic solutions of a forced relativistic pendulum via twist dynamics. Topol. Methods Nonlinear Anal. **42**(1), 51–75 (2013)
16. P. Martínez-Amores, J. Mawhin, R. Ortega, M. Willem, Generic results for the existence of nondegenerate periodic solutions of some differential systems with periodic nonlinearities. J. Differ. Equ. **91**(1), 138–148 (1991)
17. R. Martins, The attractor of an equation of Tricomi's type. J. Math. Anal. Appl. **342**(2), 1265–1270 (2008)
18. J. Mawhin, Periodic oscillations of forced pendulum-like equations. In *Ordinary and partial differential equations (Dundee, 1982)*, volume 964 of *Lecture Notes in Math.* (Springer, Berlin-New York, 1982), pp. 458–476
19. J. Mawhin, Recent results on periodic solutions of the forced pendulum equation. Rend. Istit. Mat. Univ. Trieste **19**(2), 119–129 (1987)
20. J. Mawhin, The forced pendulum: A paradigm for nonlinear analysis and dynamical systems. Expo. Math **6**(3), 271–287 (1988)
21. J. Mawhin, Seventy-five years of global analysis around the forced pendulum equation. Proceed. Equadiff **9**, 115–145 (1997)

22. J. Mawhin, From Tricomi's equation for synchronous motors to the periodically forced pendulum. Atti dei convegni Lincei-accademia nazionale dei Lincei **147**, 251–270 (1998)
23. J. Mawhin, Global results for the forced pendulum equation. Handbook Differ. Equ. **1**, 533–589 (2004)
24. J. Mawhin, Periodic solutions of the forced pendulum: classical vs relativistic. Matematiche (Catania) **65**(2), 97–107 (2010)
25. J. Mawhin, M. Willem, Multiple solutions of the periodic boundary value problem for some forced pendulum-type equations. J. Differ. Equ. **52**(2), 264–287 (1984)
26. R. Ortega, A forced pendulum equation with many periodic solutions. Rocky Mountain J. Math. **27**(3), 861–876 (1997)
27. R. Ortega, Counting periodic solutions of the forced pendulum equation. Nonlinear Anal. Theory Methods Appl. **42**(6), 1055–1062 (2000)
28. G. Sansone, R. Conti, *Non-linear Differential Equations. International Series of Monographs in Pure and Applied Mathematics*, vol. 67 (The Macmillan Company, New York, revised edition, 1964)
29. E. Serra, M. Tarallo, S. Terracini, On the structure of the solution set of forced pendulum-type equations. J. Differ. Equ. **131**(2), 189–208 (1996)
30. E. Serra, M. Tarallo, S. Terracini, Subharmonic solutions to second-order differential equations with periodic nonlinearities. Nonlinear Anal. Theory Methods Appl. **41**(5–6), 649–667 (2000)
31. J.J. Stoker, *Nonlinear Vibrations in Mechanical and Electrical Systems* (Interscience Publishers Inc, New York, N.Y., 1950)
32. G. Tarantello, On the number of solutions for the forced pendulum equation. J. Differ. Equ. **80**(1), 79–93 (1989)
33. F. Tricomi, Integrazione di un'equazione differenziale presentatasi in elettrotecnica. Annali della Scuola Normale Superiore di Pisa-Classe di Scienze **2**(1), 1–20 (1933)
34. A.J. Ureña, Using closed orbits to bifurcate many periodic solutions for pendulum-type equations. J. Math. Anal. Appl. **302**(2), 318–341 (2005)

Chapter 8
Chaos in the Duffing-Type Equation Related to Tides

8.1 Periodic and Chaotic Tides

In this chapter, following [15], we examine some of the chaotic regimes which arise in externally driven tidal systems. Interestingly, the equations modelling the tides are again of the Duffing type. Our model is based loosely on the work of Maas and Doelman [15] and the development of this chapter goes along similar lines as our recent article [2].

8.2 Introduction to the Study of Tides in a Basin

Tides have been known to influence coastal geological processes to a large extent. The reason for this is the periodic change in water levels. This leads to different parts of the coast to be subjected to wave energy throughout the day. For instance, in regions where there are large tidal ranges, the water may rise and fall up to 10 metres. Because of this, various parts of the region are exposed to erosion and deposition alternatingly. As is expected, tides affect water levels in rivers as they approach the ocean. These effects sometimes reach much farther inland, even up to hundreds of kilometres. For example, in the Amazon river, their influence can be felt up to 850 km from the mouth of the river, according to [7]. A peculiar fact is that the tides effects become even more pronounced in coastal embayments. An interesting example is in the Bay of Fundy [9, 10] where it is known that the highest tides in the world occur. The large tidal amplitudes in this region are driven by the near-resonance of the Bay of Fundy-Gulf of Maine system, which has a natural period of approximately 13 hours, close to the 12.42 hours, period of the forcing tides. Similar resonances may be observed in narrow sea straits, bays, fjords and nearly closed basins (see also the Introduction in [2]) (Fig. 8.1).

Different geometry of coasts determines different kinds of responses, which need their own description of the physics involved. Some nonlinear effects may occur

L. Burra and F. Zanolin, *The Duffing Equation*, Infosys Science Foundation Series in Mathematical Sciences, https://doi.org/10.1007/978-981-97-8301-4_8

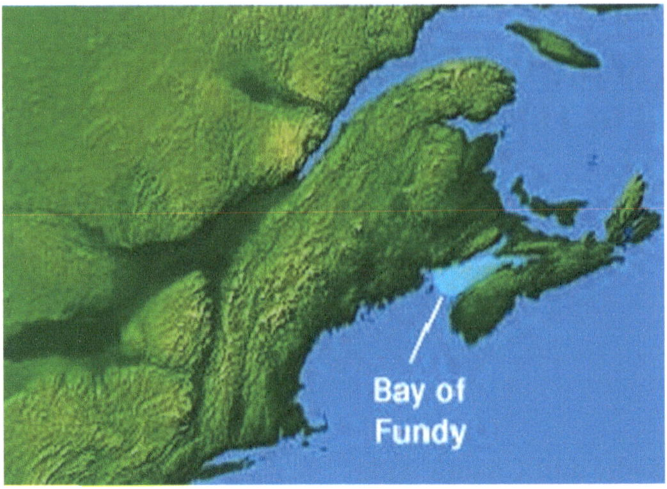

Fig. 8.1 An image of the Bay of Fundy (centred near 64 degrees west longitude and 45.5 degrees north latitude) which covers an area of 16.5 by 21 km. (Image credit Wikipedia)

in short deep basins which are connected to the sea by a thin strait [14]. Such a hypothetical configuration is depicted in Fig. 8.2, where the external tide plays the role of a periodic forcing. This is called "the Helmholtz mode". Accordingly, a periodic exchange of water between the basin and the sea is a characteristic of this kind of mode.

According to [8], the shape of the basin influences tidal elevations and the tides' amplification is usually considered as a resonance phenomenon. This is a linear phenomenon, but the governing equations are nonlinear, therefore more complicated dynamics are expected.

Fig. 8.2 A possible impression of a small basin which is connected to the open sea by a narrow strait

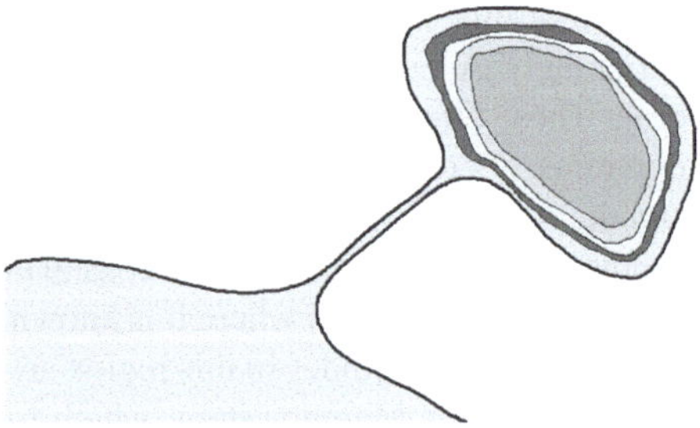

8.3 A Mathematical Formulation for the Modelling of the Tides

If $v(t)$ represents the excess volume at time t, and $\zeta(v)$ represents the surface elevation within the basin, the relation between these two quantities is given by the differential equation [8],

$$\ddot{v} + \zeta(v) = \zeta_e - \gamma(\dot{v}). \tag{8.1}$$

In Eq. (8.1), $\zeta(v)$ and $\zeta_e(t)$ denotes the surface elevation within the basin and in the exterior sea, respectively, and in [8], $\gamma(\dot{v})$ is added to represent friction effects. Typical forms of $\gamma(\dot{v})$ are usually taken as $\gamma(\dot{v}) = c\dot{v}$ with c, a constant related to the viscosity (See, for example, [15] and Eq. (8.6) later). The case of $c = 0$ is the inviscid form of the equation. The nonlinear function $\zeta(v)$ is determined by the geometry of the basin. The external forcing term $\zeta_e(t)$ is not exactly a periodic function of time. This is because the sea tide consists of both lunar and solar components, and may consist of annual, diurnal, and other harmonic frequencies. But amidst all these frequencies $\zeta_e(t)$ is at least quasi-periodic with two frequencies which are ω and ω_{sol}, the tidal undulations caused by the gravitational attractions of the moon and sun, respectively.

We take $\zeta_e(t)$, as a square wave [1, 3]. The justification for this may be found in the work of many authors. For example [4]. In their paper, the authors consider several shapes for basins connected to water bodies by channels and they show that the external tides (the forcing term) have sudden increases and decreases which may be approximated by a step function. The authors suggest that the effective geometry of the coastline influences the tides in the basins. In other words, the shape of the basin and its connection to the open water body via the channel determines the nature of the forcing term. In [16], the authors study tides entering the Strait of Magellan from the Atlantic. These are initially amplified by the geometry of the first embayment. Two narrow constrictions choke the tides and cause the tidal range to decrease in a stepwise fashion from east to west. It is concluded by them that *tidal choking* mainly affects the semidiurnal constituent tides. In [6], the authors perform sea-level and current measurements in the Mok Bay, a tidal embayment in the Dutch Wadden Sea, situated on the island of Texel, the Netherlands. They observe that the significant change in basin shape between low and high water is the cause for the existence of two frequencies of resonance. Due to its semi-enclosed nature, the basin could at both tidal phases be characterized as a Helmholtz resonator. They use a simplified semi-analytical piecewise-linear viscous Helmholtz model which captures the observed combination of tidal and eigen oscillations well. However, despite its simplicity, this model is able to display nonlinear behaviour for certain parameter values. The stepwise character of the forcing term in our model is also consistent with the numerical simulations of the Poincaré return maps of (8.1) with a choice of $\zeta_e(t) = A\cos\omega t$ and a truncated version of this function, that is

$$\zeta_e(t) = \max(-0.5, \min(0.5, \cos(\omega t))).$$

In fact they appear almost exactly the same. This is evident from Fig. 8.10 and also comparing with [15, Fig. 6].

We next consider Eq. (8.1) without both the friction and the forcing term, in which form it is integrable.

$$\ddot{v} + \zeta(v) = 0. \tag{8.2}$$

The energy integral E is given by

$$E = \dot{v}^2 + 2Z(v), \qquad Z(v) = \int_0^v \zeta(s)ds. \tag{8.3}$$

The dynamics of (8.2) is determined by the geometry of the basin given by $\zeta(v)$, the surface elevation as a function of excess volume v.

Following [15], we study a specific two-dimensional scaled basin with a shape function given by

$$z = h(x).$$

It is assumed that $h(\pm(1/2)) = 0$ with its maximum depth at $x = 0$. We also suppose that $h(x)$ is symmetric ($h(-x) = h(x)$), and an increasing function of x, for $x > 0$. In this manner, the shape function can be inverted. As a next step, we consider the wetted area, denoted by $A(z)$, which is the surface area of the water in the basin at height z. Accordingly, we have

$$A(z) = 2h^{-1}(z) \;\; (z \geq -1).$$

Then, the excess volume is

$$v = V(\zeta) := \int_0^\zeta A(z)dz = 2\int_0^\zeta h^{-1}(z)dz. \tag{8.4}$$

Maas in 1997 [14], after scaling, derives an equation for the Helmholtz resonator in terms of the excess volume, $v(t)$

$$\frac{d^2v}{dt^2} + \zeta(v) = Z_e(\sigma t) - r\frac{dv}{dt} - \gamma\left|\frac{dv}{dt}\right|\frac{dv}{dt}. \tag{8.5}$$

Where the external tidal elevation at sea $\zeta_e(\sigma_e t)$ is a given periodic function of time t, H is the maximum depth of a basin, $\zeta_e = HZ_e$. The manner in which Eq. (8.5) is derived, prompts the authors in [8] to refer to it as a "Helmholtz Oscillator".

In [15] a modified form of Eq. (8.5) is considered

$$\ddot{v} + \zeta(v) = \zeta_e(t) - c\frac{dv}{dt}, \tag{8.6}$$

where c is taken to be very small.

When the depth within the basin is uniform, the area can be taken to be $A(z) = 1$, and then it can be seen easily that $\zeta(v) = v$.
Some have considered basins with variable depth [1].

Green [12] in 1992 assumed a linear dependence on height,

$$\begin{aligned} A(z) &= 2h^{-1}(z) = 1 + z. \\ h^{-1}(z) &= \tfrac{1+z}{2} = v \\ \Rightarrow z &= 2v - 1 = h(v) \end{aligned} \tag{8.7}$$

gives the shape of the basin.

Choose $z = 2 \mid v \mid -1 = h(v)$,

$$\begin{aligned} V(\zeta) = \int_0^{\zeta} A(z)dz &= \int_0^{\zeta} (1 + z)dz \\ &= \frac{(1+\zeta)^2 - 1}{2} \\ \zeta(v) &= \sqrt{2v + 1} - 1 \end{aligned} \tag{8.8}$$

A similar system has been studied by some other authors (See [13]).

8.4 Chaotic Dynamics for the Poincaré Map Associated to Eq. (8.1)

Using the "stretching along the paths" method detailed in Chap. 4, we consider the Poincaré map associated to the system (8.1). From the fundamental theory of ODEs we have for every $t_0 \in \mathbb{R}$ and $z_0 = (x_0, y_0) \in \mathbb{R}^2$, that there exists a unique solution $\zeta(\cdot) = \zeta(\cdot; t_0, z_0)$ of (8.1), satisfying the initial condition $x(t_0) = x_0$, $y(t_0) = y_0$, which is defined on a right maximal interval of existence $[t_0, \omega[\subseteq [t_0, +\infty)$. The Poincaré map for the T-periodic problem is defined as

$$\Phi : \mathbb{R}^2 \supseteq \text{dom}(\Phi) \to \mathbb{R}^2, \quad \Phi : z_0 \mapsto \zeta(t_0 + T; t_0, z_0),$$

that is the function which maps any initial point z_0 (at the time $t = t_0$) to the point after a period T, on the trajectory of (8.1) departing from z_0. Usually, we take $t_0 = 0$ (this will be always assumed unless otherwise mentioned). The domain dom(Φ) of the Poincaré map is an open subset of the plane and Φ is a homeomorphism onto its image. Consider Eq. (8.3), in the phase-plane, this equation is

$$\begin{aligned} \dot{x} &= y \\ \dot{y} &= -\zeta(x) \end{aligned} \tag{8.9}$$

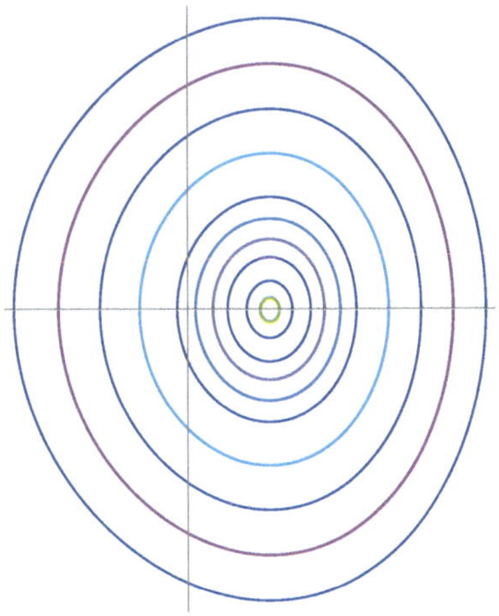

Fig. 8.3 Phase portrait of the system (8.9)

For the purpose of calculation, the variable v is replaced by x. The phase portrait of this system with $\zeta(x) = \sqrt{2x+1} - 1 + \mu$, $\zeta(x)$ being defined for $x \geq -1/2$, $\mu = 0.45$ is given in Fig. 8.3.

In Eq. (8.6), Maas and Doelman in [15] take $\zeta_e(t) = f \cos \omega t$. In our work we define $\zeta_e(t) = f \cos \omega t$. In our work we define

$$\zeta_e(t) = p_{a,b}(t) + \zeta_{per}(t),$$

with

$$p_{a,b}(t) = \begin{cases} a & \text{for } 0 \leq t < \tau_a \\ b & \text{for } \tau_a \leq t < \tau_a + \tau_b = T. \end{cases} \tag{8.10}$$

Where $\zeta_{per}(t)$ is a small perturbation. Figure 8.10 at the end of the chapter shows that the Poincaré map for a sinusoidal perturbation as in [15] and that associated with a truncated sine-function close to a square wave, as in our model, are quite similar.

Our equation now takes the form

$$\ddot{x} + \zeta(x) = p_{a,b}(t) + \zeta_{per}(t) - \gamma(\dot{x}). \tag{8.11}$$

With the definition of $\zeta_{per}(t)$ in (8.10) and treating $\zeta_{per}(t) - \gamma(\dot{x})$ for small γ as a small perturbation, we neglect it and we split Eq. (8.11) into two different systems:

$$\begin{cases} \ddot{x} + \zeta(x) = a \\ \ddot{x} + \zeta(x) = b, \end{cases} \tag{8.12}$$

so that we are led to study the alternating of the two equations:

$$\ddot{x} + \zeta(x) - a = 0 \Leftrightarrow \ddot{x} + \sqrt{2x+1} - (a+1) = 0.$$

$$\ddot{x} + \zeta(x) - b = 0 \Leftrightarrow \ddot{x} + \sqrt{2x+1} - (1+b) = 0.$$

Let $\mu \in \mathbb{R}$ be a parameter. We consider the equation

$$\ddot{x} + \sqrt{2x+1} - 1 = \mu.$$

We wish to obtain the equilibrium points of this system that is

$$\begin{aligned} \dot{x} &= \quad y \\ \dot{y} &= -\sqrt{2x+1} + 1 + \mu. \end{aligned} \tag{8.13}$$

The two equations given above are labelled as (E_a) and (E_b), respectively. The equilibrium point of the system (8.13), is given by $y = 0$, and the solution of the equation $\sqrt{2x+1} = 1 + \mu$. So it is necessary that $\mu > -1$.

Since $\mu > -1$, define the equilibrium point as $(\bar{v}_\mu, 0)$ where $\bar{v}_\mu$ is given by $\bar{v}_\mu = \frac{(\mu+1)^2-1}{2} > -\frac{1}{2}$.

Coming back to Eq. (8.2) and its first integral is given by Eq. (8.3), for our choice of $\zeta(x)$, for Eq. (8.8), the first integral is

$$y^2 + \frac{2}{3}(2x+1)^{3/2} - x(\mu+1) = \widetilde{E}. \tag{8.14}$$

We claim that there exist closed orbits around the equilibrium point $(\bar{v}_\mu, 0)$.

The energy function $\widetilde{E}(x, y)$ achieves its minimum at the equilibrium that is

$$\widetilde{E}_{min} = \widetilde{E}(\bar{v}_\mu, 0).$$

Since the energy can be defined up to an additive constant, we can modify the definition of the energy so that its minimum value at the equilibrium point is 0. Therefore, it is convenient to define the energy as follows:

$$\widetilde{E}(x, y) = \frac{y^2}{2} + \int_{\bar{v}_\mu}^{x} [(2s+1)^{1/2} - (1+\mu)]ds.$$

For the choice of the function $\zeta(x)$ we have the constraint $x \geq -1/2$. As a consequence, the maximum value of the admissible energy is $\widetilde{E}_{max} = \widetilde{E}(-1/2, 0)$.

A calculation of $\widetilde{E}_{max}$ can be easily seen to be $(\mu+1)^3$. So if $\widetilde{E}_\mu \in]0, (\mu+1)^3]$ we have that the energy level lines are closed orbits around the equilibrium point.

We fix a number q between $-1/2$ and $\bar{v}_\mu$ (excluding $\bar{v}_\mu$), and consider the energy after scaling by a suitable factor,

$$3y^2 + F_\mu(x) = \widetilde{E}_\mu(x, y)$$

where $F_\mu(x) = 2(2x+1)^{3/2} - 6(\mu+1)x + (\mu+1)^3 - 3(\mu+1))$ then $\forall q \in [-\frac{1}{2}, \bar{v}_\mu[$, $\exists$ a unique solution, $x = r$, for $x > \bar{v}_\mu$ and $F_\mu(x) = F_\mu(q)$.

We next compute the period of the orbit.

$$3y^2 + F_\mu(x) = F_\mu(q) = F_\mu(r),$$

where q and r are symmetrically on either side of $\bar{v}_\mu$.

A simple integration yields the time period of the orbit as

$$T_c = \frac{2}{\sqrt{3}} \int_q^r \frac{ds}{\sqrt{F_\mu(q) - F_\mu(r)}}.$$

8.5 The Period Map and Its Monotonicity

Thus we consider the planar differential system for $x \geq -1/2$, depending on a parameter μ :

$$\begin{cases} \dot{x} = y \\ \dot{y} = -(\sqrt{2x+1} - 1 - \mu). \end{cases} \tag{8.15}$$

The strictly increasing function $\sqrt{2x+1} - 1 - \mu$ has an absolute minimum equal to $\mu + 1$ for $x = -1/2$ and tends to infinity for $x \to +\infty$. Therefore it vanishes on $]-1/2, +\infty)$ if and only if $1 + \mu > 0$.

We assume that $\mu > -1$, so that system (8.15) has a (unique) equilibrium point $(\bar{v}_\mu, 0)$, with

$$\bar{v}_\mu := \frac{(\mu+1)^2 - 1}{2}.$$

Observe that $\bar{v}_\mu > -1/2$ if and only if $\mu > -1$.

The equilibrium point $(\bar{v}_\mu, 0)$, is a local centre surrounded by a family of periodic orbits (see Fig. 8.3). We claim that the period of these orbits is monotone decreasing as the level energy of the orbits increases, that is as the orbits move farther away from the equilibrium point.

We use a theorem of Chicone [5] (see for more details Chap. 1, Theorem 1.3) to check this claim, dealing with a planar Hamiltonian system of the form

$$\begin{cases} \dot{x} = y \\ \dot{y} = -V'(x) \end{cases} \tag{8.16}$$

with V a continuously differentiable potential having a strict local minimum at the origin (see [5, 11]). It may be recalled that, according to Chicone's theorem [5], the time mapping is monotone decreasing provided that

$$M(x) := \frac{d^2}{dt^2}\left(\frac{V(x)}{V'(x)^2}\right) \leq 0. \tag{8.17}$$

Therefore, in order to apply Chicone's theorem, we re-write system (8.15) in an equivalent form by shifting the origin to the equilibrium point. Putting also

$$\nu := \mu + 1,$$

we find that Eq. (8.15) takes the form

$$\begin{cases} \dot{x} = y \\ \dot{y} = -g(x) \end{cases} \tag{8.18}$$

with

$$g(x) := (2x + \nu^2)^{1/2} - \nu, \quad \text{with } \nu > 0.$$

System (8.18) is defined for $x \geq -\nu^2/2$. The primitive

$$G(x) := \int_0^x g(s)\, ds$$

of the function g takes the explicit form

$$G(x) := \frac{1}{3}(2x + \nu^2)^{3/2} - \nu x - \frac{\nu^3}{3}.$$

Then, for $V(x) = G(x)$ and $V'(x) = g(x)$, we compute the function $M(x)$ according to (8.17) and obtain

$$M(x) = -\frac{2}{3}\left(\frac{4\nu^4 - 4\nu^3\sqrt{2x+\nu^2} + 8\nu^2 x - 4\nu x\sqrt{2x+\nu^2} + 2x^2}{(2x+\nu^2)^{3/2}(\sqrt{2x+\nu^2} - \nu)^4}\right).$$

The sign of $M(x)$ is determined by that of

$$h(x) := -2x^2 - 8\nu^2 x + 4\nu x\sqrt{2x+\nu^2} + 4\nu^3\sqrt{2x+\nu^2} - 4\nu^4,$$

where $h : [-\nu^2/2, +\infty) \to \mathbb{R}$, with

$$h(-\nu^2/2) = -\nu^4/2, \quad \text{and } h(+\infty) = -\infty.$$

Now we show that $h(x)$ has a zero only at $x = 0$ which is the unique point of absolute maximum for $h(x)$ on its whole domain. Indeed,

$$h'(x) = 4\left(-x\sqrt{2x+\nu^2} - 2\nu^2\sqrt{2x+\nu^2} + 3\nu x + 2\nu^3\right)/\sqrt{2x+\nu^2}.$$

Notice that the sign of $h'(x)$ is the same as $(3\nu x + 2\nu^3) - (x + 2\nu^2)\sqrt{2x + \nu^2}$ and, moreover,

$$(x + 2\nu^2)^2(2x + \nu^2) - (3\nu x + 2\nu^3)^2 = 2x^3.$$

Thus we have verified that $h(x)$ is strictly increasing on $[-\nu^2/2, 0]$ and strictly decreasing on $[0, +\infty)$, with $h(0) = 0$ and then we can conclude that

$$M(x) < 0, \quad \forall x \neq 0.$$

From Chicone's theorem, we can say that the period function is strictly monotone decreasing.

8.6 Linked Twist Maps, Geometric Configuration

According to the previous definitions, we consider the system

$$\begin{cases} \dot{x} = y \\ \dot{y} = -\zeta(x) + p_{a,b}(t) \end{cases} \tag{8.19}$$

which comes from the equation

$$\ddot{x} + \zeta(x) = p_{a,b}(t). \tag{8.20}$$

and search for periodic solutions.

We split the time period T as

$$T = \tau_a + \tau_b$$

that is, with reference to the systems (E_a) and (E_b) defined below, we consider the system (E_a) for $t \in [0, \tau_a[$ and then we switch to system (E_b) for $t \in [\tau_a, T[$. At time T we get back to system (E_a) and repeat the same cycle with T-periodicity.

Such a two-state alternating behaviour can be concisely described by re-writing Eq. (8.19) as two different systems which we label as (E_a) and (E_b), respectively.

$$\begin{cases} \dot{x} = y \\ \dot{y} = -(\sqrt{2x + 1} - 1 - a) \end{cases} \tag{E_a}$$

$$\begin{cases} \dot{x} = y \\ \dot{y} = -(\sqrt{2x + 1} - 1 - b). \end{cases} \tag{E_b}$$

In order to prove that system (8.19) has very rich dynamics associated with the Poincaré map ϕ we use the linked twist map approach. Indeed, ϕ can be decomposed as

$$\phi = \phi_a \circ \phi_b ,$$

where ϕ_a is the Poincaré map of system (E_a) on the time interval $[0, \tau_a]$ and ϕ_b is the Poincaré map for (E_b) on the time interval $[\tau_a, T]$.

In the present model, we construct two annular regions $\mathcal{A}$ and $\mathcal{B}$ which cross in two disjoint topological rectangles $\mathcal{R}_1$ and $\mathcal{R}_2$. We recall from Chap. 4 that the linked twist map is defined on geometric configurations characterized by the alternation of two planar homeomorphisms which twist two overlapping annuli. Each annulus is turned onto itself by a homeomorphism which leaves the boundaries of the annulus invariant. This happens when the angular speed is monotone with respect to the radius of whether monotone increasing or decreasing (which is the case in our example). As already proved in Chap. 4, the existence of a Smale horseshoe inside $\mathcal{R}_i$ $(i = 1, 2)$ is guaranteed.

The overlapping annuli, labelled as $\mathcal{A}$ and $\mathcal{B}$, are constructed from the level lines of the first integrals associated to the system (E_a) and (E_b), respectively. The inner and outer boundaries of each annulus are closed trajectories surrounding the equilibrium points of both systems. These annuli as well as their intersections $\mathcal{R}_1$ and $\mathcal{R}_2$ are illustrated in Fig. 8.4.

Next, we provide an "orientation" to $\mathcal{R}_i$ (for $i = 1, 2$) by choosing two disjoint compact arcs in the boundary, denoted by $\mathcal{R}^-_{i,left}$ and $\mathcal{R}^-_{i,right}$. More precisely and with reference to Fig. 8.4, we orientate $\mathcal{R}_1$ and $\mathcal{R}_2$ as follows: We choose as $\mathcal{R}^-_1$ the intersection of $\mathcal{R}_1$ with the inner and outer boundaries of $\mathcal{A}$ and as $\mathcal{R}^-_2$ the intersection of $\mathcal{R}_2$ with the inner and outer boundaries of $\mathcal{B}$.

To visualize the linked twist map effect, we consider a path $\gamma : [0, 1] \to \mathcal{R}_1$ with $\gamma(0) \in \mathcal{R}^-_{1,left}$ and $\gamma(1) \in \mathcal{R}^-_{1,right}$ (See Fig. 8.5).

As can be illustrated in the figures that follow, the points of $\mathcal{R}^-_{1,left}$ move slower than those belonging to $\mathcal{R}^-_{1,right}$ under the action of system (E_a). Hence, for a choice

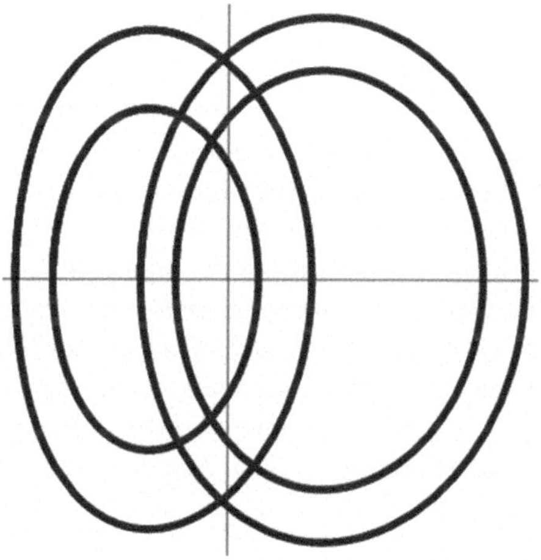

Fig. 8.4 The oriented rectangles $\mathcal{R}_1$, $\mathcal{R}_2$

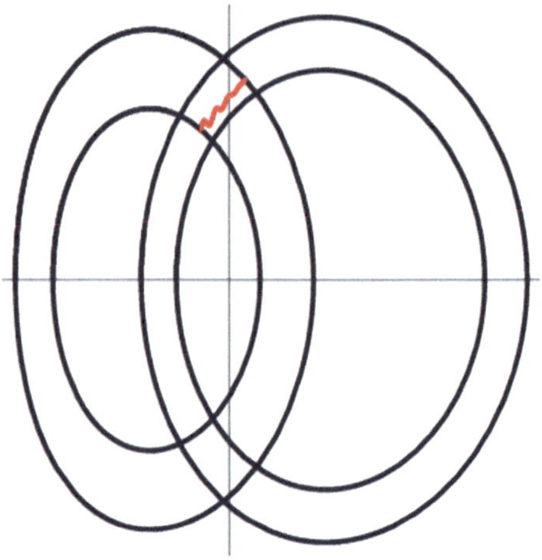

Fig. 8.5 The path γ in the region $\mathcal{R}_1$

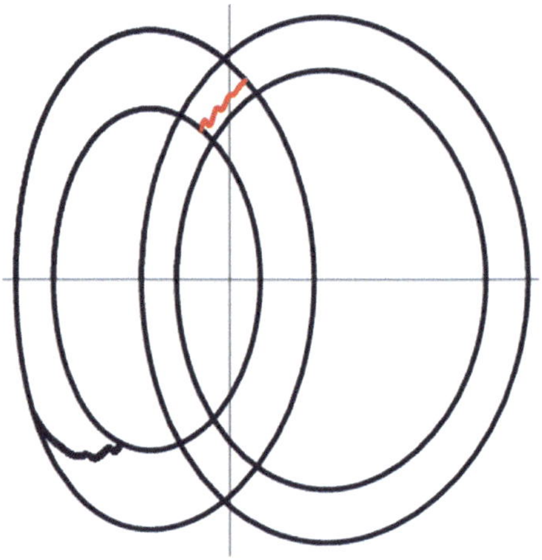

Fig. 8.6 The path γ after $t = 25$

of the first switching time τ_a large enough, it is possible to make the path $[0, 1] \ni \theta \mapsto \phi_a(\gamma(\theta))$ turn in a spiral-like fashion inside the annulus $\mathcal{A}$ and cross at least twice the rectangular region $\mathcal{R}_2$ from $\mathcal{R}^-_{2,left}$ to $\mathcal{R}^-_{2,right}$ (Figs. 8.6, 8.7 and 8.8).

Observe that the stretching of the path shows clearly that the outer orbit is of a smaller period hence points move faster on it.

We now repeat the same argument as above for the path stretched across $\mathcal{R}_2$ and under the Poincaré map ϕ_b with respect to the system E_b, and conclude that for a suitable choice of τ_b we can transform, via ϕ_b, any path in $\mathcal{R}_2$ joining the two components of $\mathcal{R}^-_2$ onto a path which crosses at least once $\mathcal{R}_1$ from $\mathcal{R}^-_{1,left}$ to $\mathcal{R}^-_{1,right}$. This stretching is shown in Fig. 8.9.

We assume that $a \neq b$. Actually, from now, on we suppose that

$$b > a > -1. \tag{8.21}$$

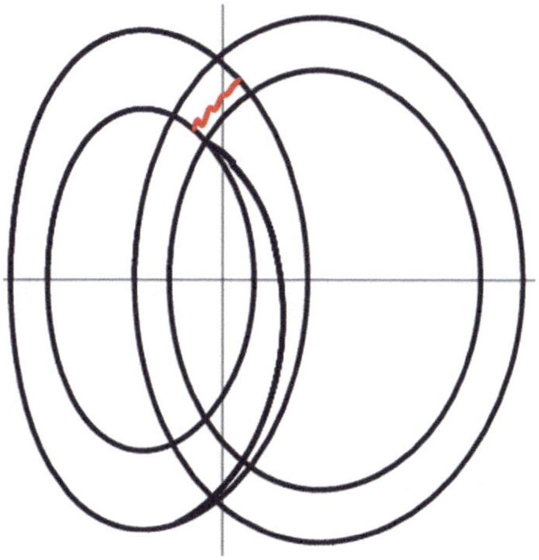

Fig. 8.7 The path γ after $t = 100$

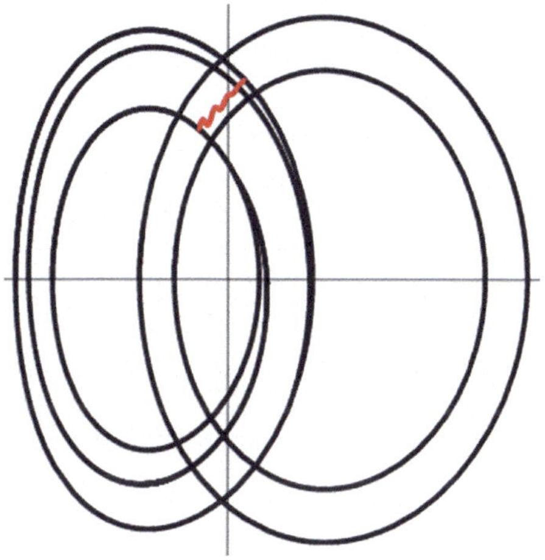

Fig. 8.8 The path γ after $t = 300$

The choice of $a, b > -1$ guarantees that the equilibrium points for both the equations of (8.12) below are in the open domain $x > -1/2$. Since $a \neq b$, we consider only the case $b > a$. The treatment of the case $a > b$ is completely symmetric and therefore it is omitted.

We are now in a position to state our main result :

Theorem 8.1 *For any choice of a and b, satisfying condition* (8.21) *and for every pair* $(\mathcal{A}, \mathcal{B})$ *of annuli linked together, the following conclusion holds:*

For every integer $m \geq 2$ *there exist two positive constants* α *and* β, *such that for each*

$$\tau_a > \alpha \quad \textit{and} \quad \tau_b > \beta,$$

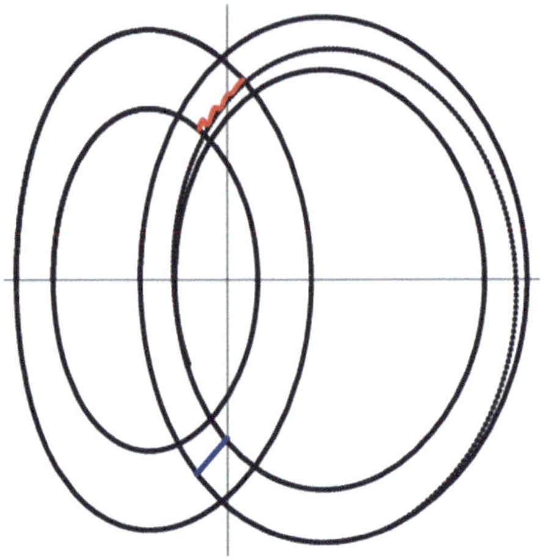

Fig. 8.9 The path γ crossing from $\mathcal{R}_2$ to $\mathcal{R}_1$

the Poincaré map associated to system (8.19) *induces chaotic dynamics on m symbols in* $\mathcal{R}_1$ *and in* $\mathcal{R}_2$ *.*

Linked twist maps (on long periods) arise as a consequence of the monotonicity of the period map which we proved in Sect. 8.5. The proof follows exactly from the same argument and estimates already derived in the previous chapters. In particular, the proof is essentially the same as that of Theorem 6.1 in Chap. 6. For the technical details, we refer also to [2].

The next Fig. 8.10 shows the Poincaré return map of systems $\dot{x}(t) = y(t)\ \ \dot{y}(t) = A\cos(\omega t)) - \sqrt{2x(t)+1} + 1$ and $\dot{x}(t) = y(t)\ \ \dot{y}(t) = A\max(-0.5, \min(0.5, \cos(\omega t))) - \sqrt{2x(t)+1} + 1$, respectively. We consider different initial points: $(0.3, .01), (.3185, .01), (\pm 0.3, 0), (0.25, 0), (\pm 0.2, 0), (-0.25, 0), (0.05, 0), (0.35, 0), (0.4, 0)$.

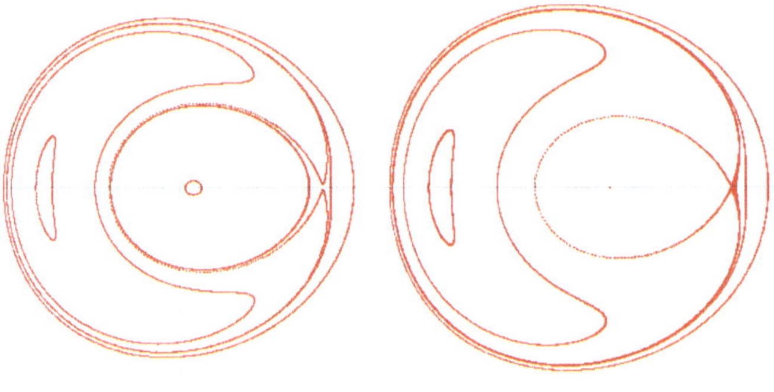

Fig. 8.10 Poincaré return maps for $\dot{x}(t) = y(t)$, $\dot{y}(t) = A\cos(\omega t)) - \sqrt{2x(t)+1} + 1$ (left) and $\dot{x}(t) = y(t)$, $\dot{y}(t) = A\max(-0.5, \min(0.5, \cos(\omega t))) - \sqrt{2x(t)+1} + 1$ $A = -.001, \omega = 1.01$ (right)

References

1. I. Boon, D. John, R.J. Byrne, On basin hyposmetry and the morphodynamic response of coastal inlet systems. Marine Geol. **40**(1–2), 27–48 (1981)
2. L. Burra, F. Zanolin, A topological approach to the problem of chaotic tides. Nonlinear Anal. Real World Appl. **68**(103699), 14 (2022)
3. E.I. Butikov, Square-wave excitation of a linear oscillator. American J. Phys. **72**(4), 469–476 (2004)
4. G.F. Carrier, R.P. Shaw, M. Miyata, *The response of narrow-mouthed harbors in a straight coastline to periodic incident waves* (J. Appl, Mech, 1971)
5. C. Chicone, The monotonicity of the period function for planar Hamiltonian vector fields. J. Differ. Equ. **69**(3), 310–321 (1987)
6. J. P. de Boer, l. R. M. Maas, Amplified exchange rate by tidal forcing of a piecewise-linear Helmholtz bay. Ocean Dyn. **61**(12), 2061–2072 (2011)
7. A. Defant, *Physical Oceanography*, vol. 1 (Pergamon, 1961)
8. A. Doelman, A.F. Koenderink, L.R.M. Maas, Quasi-periodically forced nonlinear Helmholtz oscillators. Phys. D: Nonlinear Phenomena **164**(1–2), 1–27 (2002)
9. C. Garrett, Tidal resonance in the Bay of Fundy and Gulf of Maine. Nature **238**(5365), 441–443 (1972)
10. C. Garrett, Tides and tidal power in the Bay of Fundy. Endeavour **8**(2), 58–64 (1984)
11. A. Gasull, A. Guillamon, J. Villadelprat, The period function for second-order quadratic odes is monotone. Qualit. Theory Dyn. Syst. **4**(2), 329–352 (2004)
12. T. Green, Liquid oscillations in a basin with varying surface area. Phys. Fluids A: Fluid Dyn. **4**(3), 630–632 (1992)
13. F. Mañosas, D. Rojas, J. Villadelprat, Study of the period function of a two-parameter family of centers. J. Math. Anal. Appl. **452**(1), 188–208 (2017)
14. L.R.M. Maas, On the nonlinear Helmholtz response of almost-enclosed tidal basins with sloping bottoms. J. Fluid Mech. **349**, 361–380 (1997)
15. L.R.M. Maas, A. Doelman, Chaotic tides. J. Phys. Oceanogr. **32**(3), 870–890 (2002)
16. C. Medeiros et al., Tidal characteristics of the Strait of Magellan. Continental Shelf Res. **8**(8), 947–960 (1988)
17. California NASA's Jet Propulsion Laboratory, Pasadena. The Bay of Fundy. NASA/GSFC/METI/ERSDAC/JAROS, and U.S./Japan ASTER Science Team
18. M. Pireddu, F. Zanolin, Chaotic dynamics in the Volterra predator-prey model via linked twist maps. Opuscula Math. **28**(4), 567–592 (2008)

Index

L. Burra and F. Zanolin, *The Duffing Equation*, Infosys Science Foundation Series,
https://doi.org/10.1007/978-981-97-8301-4

The manufacturer's authorised representative in the EU is Springer Nature Customer Service Centre GmbH, Europaplatz 3, 69115 Heidelberg, Germany. If you have any concerns regarding our products, please contact ProductSafety@springernature.com

Printed and bound by CPI Group (UK) Ltd, Croydon, CR0 4YY
01/10/2026
02225934-0009